The Properties of Groundwater

The Properties of Groundwater

Georg Matthess
Professor of General and Applied Geology
Kiel University, Federal Republic of Germany

Translated by
John C. Harvey, B.A., B.Sc., Ph.D. (Lond.)
Department of Environmental Sciences
Plymouth Polytechnic, England

1807 1982

175 YEARS OF PUBLISHING

A Wiley-Interscience Publication
JOHN WILEY & SONS
New York   Chichester   Brisbane   Toronto   Singapore

Library of Congress Cataloging in Publication Data:

Matthess, Georg.
  The properties of groundwater.

  Translation of: Die Beschaffenheit des
Grundwassers.
  "A Wiley-Interscience publication."
  Bibliography: p.
  Includes index.
  1. Hydrogeology. I. Title.

GB1003.2.M3713   551.49   81–7481
ISBN 0-471-08513-8   AACR2

Printed in the United States of America

10 9 8 7 6 5 4 3 2 1

Preface

Since the third edition of Konrad Keilhack's *Lehrbuch der Grundwasser –und Quellenkunde* (*Treatise on Groundwater and Springs*), published in 1935, a number of comprehensive texts on this subject and related ones have appeared in other languages, but there has not been a comparable book published in Germany. The publisher, Borntraeger Brothers, and the editor, late Prof. Wolfgang Richter, realized that instead of a revised edition of Keilhack's book, a series of specialist texts should be prepared covering the broad field of hydrogeology: *General Hydrogeology and Groundwater Balance, Groundwater Hydraulics, Karst Hydrogeology, Mineralized and Thermal Water, Geology and Water Supply, Geophysical Methods in Groundwater Exploration, Isotope Techniques in Hydrology*. The texts are printed as individual books and are obtainable separately as publications of the work of specialist authors in the respective fields.

I wrote the first contribution to this series, on the properties of groundwater. Its approval and acceptance by a broad spectrum of scientists showed that when this book was issued in 1973, it filled an important gap in German hydrogeological literature.

The reference section of the German edition has been retained in the translation because it is a record of the work that has been done on groundwater in Europe as a whole, in parallel with that published in English in other parts of the world. The original text has been preserved as well, apart from a few revisions to include the results of recently published work.

This book introduces fundamental principles that describe the geochemical mechanisms that control the properties of groundwater, the occurrence of the various dissolved substances, and their natural variations present in groundwater. The relationship between these components and groundwater is described. The physical, chemical, biological, and hygienic properties of groundwater determine its usefulness for human purposes, namely agriculture, industry, and domestic use. Furthermore, groundwater properties give important indications of the nature of aquifers, and supply valuable information about the origin, flow velocity, and direction of groundwater. Modern geochemical prospecting techniques in which substances are carried by groundwater are used for detecting concealed mineral ores and deposits of oil and natural gas. Spas use groundwater in which unusual dissolved mineral matter makes it of therapeutic value to the sick. A high concentration of elements important to agriculture—potassium, bromine, iodine, and others—can be considered to be useful. Water quality is subject to much spatial, and sometimes also periodic variations, the causes of which can very often be discovered only with difficulty, particularly because the chemistry of water itself can be very complex. Many of the physical and chemical principles have been known for a long time, but the rapid increase in the use of water and the development of geochemistry generally, as well as the application of analytical techniques in recent years, have made possible rapid progress in furthering knowledge of groundwater chemistry.

This book will be useful not only to earth scientists, but also to scientists in general, sanitary and civil engineers, and public health experts, to whom a knowledge of the properties of groundwater is important.

I gratefully acknowledge the careful and difficult work of the translator, John C. Harvey, and the advice and help given by Richard E. Jackson of the Inland Waters Directorate of Canada, Richard Buschner, Wiesbaden, and Wilhelm Schneider, Institut Fresenius, Neuhof, near Wiesbaden.

GEORG MATTHESS

Kiel, West Germany
October 1981

Contents

UNITS AND ABBREVIATIONS

1 PHYSICAL AND CHEMICAL PRINCIPLES   1
  1.1.   Physical properties of water, 1
      1.1.1.   The water molecule, 1
      1.1.2.   The isotopic composition of water, 3
      1.1.3.   Physical properties of pure water, 12
           Physical states and latent heats of water 13, Density 16,
           Compressibility 16, Coefficient of cubic expansion 17,
           Viscosity 18, Surface tension 18
      1.1.4   Electrolytic dissociation, solubility product, pH value,
           specific electrical conductance, 18
  1.2.   Solutions, 19
      1.2.1.   Solubility of gases, 19
      1.2.2.   Solubility of solids and liquids, 22
           Colloidal solutions 26, Solubility of organic substances 27
      1.2.3.   Factors causing solublity changes, 29
           1.2.3.1.   Ionic activity, 29
           1.2.3.2.   Solubility in aqueous solutions, 44
           1.2.3.2.1.   Carbonate–carbon dioxide equilibria, 45
                Calcium carbonate–carbon dioxide equilibrium, 45
                $CaCO_3$ precipitation, Influence of pressure on carbonate,
                equilibrium 58, Solubility of other carbonates, 59
           1.2.3.3.   Importance of pH with regard to solubility, 59
           1.2.3.4.   Importance of redox potential with regard to solubility, 61
           1.2.3.5.   Stability field diagrams to show dependence of
                solubility on ph and Eh, 64
      1.2.4.   Specific electrical conductance of aqueous solutions, 70

2 GEOCHEMICAL PROCESSES AND WATER   73
  2.1.   Dissolution, hydrolysis, and precipitation, 73
      Role of contract surface and contact time between water and rock,
      and temperature 74, Dissolution of salts 75, Hydrolysis 75.
      Changes in water quality caused by dissolution and precipitation 78,
      Dependence of dissolution and precipitation on climate 80, Degree of
      concentration and precipitation 83, Dissolution and precipitation
      during mixing processes 85, Coprecipitation 85, Complexing 86,

2.2. Adsorption and ion exchange, 86
Ion exchange 88, Exchange capacity and ion species 91, Exchange
rates 92, Influence of temperature 92, Exchange capacity and pH
value of the solution 92, Ion selectivity 92, Adsorption processes in
the unsaturated zone 103, Exchange processes in the groundwater
zone (saturated zone) 103, Sorption and filter effect 105, Ion filtration
by osmosis, 105
2.3. Oxidation and reduction, 107
2.4. Gas exchange between groundwater and atmosphere, 114
2.5. Biological processes, 119
Microbial metabolism 119, Sulfur and sulfur compounds 123,
Nitrogen compounds 125, Iron and manganese 126, Organic
substances 127, Control of microbiological activity by poisons 130,
Organisms occurring in groundwater 134, Influence of higher plants
on groundwater quality 135
2.6. Man-made factors, 139
Man-made pollution of groundwater 139, Pollution by gases 143,
Pollution by liquids 145, Pollution by solids 146, Effect of sewage
and wastewater 146, Discharge to surface runoff 147, Infiltration
from septic tanks and drains 148, Irrigation and spray irrigation 148,
Ground disposal into deep strata 149, Disposal of sewage (drying and
burning) 149, Disposal of radioactive wastewater 149, Consequence
of solid waste disposal on groundwater quality 151, Effects caused by
fertilizers 156, Effects caused by pesticides 157, Pollution caused by
the use of salt on roads 158, Migration of pollution, 158

3 GROUNDWATER   164
3.1. Origins of groundwater, 164
3.1.1. Atmospheric precipitation, 164
Dissolved and suspended solids 164, Aerosols from gas
reactions, 170
3.1.2. Percolation water, 172
Residence time in the unsaturated zone 173, Dissolution of
the ground air 174, Dissolution of soil material 175, Soil
solutions, 178, Lysimeter measurements 178, Periodic
changes in the composition of soil solutions 184, Comparisons
between percolate and groundwater quality, 185
3.1.3. Inland surface waters, 188
3.1.4. Seawater, 195
3.2 Groundwater properties and constituents, 196
3.2.1. Groundwater temperatures, 197
Spring water temperatures, 206
3.2.2. Substances in groundwater, 210
3.2.2.1. Dissolved constituents: Total dissolved solids, 215
3.2.2.1.1. Major constituents, 215
Sodium 215, Potassium 216, Calcium 236, Magnesium 239,
Iron 240, Manganese 243, Carbon dioxide, carbonate,
hydrogen carbonate, alkalinity, and carbonate hardness 245,

Nitrate, nitrite, ammonia, and other nitrogenous compounds
249, Sulfate 251, Chloride 254, Silica 256
3.2.2.1.2. Minor constituents and trace elements, 257
Lithium 258, Rubidium 258, Beryllium 258, Strontium 259,
Barium 260, Aluminum 261, Vanadium 261, Chromium 262,
Molybdenum 262, Cobalt 262, Nickel 263, Copper 263, Silver
264, Zinc 264, Cadmium 265, Mercury 266, Germanium 266,
Lead 266, Boron 267, Phosphorus 267, Arsenic 268, Selenium
268, Fluorine 269, Bromine 270, Iodine 271, Uranium 271,
Radium 272. Other trace elements 273, Noble gases 274
3.2.2.1.3. Organic substances, 276
3.2.2.2.   Insoluble constituents, 278
3.2.2.2.1. Suspended matter, 278
3.2.2.2.2. Organisms in groundwater, 279
3.3 Influence of aquifer materials on groundwater quality, 281
Igneous and metamorphic rocks 281, Groundwater in granite,
rhyolite, gneiss, and similar rocks 282, Groundwater in gabbro,
basalt, and similar crystalline rocks 283, Groundwater in sedimentary
rocks 284, Groundwater in sandstones and other psephitic-psammitic
hard rocks 284, Groundwater in unconsolidated aquifer materials
284, Groundwater in carbonate rocks 285, Groundwater in clayey,
marly, and silty rocks 287, Groundwater from gypsum-, anhydrite-,
and salt-bearing rocks 287, Groundwater in contact with
caustobiolites 288

4 CLASSIFICATION AND ASSESSMENT OF GROUNDWATER   291
4.1. Reporting and handling of physical and chemical data; methods of
illustrating, 291
4.1.1. Numerical presentation of analyses, 292
4.1.2. Graphical presentation of analyses, 297
4.1.2.1. Pictorial diagrams, 299
Bar graphs 299, Circular diagrams 301, Radial diagrams, 304
4.1.2.2. Multivariate diagrams, 307
Trilinear diagrams 307, Four-coordinate diagrams 312,
Two-coordinate diagrams 314, Parallel scale
diagrams 314, Horizontal scale diagrams 315,
Vertical scale diagrams 315
4.1.2.3. Plotting data on maps, 317
4.1.3. Data processing and storage, 320
4.2. Groundwater classification, 321
4.2.1. Classification on basis of origin, 321
4.2.2. Classification on basis of dissolved constituents, 325
4.2.3. Classification on basis of potential use, 335
4.2.3.1. Potable water, 336
4.2.3.2. Water for agriculture, 343
4.2.3.3. Industrial water, 350

REFERENCES   353
INDEX   399

Units and Abbreviations

Prefixes for Multiples and Fractions

G	= giga	= 10^9	m = milli	= 10^{-3}
M	= mega	= 10^6	μ = micro	= 10^{-6}
k	= kilo	= 10^3	n = nano	= 10^{-9}
c	= centi	= 10^{-2}	p = pico	= 10^{-12}

In general chemical nomenclature follows Nomenclature of Inorganic Chemistry, American version, IUPAC Inorganic Rules as published in *J. Am. Chem. Soc.* **82**, 5525 (1960).

Units of Measurement

Length	m	metre
Area	m²	square metre
	ha	hectare = 10^4 m²
Volume	m³	cubic metre
	l	litre = 10^{-3} m³
Mass	g	gram
	kg	kilogram
Time	s	second
	h	hour
	d	day
	a	year
Temperature	K	Kelvin (SI)
	°C	Celsius ($T = t = T_0$; T in K, t in °C, $T_0 = 273.15$ K)
Standard temperature		25°C = 298.15 K
Pressure	Pa	pascal (SI) = newton per metre² (N·m⁻²)
	bar	bar = 1000 mbar = 10^5 Pa
Standard atmospheric pressure	atm	atmosphere = 1013.24 mbar − 760 torr

Specific electrical conductance	S/cm	siemens per centimetre (SI)
Electrical charge	C	coulomb = 1A · s
Radioactivity	Bq	becquerel (SI) = s^{-1} = 27 picocuries (pCi)
Concentration	mg/l	milligram per litre
	mg/kg	milligram per kilogram (SI) (formerly ppm)
	mol/m^3	mole per cubic metre (SI) = 1 mmol/l
	meq/l	milliequivalent/l
	meq/kg	milliequivalent/kg
Circle	g	grad (400 grads = 360°)
Groundwater permeability	darcy	darcy = 9.87×10^9 cm^2

The Properties of Groundwater

Chapter 1 Physical and Chemical Principles

1.1 PHYSICAL PROPERTIES OF WATER

1.1.1 The Water Molecule

Water occurs in large quantities in the hydrosphere, the atmosphere, and the lithosphere (rocks), in solid, liquid, and gaseous states. Water as a liquid is the most important phase; being the most powerful solvent, it is involved in the distribution of geochemical materials around and through the Earth's surface. This capability is the result of a number of specific properties of water: its unusually high surface tension (71.97×10^{-3} N·m^{-1} at 25°C), dielectric constant (78.25 at 25°C), latent heat of vaporization (2282.2 J·g^{-1} at 1013. 24 mbar), and latent heat of fusion (333.73 J·g^{-1}). Its maximum density, which occurs above the freezing point, occurring at 4°C, is also unusual as is the reduction in coefficient of volume compressibility with increasing temperature (up to 50°C). The viscosity decreases with increase in pressure up to 32°C. Finally there is the phenomenon that among the substances that are liquid at standard temperature and pressure (25°C, 1013.24 mbar), water has the lowest molecular weight. These properties are the result of the peculiar structure of the water molecule, a combination of two hydrogen atoms of atomic weight, 1.00797 with an oxygen atom of atomic weight 15.9994; its molecular weight is therefore 18.01534. The radius of a water molecule is 138 pm.

The hydrogen nuclei are asymmetrically placed with respect to the electrons and to the oxygen nucleus (Fig. 1). The straight lines joining both hydrogen atoms to the center of the oxygen atom have a length of 96 pm and form an angle of about 104.5°. The molecular structure can be described as a tetrahedron, with the oxygen atom located at its center, while the centers of mass of the two hydrogen atoms occupy at any given moment one corner each, and the centers of charge of the two electron pairs at the same moment occupy the other two corners. Thus the four electrons are located as far away as possible from the oxygen nucleus and the hydrogen nuclei are bonded to the oxygen nucleus. Four of the other six electrons of the water

1

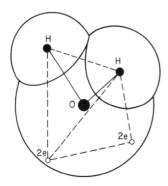

Fig. 1 Model of a water molecule. The mean positions of the two hydrogen nuclei and each of two pairs of electrons form a tetrahedron. The nucleus of the oxygen is in the center of the tetrahedron.

molecule form covalent bonds between the oxygen nucleus and the hydrogen nuclei, while the two remaining electrons remain close to the oxygen nucleus. Because of this asymmetric arrangement of the centers of gravity of the positive and negative charges, the water molecule is a dipole; the dipole moment (μ) is defined as the charge (e) multiplied by the distance (l) $= 1.84$ debye units; this comparatively high dipole moment is the cause of the much greater internal cohesion of water compared with other liquids.

The arrangement of the atomic nuclei and the electrons also influences the bonding of water molecules, which have a bonding network held together by hydrogen bonds. The hydrogen and oxygen atoms thereby arrange themselves tetrahedrally in the same way that silicon and oxygen combine in the quartz crystal lattice. This holds ideally for ice, but water in the liquid state also shows adjacent molecules forming into groups of various dimensions. The groups or clusters consist of an average of 90 molecules at 0°C, and about 25 at 70°C. Many observations also show an average group size of a few hundred water molecules (about 650 molecules at 0°C, approximately 74 molecules at 100°C). They become larger and more stable at lower temperatures, until at 4°C a maximum density is reached. At this temperature the structure is not rigid, but contains many unattached molecules alongside the larger aggregates of molecules. Further cooling reduces the kinetic energy of the molecules until it becomes too low to prevent orientation of molecules in a hydrogen bond structure; the structure becomes more open and there is a decrease in density.

When freezing occurs there is a final increase in the orientation to a maximum, the distances between molecules are greatest, and they all become more or less rigidly fixed in a crystal lattice as ice. One can therefore consider water to be a polymerized liquid with the formula $(H_2O)_n$, in which the molecules have various degrees of polymerization depending on the temperature and pressure. These polymerized components in water, which correspond structurally to ice, make up about 25% at 0°C, but in contrast only 5% at 100°C. Further structural elements consist of twofold to sixfold molecular aggregates with nontetrahedral hydrogen bonds, which allow a denser packing. The juxtaposition of individual molecules, tetrahedral and nontetrahedral molecular aggregates, explain the inherent physical behavior of water at different temperatures.

The breaking of the intermolecular bonds in water demands a relatively high amount of energy, which explains the high values of the surface tension and the latent heat of vaporization. The tendency toward hydrogen bonding and the dipole character of the molecule explain the unusual solvent action of water, which can dissolve more substances and in greater concentrations than other liquids. Some compounds (e.g., sugar and alcohol) are held in solution through hydrogen bonds. The high dielectric constant of water favors the solution of ionized substances, such as sodium chloride, because the ions of opposite charges are separated by "shells" of structured water molecules that form around ions in solution (Dorsey 1940, Hendricks 1955, Hodgman et al. 1958, Hutchinson 1957, Krauss 1961, Küster et al. 1969, Luck 1965, 1974, Pauling 1968, Schoeller 1962, Wells 1950).

1.1.2 The Isotopic Composition of Water

Natural water contains, in addition to hydrogen (H) of mass 1 (^{1}H) and oxygen (O) of mass 16 (^{16}O), small amounts of the stable isotope of hydrogen (^{2}H, deuterium D), the stable isotopes of oxygen (^{17}O and ^{18}O), and the radioisotope tritium (^{3}H, T) (Table 1). The unstable oxygen isotopes (^{14}O, ^{15}O, and ^{19}O) have such short half-lives (77, 118, 30 s, respectively) that they are of no importance in natural waters.

Table 1 Relative Distribution of Hydrogen and Oxygen Isotopes in Natural Water (Rankama 1954, Jacobshagen 1961)

Isotope	Abundance	Isotope	Abundance
^{1}H	99.9844%	^{16}O	99.76%
D	D/H = 1.49×10^{-4}	^{17}O	^{17}O/^{16}O = 4×10^{-4}
T	T/H = 1.3×10^{-18}	^{18}O	^{18}O/^{16}O = 2.04×10^{-3}

The isotopes ^{1}H, D, T, ^{16}O, ^{17}O, and ^{18}O combine to form 18 water molecules and 12 ion species. The rare isotopes D, T, ^{17}O, and ^{18}O occur in greatest abundance in combination with the commonest isotopes ^{1}H and ^{16}O, hence as HD(^{16}O), HT(^{16}O), H$_2$(^{17}O), and H$_2$(^{18}O). Molecules in which two or even three scarce isotopes appear are, on the other hand, very rare.

Water is therefore to be considered to be a mixture of these isotope combinations, having molecular weights of 18 (H$_2$O) up to 24 [T$_2$(^{18}O)], and with 99.8% of the occurring water molecules possessing the molecular weight 18 (Alekin 1962). The proportion of heavy water (HDO) in natural water is very low, usually 0.3‰. The physical properties of isotopically divergent water types are different from those of normal water (Table 2).

Table 2   Physical Data of Normal and Heavy Water (After Schatenstein 1960)

Constant	H_2O	D_2O
Maximum density (g·cm^{-3})	1.00000	1.10597
Temperature for maximum density (°C)	3.98	11.23
Viscosity at 25°C (10^{-3} Pa·s)	0.893	1.101
Melting point (°C)	0.000	3.813
Latent heat at melting point (J/mole)	6,012	6,343
Boiling point (°C)	100.000	101.437
Latent heat at boiling point (J/mole)	40,692	41,562
Surface tension at 25°C (10^{-3} N·m^{-1})	71.97	71.93

The stable isotope content is determined by mass spectrometer in which H_2 is used as the gas specimen for the deuterium analysis, and CO_2 for the ^{18}O isotope. To obtain the required accuracy of measurement, the isotope content of the specimen R is compared with a standard R_{st} and is given as a relative deviation (in %o) from the standard content:

$$\delta = \frac{R - R_{st}}{R_{st}} 1000$$

Following a proposal by Craig (1961b), the D and ^{18}O values are generally expressed on the basis of Standard Mean Ocean Water (SMOW). The accuracy of measurement is of the order of ±0.3%o for ^{18}O and ±2%o for D.

Tritium may be detected after a preliminary isotope enrichment, by counting particles ($E_{max} = 18$ keV) in a Geiger counter, into which the test material has been inserted in a gaseous state. Liquid scintillation counters are now generally used. The concetration of tritium is usually expressed in tritium units (TU), where 1 TU corresponds to a tritium content of [T]/[H] = 10^{-18}, and a specific tritium activity of 2.22 mβq/mole H_2O. The accuracy of measurement is of the order of about 5%, and the limit of detection today is 0.1TU, that is, one-tenth of the naturally occurring average global concentration of tritium on the Earth's surface.

As a consequence of the varying amounts in which the isotopes are distributed in the water molecule, there are isotope separating processes taking place during the hydrologic cycle. The variations in the content of the heavy isotopes deuterium and ^{18}O in natural waters are almost entirely attributable to isotope fractionation, which arises from partial evaporation and condensation. Moreover, there is an additional kinetic fractionation possible during evaporation under thermodynamic equilibrium, which is determined by the ratio of the vapor pressures p_{H_2O}/p_{HDO} relative to $p_{H_2}(^{16}o)/p_{H_2}(^{18}o)$ because of the variable diffusion velocity of $H_2(^{16}O)$, HDO, and $H_2{}^{18}O$ in the adjacent moving air mass (Ehhalt & Knott 1965, II. Phys. Inst. Heidelberg 1971):

$$\frac{p_{H_2O}}{p_{HDO}} = 1.0852 \pm 0.5\%o$$

$$\frac{p_{H_2}(^{16}O)}{p_{H_2}(^{18}O)} = 1.0088 \pm 0.4\%o$$

The following fractionation factors hold for diffusion from water vapor in air α_D:

$$\frac{D_{H_2O}}{D_{HDO}} = 1.015 \pm 3\%o$$

$$\frac{D_{H_2}(^{16}O)}{D_{H_2}(^{18}O)} = 1.030 \pm 1\%o$$

The D-values are coefficients of diffusion of the isotope water molecules in air (20°C, 1013.24 mbar) (Ehhalt & Knott 1965). The higher freezing point of HDO relative to $H_2(^{18}O)$ for water leads to an enrichment of the heavy isotope in the solid phase. Finally hydrological and geochemical processes, such as the mixing of waters of varying isotopic composition under natural conditions, can give rise to observed isotope ratios.

The concentrations of deuterium and ^{18}O in water may vary by about 40 and 6%, respectively, and have a linear relationship (Friedman 1953, Epstein & Mayeda 1953, Dansgaard 1953, 1961, Craig 1961a). The concentration of heavy isotopes in atmospheric precipitation is normally diminished progressively. This atmospheric Rayleigh process by liquid-gas equilibrium also explains why with decreasing mean annual air temperature the precipitated water and fresh water become increasingly lighter (Fig. 2), whereas water samples from tropical regions exhibit only very small deviations from standard mean ocean water (SMOW) (Degens & Chilingar 1967). Furthermore, the density of precipitated water is seen to be reduced as one proceeds from the coast to the interior of continents. These processes lead to the conclusion that except for the concentrated water of seas with internal drainage (Sofer & Gat 1975) compared with normal seawater, a diminution of the heavy isotope compared with its content in seawater always occurs.

The isotope composition of groundwater varies from the mean isotopic composition of rainwater because of the annual variations that play a part in periodic groundwater recharge, and because of isotope fractionation, during interception, evaporation, and transpiration, and surface runoff, infiltration, and vaporization in the soil. Observations in Israel by Gat & Tzur (1967) have shown that ^{18}O enrichment up to 0.5% and a corresponding increase in deuterium can be caused by these processes in combination with groundwater recharge. The tendency toward reduction in heavy isotope content at increasing distances from the sea is also recognizable in groundwater (Table 3): the geyser at Mammoth Hot Springs is 1200 km from the Pacific Ocean.

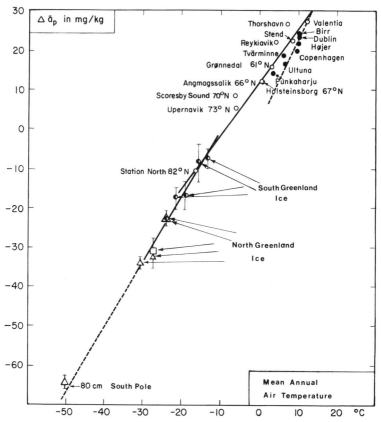

Fig. 2 Relationship between mean annual temperature and yearly average of the ^{18}O content in precipitation $\Delta\bar{a}_p$. Open circles indicate stations at sea level and near the sea in the North Atlantic area; solid circles show low altitude stations with more or less continental climate; other symbols ($\bigcirc$, $\triangle$, and $\square$) show nine stations on the Greenland icecap and one station at the South Pole (after Dansgaard 1961).

Isotope exchange between groundwater and chemically or physically bonded water molecules in aquifers is possible; for example kaolinite, illite, and montmorillonite can exchange H_2O adsorbed between the lattice layers for HDO (Jacobshagen 1961). There is reason to believe that if a complete exchange occurs, the influence on the oxygen isotope content of the water could be greater than that affecting hydrogen, because there is more oxygen than hydrogen stored in the mineral phase. The subsurface water movement, however, does not appear to have a marked influence on the oxygen isotope content of water. For example, the oxygen isotope ratios in groundwater samples taken at intervals of 30–160 km over a distance of 1120 km and at depths of 150–600 m remained constant within 1‰, whilst the chemical composition fluctuated strongly in relation to the flow path and the state of diagenesis of the rock material in the aquifer (Nubian Series, Western Desert in Egypt) (Knetsch et al. 1962, Degens 1962).

The oxygen isotope composition of some strongly saline oil field water in Cambrian to Tertiary strata (Fig. 3) does not vary appreciably from that of

Table 3 Deuterium Content δD of Mineralized Spring Water (After Holl 1965)

	δD (%)
Queen Spring at Spa, Belgium	− 3.0
Thermal Spring, Agnano, Naples	− 3.1
Boniface Spring, Bad Salzschlirf	− 5.8
Jordansprudel, Bad Oeynhausen	− 6.0
Grosser Sprudel, Bad Nauheim	− 6.8
Kochbrunnen, Wiesbaden	− 7.0
Mühlbrunnen, Karlsbad	− 7.0
Cachat Spring, Evian, Lake Leman	− 7.9
St. Peter Spring, Vals, Graubünden	− 9.7
Mammoth Hot Springs, Rocky Mountains	−14.3

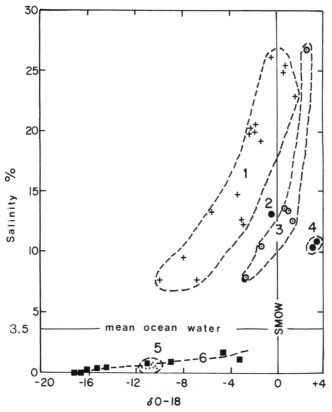

Fig. 3 Comparison between salt content and oxygen isotope composition in oil field waters from marine and lacustrine oil deposits (after Degens et al. 1964). 1, Cambrian-Ordovician (Oklahoma); 2, Devonian (Oklahoma); 3, Upper Carboniferous (Oklahoma); 4, Tertiary (Texas); 5, Chalk (Colorado); 6, Tertiary (Utah) (after Degens et al, 1964).

present day seawater. The variations in the negative range of $\delta^{18}O$ are clearly connected with a decrease in the salt content, which in turn has been caused by dilution with circulating waters during changes in the hydrogeological conditions during uplift, denudation, or other geological phenomena (Degens et al. 1964). Small deviations of the $\delta^{18}O$ values of some samples into the positive range may be caused by evaporation at the Earth's surface or by isotope equilibration during a long contact time (Degens & Epstein 1962).

Natural water also contains radioactive tritium, which by β-radiation changes into ^3He, having a half-life of about 12.35 years. In nature it forms HTO, the tritiated water molecule, which is an ideal tracer for the study of groundwater circulation. One must, however, remember that in groundwater studies isotope exchange can often occur between the tritium in the water and that which is physically or chemically bound in minerals, for example in montmorillonite and bentonite (as hydrogen or hydroxyl) (Knutsson & Forsberg 1967). Tritium is produced in the atmosphere by cosmic radiation at a rate of about 0.25 atom·cm^{-2}·s^{-1}. This production rate and the following radioactive disintegration results in an equilibrium state for the Earth as a whole at about 1.2×10^3 moles of HTO, which, divided by the total exchangeable water available, yields a mean surface tritium concentration T/H of approximately 10^{-20}. The oceans therefore contain more than 95% of the terrestrial reservoir of this substance (Begemann 1961, Eriksson 1965, Haxel & Schumann 1962, Münnich 1963).

The natural distribution of tritium has been greatly disturbed since 1950 because of fallout of large amounts of the isotope from surface tests of atomic weapons (Eriksson 1965, Fergusson 1965). The overall result of atomic weapons testing was to increase the tritium amount by some 50 times that present under natural equilibrium (Roether 1967a), and the maximum tropospheric concentration was raised to 500 times the natural level. Natural tritium and tritium due to atomic weapons occur in the upper atmosphere and take part in the hydrologic cycle. Tritium oxidizes to water (HTO), as soon as it is formed, and it reaches the troposphere after about 1 year's delay, mostly arriving during the spring in higher latitudes.

Tritium that has been precipitated from the atmosphere over continents remains to a large extent on the surface and reverts, particularly during the warm period of the year, to the atmosphere, with the result that the net rate of tritium precipitation on continents approaches an order of magnitude less than that on the oceans. Tritium precipitation proceeds so rapidly that the horizontal air mass movements can equalize differences in tritium content within the troposphere to a limited extent only. The tritium content of precipitation therefore varies greatly from one rainstorm to another, because at any given time it depends on the arrival of a new supply from higher layers of the atmosphere and on dilution within the troposphere. There is an average content, with superimposed deviations that have a recognizable statistical distribution.

For hydrologic studies in general the tritium content of the precipitation in the research area concerned is not known directly but must be interpolated over a very wide mesh net of stations. For monthly precipitation tests this is possible with a standard error of about ± 30 (Roether 1967a). The variations in the air mass exchange between stratosphere and troposphere accordingly produce annual changes in the tritium content, with a maximum in spring or early summer. The tritium content at the annual maximum is about five times that at the annual minimum (Roether 1967a). Tritium is, as it were, dammed up in the troposphere above the continents. Therefore for example, the tritium concentration rises from the coast of western Europe to the interior of the continent by a factor of 3 approximately. This continental effect is shown in Fig. 4. The figures given can be used for estimating the tritium content in precipitation for chosen research areas in central Europe.

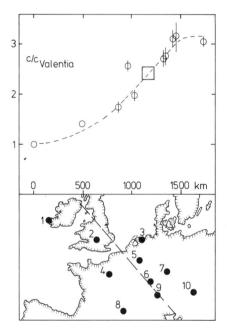

Fig. 4   Continental effect on the tritium content in rainwater in Europe (after Roether 1967a). Horizontal scale: distances from the reference station at Valentia on the west coast of Ireland (station 1). Projections perpendicular to the transect line on the map mark off the distances plotted along the abscissa on the graph above to give station locations measured from station 1. The square indicates a mean reference level for central Europe used in the concentration-time graph (Fig. 5).

Tritium has been released by degrees during the various atomic weapons tests carried out since 1950, and besides causing a general increase in the amount, this release has given rise to marked variation in concentration from one year to another (Fig. 5). In particular the peak tritium content in rainwater during the period 1963–1964 should be noted (Roether 1970).

Tritium content can be used for groundwater dating (Münnich 1968). To this end the measured isotope concentration (C) of a groundwater sample is compared with the estimated (or measured) initial isotope concentration

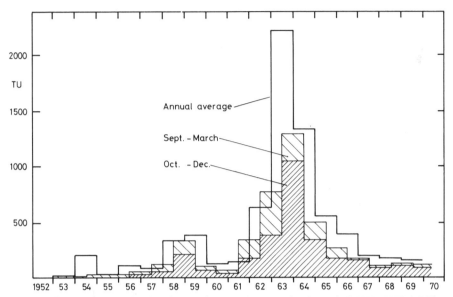

Fig. 5   Tritium concentration in rainwater in central Europe for the period 1952–1970 (cf. Fig. 4) (after Roether 1970). Prior to 1953 the annual mean was about 6 TU, and the mean September–March concentration was 4 TU (Roether 1970).

(C^0) of the water entering the aquifer. The decrease in concentration and the known half-life of the isotope ($\tau_{1/2}$) are used to calculate the age of the water sample in the aquifer, following the law for radioactivity dating: $t = \ln (C_0/C) \cdot \tau_{1/2}/(\ln 2)$. The annual variations in tritium content of the precipitation are taken into account when determining the initial tritium concentration. In this connection one has to remember that the precipitation in the cooler part of the year makes the major contribution to groundwater recharge. As an approximation, the tritium content of the groundwater recharge may be taken as the average tritium content of the precipitation over the months September to March in the Northern Hemisphere (Roether 1967b, Münnich et al. 1967). These mean values (Fig. 5) are without exception smaller than the mean annual value. The natural tritium concentration in the groundwater recharge in central Europe is about 4 TU (Roether 1967a) (Figs. 4 and 5).

For groundwater that is more than 20 years old, radioactive dating based on the natural tritium content is possible. The method is useful for ages up to 50 years. Minor impurities in younger waters that contain tritium released from nuclear weapons can indeed greatly increase the tritium content, and this fact makes tritium dating of groundwater a useful tool for finding out whether and to what extent younger water is mixed with older (Matthess et al. 1968, 1975).

A distinct age sequence in groundwater has been demonstrated by tritium dating in the region of the unconsolidated sand and gravel sediments near

Batsto (Wharton Tract Well Field, New Jersey). That area has approximately 60 m of Tertiary fine to coarse sand deposits with interspersed clay lenses underlying about 1 m of Quaternary river deposits of fine to coarse sands. The horizontal hydraulic conductivity is about 0.03 cm/s, the average porosity is about 40%, and the hydraulic gradient is approximately 2 m/km. The mean flow velocity through a given cross section is approximately 300 m/year.

The vertical distribution of the tritium content in the area near Batsto shows four age layers in the groundwater (Fig. 6). The oldest water (more than 25 years) is 30 m deep and contains 0–2 TU; the second oldest water, containing 5–8.3 TU (7–8 years) is about 15 m deep. The water between 7.5 and 15 m deep has such a high tritium content (21–48 TU) that it must have infiltrated into the ground after the thermonuclear explosions (from March 1954). This water was therefore between 4 years 8 months and 8 months old at the time of the fieldwork (March and November 1958). The topmost layer gave values of 82–147 TU directly below the groundwater surface and evidently originated from precipitation that took place at the time of the H-bomb tests. This water was therefore at the most 8 months old (Carlston et al. 1960).

This first proof of groundwater layering was corroborated in Europe during research in unconsolidated sand and gravel aquifers in northern Germany (Fuhrberge) (Geyh & Kuckelkorn 1969) and in the Upper Rhine Plain (Münnich et al. 1967).

Quantitative statements can be made on the basis of measurement of the tritium content over a series of years. Hence the time factor in the growth of tritium content of groundwater recharge since 1950 is worth analyzing (Figs. 5, 7). Ideally a time series record of the groundwater as far as the observation point can be determined (Brown 1961, Münnich 1963, Davis et al.

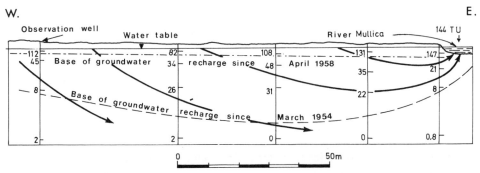

Fig. 6 Distribution of tritium content in a groundwater occurrence at Batsto, NJ (Wharton Tract Well Field). The numbers beside the well positions give the tritium concentrations in TU. The broken lines separate the ages of groundwater layers (after Carlston et al, 1960).

1967), and the water budget of larger catchment areas studied (Eriksson 1963, Buttlar & Wendt 1958). For these purposes one makes use of the fact that the precipitation worldwide has been marked with tritium in various amounts with respect to time, as shown in Fig. 5. The half-life comes into the calculations as a correcting factor. Finally, in areas in which spatially homogeneous groundwater recharge occurs, one may calculate the vertical level of the recharge from the tritium profile. This is done by adding up the amounts of tritium in the groundwater (Fig. 7), then referring to the concentration distribution in Fig. 5 to find the mean groundwater recharge (Münnich et al. 1967).

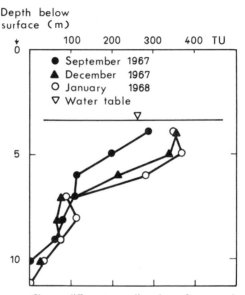

Fig. 7 Vertical tritium profile on different sampling dates for groundwater at Sandhausen, south of Heidelberg. Comparison with Fig. 5 shows that the water at 10 m depth is more than 10 years old (after Thilo 1969, Roether 1970, Münnich et al. 1967).

When the geochemistry of an isotope is known, investigation of the concentration can contribute to research work into the history of groundwater. However, isotope analysis is generally too expensive for use in routine measurements in groundwater research and other methods have to be used.

1.1.3 Physical Properties of Pure Water

Some important properties of water are temperature dependent: phase state, vapor pressure, density, viscosity, and compressibility (Fig. 8). These variables are important in the movement, extraction, and use of groundwater. The water temperature controls the solvent action and therefore the chemi-

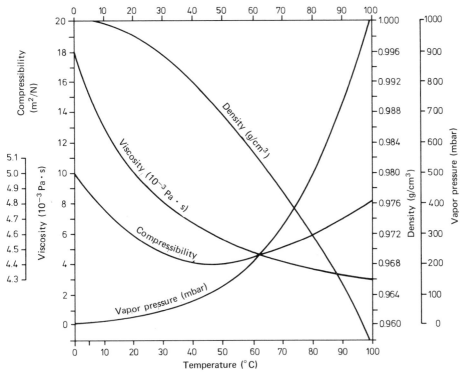

Fig. 8 Temperature-dependent water characteristics (density, vapor pressure, viscosity, and isothermal compressibility).

cal quality of the water and its general usefulness for drinking and industrial purposes. The viscosity controls the mobility of groundwater. Changes of state are most important: water can be immobile in the form of ice lenses below the surface, in the liquid state water moves relatively slowly under the influence of gravity, and water vapor can migrate under the influence of concentration and temperature gradients.

Physical States and Latent Heats of Water. Water occurs in nature in solid, liquid, and gas phases, but these three phases can coexist with respect to temperature and pressure only at the triple point, which lies at 1013.24 mbar pressure and 0°C. At temperatures above the triple point the liquid and gas phases coexist; at lower temperatures the solid and gas phases are in equilibrium. When water vapor pressure at a given temperature lies below this curve, only water vapor can exist; water or ice in this condition must vaporize because of the low pressure. The vapor pressure of water increases from 6.105 mbar at 0°C to 1013.24 mbar at 100°C (Table 4). Above the critical pressure ($p = 221,423.3372$ mbar) and critical temperature ($t = 374.2 \pm 0.1$°C) water vapor cannot be liquefied by increased pressure.

Table 4 Physical States of Water

Temperature (°C)	Vapor pressure (mbar)	Density (g · cm⁻³)	Isothermal compressibility at 1013.24 mbar, X_T (10^{-10} m² · N⁻¹)	Temperature dependent constant, L (10^7 N · m⁻²)	Dynamic viscosity (10^{-3} Pa · s)	Surface tension (10^{-3} N · m⁻¹)
0	6.105	0.99984	5.01-5.04		1.7921	75.64
1	6.567	0.99990			1.7313	
2	7.058	0.99994			1.6728	
3	7.579	0.99996			1.6191	
4	8.134	0.99997			1.5674	
5	8.723	0.99996			1.5188	74.92
6	9.350	0.99994			1.4728	
7	10.016	0.99990			1.4284	
8	10.726	0.99985			1.3860	
9	11.478	0.99978			1.3462	
10	12.278	0.99970	4.78		1.3077	74.22
11	13.124	0.99960			1.2713	74.07
12	14.023	0.99949			1.2363	73.93
13	14.973	0.99937			1.2028	73.78
14	15.981	0.99924			1.1709	73.64
15	17.049	0.99910			1.1404	73.49
16	18.177	0.99894			1.1111	73.34
17	19.372	0.99877			1.0828	73.19
18	20.634	0.99859			1.0559	73.05
19	21.967	0.99840			1.0299	72.90

Temperature (°C)	Vapor pressure (mbar)	Density (g · cm⁻³)	Isothermal compressibility at 1013.24 mbar, X_T (10^{-10} m² · N⁻¹)	Temperature dependent constant, L (10^7 N · m⁻²)	Dynamic viscosity (10^{-3} Pa · s)	Surface tension (10^{-3} N · m⁻¹)
20	23.378	0.99820	4.58		1.0050	72.75
25	31.672	0.99704	4.57	29.96	0.8937	71.97
30	42.428	0.99564	4.46		0.8007	71.18
35	56.229	0.99403	4.48	30.55	0.7225	70.38
40	73.759	0.99221	4.41		0.6560	69.56
45	95.83	0.99022	4.44	30.81	0.5988	68.74
50	123.34	0.98804	4.40		0.5494	67.91
55	157.37	0.98570	4.44	30.78	0.5064	67.05
60	199.16	0.98321	4.43		0.4688	66.19
65	250.03	0.98056	4.48	30.52	0.4355	
70	311.57	0.97778	4.49		0.4061	64.42
75	385.43	0.97486	4.55	30.05	0.3799	
80	473.42	0.97180	4.57		0.3565	62.61
85	578.08	0.96862	4.65	29.39	0.3355	
90	700.95	0.96531	4.68		0.3165	60.75
95	845.13	0.96189			0.2994	
100	1013.24	0.95835	4.80		0.2838	58.85
	Vogel 1956	Hodgman et al. 1958				Dorsey 1940

15

   The melting and freezing points are pressure dependent. At 1013.24 mbar the melting temperature is 0°C. The vapor pressure of a solution is reduced by the presence of the solute within it, which leads to a rise in the boiling point and to a reduction of the melting point proportional to the molar concentration. The reduction in freezing point amounts to 1.867°C · mole^{-1} and the increase in boiling point to 0.513°C · mole^{-1}. The specific heat, which is the amount of heat necessary to raise a unit mass of water through a temperature rise of 1°C, varies slightly over the range between 1 and 100°C (by 1% at the most). The median value, 4.1868 J, is 1/100 of the heat necessary to raise the temperature of 1 g of water from 0 to 100°C. The specific heat at constant volume is not appreciably different from the specific heat at constant pressure.
   The latent heat of fusion is the amount of heat that is necessary at constant pressure for a change in state from solid to liquid. It is identical for the reverse process. The value is 333.73 J · g^{-1} at 1013.24 mbar pressure. Analogously, the latent heat of vaporization is the amount of heat necessary to change the state of a unit mass of a liquid to a gaseous state. The value is 2282.2 J · g^{-1} at 1013.24 mbar. This is identical to the heat of condensation for the reverse process.
   The thermal conductivity is defined as the number of joules that flow in 1 s from an outside surface of a cube having 1 cm edges to the opposite surface, which has a temperature 1°C lower than that of the first. The value is 5.69×10^{-3} J · cm^{-1} · s^{-1} · °C^{-1} (at 25°C), and is very low compared with that for other substances; that is, water is a poor conductor of heat. The thermal conductivity varies with temperature.

Density.   The density (ρ) of a body is the mass per unit volume of that body (g · cm^{-3}). The density of pure water at its maximum value (at 3.98°C and 1013.24 mbar pressure) serves as the unit and reference value. The density of water decreases with temperature (Table 4) and increases with pressure. The temperature for maximum density decreases with increasing pressure. At 1013.24 mbar this maximum value (defined as 1) occurs at 3.98°C.

Compressibility.   Water can be compressed by only a very small amount. For the temperatures occurring in normal groundwater practice the isothermal coefficient of compressibility K_T is defined as the relative reduction in volume $-\Delta V/V$ for an increase in pressure Δp.
   The isothermal compressibility of water X_T is calculated from the relationship

$$X_T = \frac{C}{L + p} \tag{1}$$

in which C is a constant that varies with the material (for water $C = 0.1368$), L is a temperature-dependent constant, and p is the pressure. Table 4 gives values for X_T at 1013.24 mbar, and for L for different values of pressure.

Coefficient of Cubic Expansion (Dilatation). Changes in the volume of water occur corresponding to the variation in density at different pressures and temperatures. The coefficient of cubic expansion α at constant pressure is the relative increase in volume $\Delta V/V$ for a temperature change of ΔT. There is also a coefficient β to measure the relative change in pressure at constant volume $\Delta p/p$ per unit temperature change. The expansion V for temperatures T from -10 to $+110°C$ can be calculated from equation 2, due to Schoeller (1962).

$$V = 0.991\ 833 + 2.25208 \times 10^{-4}T + 2.84475 \times 10^{-6}T^2$$
$$+ 0.539\ 777 \times 10^{-6} \times 10^{1143/(273+T)} \qquad (2)$$

For temperatures between 0 and 25°C the approximate value can be found from equation 3 (Schoeller 1962).

$$V = 0.43668 + 0.002\ 005T + \frac{153{,}820}{273 + T} \qquad (3)$$

The coefficient of cubic expansion α is almost the same for all pressures at 48°C. Below 48°C it increases with pressure, and above 48°C it decreases with increasing pressure (Fig. 9). For an increase of about 1°C water temperature per 33 m (average geothermal gradient), starting from a temperature of

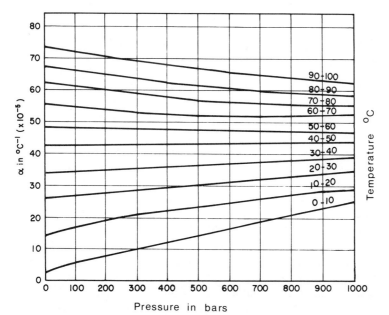

Fig. 9 Coefficient of cubic expansion α of liquid water at constant temperature (after S. Chenais & J. Chenais 1939).

10°C in the soil, the coefficient of expansion increases accordingly; hence the density also increases steadily with depth.

Viscosity. The internal resistance of a liquid to motion is known as its viscosity, which is usually measured in millipascal-seconds (mPa·s). There is a distinction between the dynamic viscosity μ and the kinematic viscosity $\nu = \mu/\rho$. The viscosity depends on intermolecular attraction, and is greatly influenced by temperature (Table 4 and Fig. 8). Small changes in pressure have only a negligible effect on viscosity.

Surface Tension. The surface separating water from air behaves as an elastic membrane that is fixed at the edges. This phase boundary between water and gas exerts a constant tension on the limiting edge and acts like a wall. The surface tension is that force which is exerted outward per unit length. Surface tension varies with the characteristics of the liquid and those of the phase boundaries, and is highly dependent on temperature (Table 4). The surface tension of water at 1013.24 mbar pressure and at 25°C is 71.97×10^{-3} N · m^{-1}. With increasing temperature it decreases rapidly, and it disappears completely at the critical temperature (Table 4 and Fig. 8).

1.1.4 Electrolytic Dissociation, Solubility Product, pH Value, Specific Electrical Conductance

Pure water dissociates to a small extent into H^+ and OH^- ions. At 24°C the product of the concentrations is 10^{-14}. The degree of dissociation is usually designated by the symbol pH, which is a measure of the hydrogen ion concentration expressed in terms of the logarithmic value of this concentration: $pH = -\log_{10} [H^+]$. The pH value of pure water at 24°C is therefore 7.00. A concentration of 10^{-5} mole H^+ per litre corresponds to a pH value of 5. The H^+ ion cannot exist in the free state as a subatomic particle, as a proton, but joins with water to form hydronium or hydroxonium ion H_3O^+. The dissociation is dependent on temperature (Table 5).

The electrical conductance is the reciprocal of electrical resistance. The unit is the siemens (S), formerly mho. The specific electrical conductance of water is the ability of a cube of the substance, with side measuring 1 cm, to conduct electrical current. It is dependent on temperature and on the type and concentration of the dissolved ions. It is usually defined at 25°C, so that differences in conductance are a function of the concentration and type of dissolved ions only. The specific electrical conductance permits a rapid evaluation of the chemical quality of the water sample (i.e., its total dissolved electrolyte content).

The specific electrical conductance of water in its purest state is $K = 4.2 \times 10^{-2}$ μS · cm^{-1}. This degree of purity cannot be maintained. Trace impurities − for example some CO_2 − raise the conductance so that the specific

Table 5 Temperature Dependence of pH Value of Pure Water
(Haberer 1969a)

Temperature (°C)	pH
0	7.47
5	7.37
10	7.27
15	7.17
20	7.08
24	7.00
25	6.99
30	6.92
35	6.84
40	6.77
45	6.70
50	6.63
55	6.57
60	6.51

electrical conductance of the purest water obtainable in practice is $K = 7 \times 10^{-1} \mu S \cdot cm^{-1}$, and that of common distilled water is $K = 0.5-5 \mu S \cdot cm^{-1}$.

1.2 SOLUTIONS

Water is the commonest, and because of its dipole nature, the most effective solvent for many types of solid, liquid, and gaseous substances. According to the particle size of the dissolved substance, solutions can be divided into "true" and "colloidal" types. In "true" solutions polar compounds mostly dissociate into ions, whereas nonpolar compounds occur as molecules. The dissolution process proceeds faster as the temperature increases, the greater the contact surface at the phase boundaries, and the greater the difference between the highest possible solution concentration and the actual concentration at the time. Finally the solution process is influenced by the type of removal and dispersion of the dissolved particles in the solution. If in practice it is dependent only on the slow effect of diffusion, then it is less than in stirred water, in which the solvent at the phase boundary is constantly being renewed.

1.2.1 Solubility of Gases

The solubility of gases in water depends on their type, the water temperature, the gas pressure or the partial pressures for mixtures of gases, and the total solute content. Solubility is governed by Henry's law; that is, at a given temperature the weight of gas dissolved in a given quantity is proportional to the pressure or partial pressure:

$$\lambda = K \cdot p \tag{4}$$

where λ is the solubility of the gas in litres per litre, p the pressure or partial pressure, and K is a constant dependent on temperature; investigators usually use the Bunsen absorption coefficient α, which refers to the volume of dissolved gas. Tables 6 and 7 give concentrations of various gases dissolved in water. The solubility of a gas at a given temperature is directly proportional to the pressure, and that of a gas mixture to its partial pressure.

Table 6   Solubility of Gases in Water at 0°C and 1013.24 mbar Partial Pressure (Alekin 1962, Haberer 1969a, Landolt–Börnstein 1923)

Gas	Solubility	
	ml/l	mg/kg
He	9.7	1.7
H_2	21.48	1.92
N_2	23.59	28.8
O_2	49.22	69.45
CH_4	55.63	39.59
Ar	57.8	100.7
NH_3	1300	1000
CO_2	1713	3346
H_2S	4690	7100

Henry's law does not hold for gases that react with water, such as ammonia and carbon dioxide. Because CO_2 becomes transformed to carbonic acid only to the extent of 1%, in this instance the deviation from Henry's law is quite small. Solubilities of gases in water are given in Table 6.

Gay-Lussac formulated an expression for the relationship between solubility λ and temperature of a gas:

$$\lambda = \frac{\alpha(1-T)}{273} \tag{5}$$

in which T is given in K. Therefore the solubility of gases generally decreases with increasing temperature.

Table 7 compares solubility coefficients of atmospheric nitrogen, oxygen, and carbon dioxide at different temperatures.

The solubility of gases in water is reduced in the presence of dissolved

Table 7   Solubility of Gases in Water (Landolt-Börnstein 1923)

Temperature (°C)	Atmospheric nitrogen (98.815 vol % N_2 + 1.185 vol % Ar)		Oxygen		Carbon dioxide	
	α^a	q^b	α	q	α	q
0	0.02354	29.42	0.04922	69.92	1.713	3346
1	0.02297	28.69	0.04788	67.98	1.646	3213
2	0.02241	27.98	0.04661	66.17	1.584	3091
3	0.02187	27.30	0.04540	64.39	1.527	2978
4	0.02135	26.63	0.04426	62.73	1.473	2871
5	0.02086	26.00	0.04317	61.15	1.424	2774
6	0.02037	25.37	0.04214	59.66	1.377	2681
7	0.01990	24.77	0.04115	58.22	1.331	2589
8	0.01945	24.19	0.04020	56.84	1.282	2492
9	0.01902	23.65	0.03929	55.52	1.237	2403
10	0.01861	23.12	0.03842	54.25	1.194	2318
11	0.01823	22.63	0.03759	53.04	1.154	2239
12	0.01786	22.16	0.03679	51.87	1.117	2165
13	0.01750	21.70	0.03603	50.75	1.087	2098
14	0.01717	21.26	0.03530	49.68	1.050	2032
15	0.01685	20.85	0.03459	48.63	1.019	1970
16	0.01654	20.45	0.03391	47.63	0.985	1903
17	0.01625	20.06	0.03326	46.67	0.956	1845
18	0.01597	19.70	0.03263	45.74	0.928	1789
19	0.01570	19.35	0.03203	44.85	0.902	1737
20	0.01545	19.01	0.03145	43.98	0.878	1688
21	0.01522	18.69	0.03090	43.15	0.854	1640
22	0.01498	18.38	0.03036	42.34	0.829	1590
23	0.01475	18.09	0.02985	41.57	0.804	1540
24	0.01454	17.80	0.02935	40.61	0.781	1493
25	0.01434	17.51	0.02887	40.08	0.759	1449

[a] α = Bunsen absorption coefficient, the dissolved volume of a gas within a certain volume of solvent at the given temperature (reduced to 0°C and 1013.24 mbar) when the partial pressure of the gas is 1013.24 mbar.
[b] q = Amount of gas (mg) which 1 kg of the pure solvent at the reference temperature will dissolve when the total pressure (partial + saturation vapor pressure of the liquid at the temperature of solution) is 1013.24 mbar.

solid substances (Table 8). According to Alekin (1962), the solubility of oxygen in a 4% salt solution is reduced by about 25%.

When water comes into contact with air, as occurs during precipitation and infiltration, and during contact between groundwater and ground air, equilibrium occurs, or is approached, corresponding to the partial pressures of (a) the gases in the pores and (b) the dissolved gases (Table 9). Wa-

Table 8 Solubility of Oxygen in Contact
with Air in Water of Varying Cl^-
Content at 15°C and 1013.24 mbar
(Haberer 1969a)

Chloride content (mg/1)	Dissolved oxygen (mg/1)
0	10.15
5,000	9.65
10,000	9.14
15,000	8.63
20,000	8.14

ters, which under high pressure are saturated with a gas (e.g., CO_2), give up this gas until equilibrium with the atmosphere is reached when the pressure is reduced. This holds especially for gases such as H_2S and H_2, also CO_2, which in the atmosphere have relatively very low partial pressures. Conversely, distilled water in contact with the atmosphere absorbs gases until equilibrium is reached.

Table 9 Solubility of Gas Mixtures (i.e., the atmosphere) and of Pure Gases in Water (Landolt-Börnstein 1923)

Temperature (°C)	Atmospheric gases (mg/1) in equilibrium with atmosphere at 1013.24 mbar			Atmospheric gases (mg/1) in equilibrium with pure gas concerned at 1013.24 mbar		
	N_2	O_2	CO_2	N_2	O_2	CO_2
0	23.4	15.4	1	29.42	69.92	3346
10	18.3	11.9	0.7	23.12	54.25	2318
20	15.1	9.6	0.5	19.01	43.98	1688
30	12.9	8.1	0.4	16.24	36.75	1257
50	9.7	5.9	0.2	12.16	26.68	761

1.2.2 Solubility of Solids and Liquids

In "true" solutions the dissolved substances are in the state of separated single molecules or ions, their dimensions ranging from 10^{-6} to 10^{-8} cm, and the solute is therefore quite transparent to transmitted light.

Substances that dissolve in water can be divided into electrolytes and nonelectrolytes. Examples of nonelectrolytes include sugar, urea, and glycerine, that is, substances that cannot conduct electric currents when in solution. Electrolytes, a class to which acids, bases, and salts belong, can dissociate into ionic form in dilute solutions, which carry the equivalent positive and negative charges corresponding to the elementary charge 1.6016×10^{-19} C or whole number multiples of it.

The number of the ionic charge on an ion corresponds to the ionic valence. Substances that in the solid state are built up from ions (e.g., salts) are held together by electrostatic forces between positively and negatively charged particles (ionic bonding). This force K is equal to the product of the charge of the cation e^+ and that of the anion e^-, divided by the square of the distance apart a:

$$K = \frac{e^+ \cdot e^-}{a^2} \tag{6}$$

A solvent decreases this force, and the greater its dielectric constant D, the more pronounced the decrease. The bonding force is therefore smaller:

$$K = \frac{1}{D} \cdot \frac{e^+ \cdot e^-}{a^2} \tag{7}$$

The unusually high dielectric constant of water (78.25 at 25°C, 81.5 at 17°C) explains its powerful solvent action. The dissolved molecules, above all the ions, form hydrates with water, in which water as a dipole bonds the ions or the positive or negative particles in the solution. Nonelectrolytes can also take part in the formation of hydrates if they are polarized (e.g., OH groups in organic compounds like alcohols and cane sugar). Unpolarized molecules, such as the hydrocarbons benzine, benzol, naphtha, and anthracene, are not truly soluble in water.

The solubility of individual substances is extraordinarily variable. Table 10 gives solubility data for a number of compounds commonly dissolved in water. The solubility of solid substances, with few exceptions, increases with increase in temperature, and it is practically independent of pressure.

The solubilities given in Table 10 for substances in pure water are approximations only, considering natural waters, because their solubilities are greatly influenced by other ions also present in the solution. The effect is most marked in the range of the respective saturation concentrations. In simple terms, the concept of the solubility product K_{SP} is derived from the law of mass action as follows. For the dissociation reaction

$$AB \rightleftharpoons \nu_A \, A^{z_A-} + \nu_B \, B^{z_B+}$$

the law of mass action is written

$$K_{AB} = \frac{[A^{z_A-}]^{\nu_A} \, [B^{z_B+}]^{\nu_B}}{[A \, B]} \tag{8}$$

where the brackets enclose the molar concentrations. In the general case,

Table 10 Solubilities of Some Inorganic Substances in Water at Different Temperatures (After Vogel 1956)[a]

Substance (Solid)	0°C	10°C	20°C	30°C	40°C	50°C	60°C	70°C	80°C	90°C	100°C	T.P.
$CaCl_2 \cdot 6\ H_2O$	603	650	745									29.8°C 1006
$CaCl_2 \cdot$				1020								
$CaCl_2 \cdot 2\ H_2O$						1323	1368	1417	1470	1527	1590	453°C 1302
$CaSO_4 \cdot 2\ H_2O$	1.76	1.925	2.036	2.10	2.122							42°C 2.1
$CaSO_4$											0.67	
$Ca(HCO_3)_2$	161.5		166		170.5		175		179.5		184.0	
$CaNO_3 \cdot 4\ H_2O$	1010	1153	1270									
$Ca(NO_3)_2 \cdot 3\ H_2O$					196	281.5						42.6°C 2390
KCl	281.5	313	343.5	373	403	431.0	456	483	510	534	562.0	
K_2SO_4	73.3	92	111.5	129.1	147.9	165	182	197.5	212.9	228	241.0	
$K_2CO_3 \cdot 1,5\ H_2O$	(1073)	(1090)	1115	1140	1170	1212	1270	1331	1400	1475	1560	
KNO_3	132.5	215	315	456	639	857	1099	1380	1680	2020	2452	
$MgCl_2 \cdot 6\ H_2O$	535.0		542.5	553	565	587	607		658.7		727	−3.4°C 523
$MgSO_4 \cdot 7\ H_2O$		300.5	356	408	454							1.8°C (267)
$MgSO_4 \cdot 6\ H_2O$							544					48.2°C (492)
$MgSO_4 \cdot 1\ H_2O$										515	(480)	67.5°C 58.7
$Mg(NO_3)_2 \cdot 6\ H_2O$	639		705		818		937		1109			−14.7°C 589
$Mg(NO_3)_2 \cdot 2\ H_2O$							2145		2330		2640	52.7°C (2080)
$NaCl \cdot 2\ H_2O$	356.0											0.2°C (356.0)
$NaCl$		357.0	358.5	361.5	364.2	367.2	370.5	375	385	387	392	
$Na_2SO_4 \cdot 10\ H_2O$	45.6	91.4	190.8	408.7								−1.25°C 42.6
$Na_2SO_4 \cdot (10\ H_2O)$					481							32.383°C 497
Na_2SO_4						466	452.6		430.9		423	
$Na_2CO_3 \cdot 10\ H_2O$	68.6	119.8	215.8	397	489	474						−2.05°C 60.6
$Na_2CO_3 \cdot 1\ H_2O$							462	452	445	445	445	35.37°C 495
$NaNO_3$	707		880		1049		1247		1480		1760	

[a] (In grams of anhydrous substance in 1000 g of water; T.P. = transition point for hydration in terms of temperature and concentration).

the concentration [AB] is variable. In the special case of a saturated solution, in which dissolved AB is in equilibrium with crystalline AB, the concentration of the undissociated molecule $[AB]_{sat}$ is constant for a given temperature. Fort this case equation 8 simplifies to (Wiberg 1971)

$$K_{SP} = K_{AB} [AB]_{sat} = [A^{z_{A}-}]^{\nu_{A}} \cdot [B^{z_{B}+}]^{\nu_{B}} \tag{9}$$

The solubility product as the product of the molar ionic concentrations for

Table 11   Solubility Products of Solids in Water at 25°C

Solubility product		Source[a]
$[Ag^+] \cdot [Cl^-]$	1.7×10^{-10}	(3)
$[Ag^+]^2 \cdot [S^{2-}]$	1×10^{-51}	(3)
$[Ba^{2+}] \cdot [CO_3^{2-}]$	8.1×10^{-9}	(4)
$[Ba^{2+}] \cdot [F^-]^2$	1.73×10^{-6}	(4)
$[Ba^{2+}] \cdot [SO_4^{2-}]$	1.08×10^{-10}	(4)
$[Ca^{2+}] \cdot [CO_3^{2-}]$	4.82×10^{-9}	(3)
$[Ca^{2+}] \cdot [F^-]^2$	2.95×10^{-11} (26°C)	(4)
$[Ca^{2+}] \cdot [SO_4^{2-}]$	6.1×10^{-5}	(3)
$[Ca^{2+}] \cdot [OH^-]^2$	7.9×10^{-6}	(3)
$[Ca^{2+}] \cdot [Mg^{2+}][CO_3^{2-}]^2$	4.7×10^{-20}	(2)
$[Cd^{2+}] \cdot [CO_3^{2-}]$	2.5×10^{-14}	(1)
$[Cu^{2+}] \cdot [CO_3^{2-}]$	1.37×10^{-10}	(1)
$[Cu^+]^2 \cdot [S^{2-}]$	2×10^{-47} (18°C)	(1)
$[Fe^{2+}] \cdot [CO_3^{2-}]$	2.11×10^{-11}	(5)
$[Fe^{2+}] \cdot [OH^-]^2$	1.65×10^{-15}	(3)
$[Fe^{3+}] \cdot [OH^-]^3$	4×10^{-38}	(3)
$[Fe^{2+}] \cdot [S^{2-}]$	4×10^{-19}	(5)
$[Hg^+]^2 \cdot [CO_3^{2-}]$	9×10^{-17}	(1)
$[Hg^+] \cdot [Cl^-]$	2×10^{-18}	(1)
$[Mg^{2+}] \cdot [CO_3^{2-}]$	1×10^{-5}	(3)
$[Mg^{2+}] \cdot [OH^-]^2$	5.5×10^{-12}	(3)
$[Mn^{2+}] \cdot [OH^-]^2$	7.1×10^{-15}	(3)
$[Ni^{2+}] \cdot [OH^-]^2$	1.6×10^{-14}	(1)
$[Ni^{2+}] \cdot [S^{2-}]$	1×10^{-26} (20°C)	(1)
$[Pb^{2+}] \cdot [CO_3^{2-}]$	1.5×10^{-13}	(3)
$[Pb^{2+}] \cdot [S^{2-}]$	3.4×10^{-28} (18°C)	(1)
$[Ra^{2+}] \cdot [SO_4^{2-}]$	4.25×10^{-11} (20°C)	(1)
$[Sb^{3+}] \cdot [OH^-]^3$	4×10^{-42}	(1)
$[Sn^{2+}] \cdot [OH^-]^2$	5×10^{-26}	(1)
$[Sn^{4+}] \cdot [OH^-]^4$	1×10^{-56}	(1)
$[Sr^{2+}] \cdot [CO_3^{2-}]$	1.6×10^{-9}	(4)
$[Sr^{2+}] \cdot [SO_4^{2-}]$	2.8×10^{-7}	(1)
$[Zn^{2+}] \cdot [CO_3^{2-}]$	6×10^{-11}	(1)
$[Zn^{2+}] \cdot [S^{2-}]$	1.1×10^{-24}	(1)

[a] D'Ans & Lax 1967 (1), Garrels et al. 1962 (2), Haberer 1969a (3), Hodgman et al, 1958 (4), Latimer 1953 (5).

saturated solutions of low solubility salts at a given temperature is constant. The solubility product generally increases with temperature. For substances that exhibit phase changes at certain temperatures, the product decreases at the inversion temperature, for example when gypsum $CaSO_4 \cdot 2H_2O$ inverts to anhydrite $CaSO_4$ at 60°C. Solubility product also increases with rising concentration of the solution. This effect is especially noticeable in highly concentrated mineralized water and NaCl brines, in which the concentration can reach very high values.

Table 11 lists the solubility products of important major and trace constituents of natural water. The values given allow calculations of equilibrium concentrations of ions in terms of dependence of the concentration on other ions in solution to be made.

For example, iron(II) hydroxide solution has twice as many hydroxyl ions as iron(II) ions:

$$[Fe^{2+}] = 2[OH^-]$$

Hence for the solubility product

$$L = [Fe^{2+}]\left[\frac{1}{2} Fe^{2+}\right]^2 = 1.65 \times 10^{-15}$$

$$[Fe^{2+}] = \sqrt[3]{\frac{1.65 \times 10^{-15}}{4}} = 0.744 \times 10^{-5}$$

The concentration (mg/1) is obtained by multiplying by the molecular weight and by 1000.

$$Fe^{2+} = 0.744 \times 10^{-5} \times 55.85 \times 1000 = 0.42 \text{ mg/l}$$

The addition of another iron salt or OH^- ion decreases the solubility of the salt because of the phenomenon known as common ion effect (Le Chatelier's principle). For example if the hydroxyl content rises to 10^{-4} (pH 10) the iron content will decrease:

$$[Fe^{2+}][10^{-4}]^2 = 1.65 \times 10^{-15}$$

$$[Fe]^{2+} = \frac{1.65 \times 10^{-15}}{10^{-8}} = 1.65 \times 10^{-7}$$

$$Fe^{2+} = 1.65 \times 10^{-7} \times 55.85 \times 1000 = 9.25 \times 10^{-3} \text{ mg/l}$$

Colloidal Solutions. Colloidal solutions contain particles in the size range 10^{-6}–10^{-4} cm that can be observed in an ultramicroscope because of the diffraction patterns they cause. An incident light ray stands out visibly in the dispersion (Tyndall effect). The colloids in these dispersons bear an electric charge that originates through three principal causes. (1) Individual molecules on the surface of a particle can dissociate; thus H^+ or OH^- ions can be separated off and the colloids are left with positive or negative charges, respectively. (2) Surface charge may be caused by lattice imperfections at the particle surface or by isomorphous replacement within the lattice. (3) Positive or negative ions can be adsorbed from the liquid, and these will charge the particles accordingly (Stumm & Morgan 1970).

In natural waters (i.e., pH 5–9), colloids of ferric hydroxide are positively charged and occur as relatively stable colloidal dispersions in surface waters with concentrations of a few hundredths of a milligram per litre. Examples of colloids that are generally negatively charged are silica, sulfur, sulfide, MnO_2, and clay minerals.

The occurrence of highly charged ions generally leads to the breakdown of the colloidal state and causes flocculation of the colloids. In the opposite sense, the effect of charged substances is known as peptization.

Loss of water can turn a colloidal dispersion into a gel, which in respect of many physical properties (e.g., constancy of state) stands between solids and liquids, as seen in gels of metal hydroxides and silica.

Emulsions and suspensions contain larger, more widely dispersed particles, which occur within the water as a few large molecules. Thus emulsions and suspensions are transitional to colloidal dispersions.

Solubility of Organic Substances. Whereas organic acids and bases can form true solutions as electrolytes, many nonelectrolytic organic substances occur in nature in contact with water. Thus, for example, there are the crude oils and the products derived from them, especially paraffin and naphtha hydrocarbons. For mixtures of this type, which include a greater or smaller number of individual chemical substances, no clearly defined solubility can

Table 12   Saturation Concentrations[a]

Type of mineral oil product	Saturation concentration in water (mg/l)	
	Based on Committee "Wasser und Mineralöl" (1969)	Based on Steck (1971)
Benzole	1700	—
Light benzene	—	60
Gasoline	50–500	—
Gasoline, normal	—	164
Gasoline, super	—	380
Diesel fuel	10–50	4
Heating oil, extra light	10–50	—
Kerosene	0.1–5	9
Petroleum	0.1–5	—

be given. This is because the experimentally obtained solubility values always depend on the relative amounts of the components, water and substance, under test; the various true solubilities of the individual compounds in the mineral oil phase interact because of the experimental conditions. Hence the concept of saturation concentration is useful; that is, we consider that the mineral oil is brought into contact with water in such a large quantity that its composition is not significantly altered. The quantities in Table 12 are mean values for several similar substances at 20°C.

Figure 10 shows the true solubility values of various organic compounds derived from mineral oil. With increasing molecular volume, the solubility of all groups of compounds decreases. The aromatics have the highest solubility and paraffin the lowest. The solubility increases with temperature.

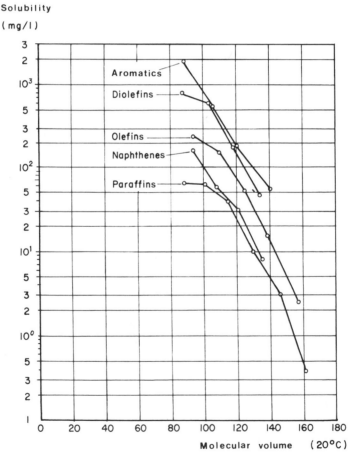

Fig. 10 Relationship between solubility and molecular volume (molecular weight divided by density) (after McAuliffe 1966, and Committee "Wasser und Mineralöl" (1969).

1.2.3   Factors Causing Solubility Changes

A state of chemical equilibrium exists or is approached between the substances dissolved and those still in the solid state, but this can be shifted by variations in the concentration of the solutes or by the addition of other soluble substances.

1.2.3.1   Ionic Activity.

In nonideal solutions the ions present in not infinitely dilute solution influence other ion species through the electric fields surrounding them, so that the latter appear to be present in smaller amounts. This factor is taken into consideration in the concept of ionic strength, a quantity proposed by Lewis & Randall (1921) for ideal solutions.

The effect of dissolved electrolytes on the solubility of a solid is remarkably uniform considering the charge and molar concentrations of the ions concerned. In sufficiently dilute solutions the reciprocal interionic interaction is controlled by the number of ions and the charge on them.

For the ionic strength I the empirical equations holds:

$$I = \frac{1}{2} \sum_i C_i \cdot z_i^2 \tag{10}$$

The molar ionic strength is calculated in terms of molar concentration C_i (in mole/l (litre-molarity), in mole/kg solvent (kilo-molarity) and valency z_1. The calculation is valid only for dilute solutions, and a good approximation is achieved by using the concentrations of the most important ion species found by analysis. The analysis of groundwater from the Friedberg area in Upper Hesse serves as an example of this type of calculation (Tables 13, 79) (Scharpff 1972).

Table 13   Analysis of Groundwater from Upper Hesse

Concentration	Ions						
	Na^+	K^+	NH_4^+	Mg^{2+}	Ca^{2+}	Mn^{2+}	Fe^{2+}
mg/kg	8.26	1.17	0.12	25.5	84.3	0.03	0.24
mmole/kg	0.3593	0.0299	0.0066	1.0485	2.1033	0.0005	0.0043
	Cl^-	NO_3^-	HCO_3^-	HPO_4^{2-}	SO_4^{2-}		
mg/kg	12.8	2.61	357.5	0.02	22.2		
mmole/kg	0.3610	0.0421	5.858	0.0002	0.2311		

The values given in the first row are divided by the respective molecular weights to give the kilo-molarity values in the second row. An error, though very small, enters because of the difference between the kilo-molarity (mole/

kg solvent) and the analysis results (parts by weight/kg solution). The ionic strength is calculated thus:

$$I = \frac{1}{2}(0.3593 \times 1^2 + 0.0299 \times 1^2 + 0.0066 \times 1^2 + 1.0485 \times 2^2 + 2.1033 \times 2^2$$
$$+ 0.0005 \times 2^2 + 0.0043 \times 2^2 + 0.3610 \times 1^2 + 0.0421 \times 1^2 + 5.858 \times 1^2$$
$$+ 0.0002 \times 2^2 + 0.2311 \times 2^2) \times 10^{-3}$$
$$= 10.105 \times 10^{-3}$$

In waters that contain only mono- or divalent ions, the ionic strength is calculated to sufficient accuracy from the equation

$$I = 0.5C_1 + 2C_2 \tag{11}$$

Where C_1 is the sum of the concentrations of all the monovalent ions and C_2 that of all the divalent ions in mole/kg. In the example above, based on Table 13, the values are: $C_1 = 6.6569 \times 10^{-3}$ mole/kg and $C_2 = 3.3879 \times 10^{-3}$ mole/kg. Hence the ionic strength is $I = (3.328 + 6.775) \times 10^{-3} = 10.103 \times 10^{-3}$.

Axt (1962) gives another approximation for waters in which the solutes consist of mono- and divalent ions:

$$I = 2.5 \times 10^{-5} \times R \tag{12}$$

in which R is the residue after evaporation in mg/l. The error involved in the application of this expression is generally less than 20%. When applied to the analysis above (Table 13), the calculated ionic strength is:

$$I = 2.5 \times 10^{-5} \times 350.5 \cong 8.8 \times 10^{-3}$$

Garrels (1960) found an ionic strength of 0.1082 for a sample of mineral water with 7130 mg/l dissolved solids, corresponding approximately to the average ionic strength of the subsurface water. Normal mineralized groundwater, rivers, and lakes usually reach approximately one-tenth of this value, while the ionic strength of seawater is about 10 times higher.

For application of the laws of solubility derived from the law of mass action for ideally dilute solutions, for example natural water, the activities are used instead of the measured concentrations.

The activity of a dissolved substance is a relative quantity that is derived from consideration of a solution approaching infinite dilution, for which the ratio of the activity to the kilo-molarity is equal to unity. The activity of solids and pure liquids is unity, that of water in dilute aqueous solutions is approximately unity. The activity of a gas in the lower pressure region (a few bars) and at 25°C, where it behaves as an ideal gas, is equal to the partial pressure; expressed in terms of mole fraction X_{gas}; the mole fraction is the quotient of the partial pressure P_1, and total pressure P_t:

$$X_{gas} = \frac{P_1}{P_t} \tag{13}$$

For ideal solutions the activity of a dissolved species is proportional to its concentration expressed in terms of the mole fraction (equations 13, 14). For the mole fraction X_i of a solution

$$X_i = \frac{n_1}{n_1 + n_2 + n_3 + \cdots + n_i} = \frac{n_1}{\sum_i n_i} \tag{14}$$

dissolved substances and the solvent are included. The mole fraction of NaCl in a $1M$ solution is:

$$X_{NaCl} = \frac{1}{1 + 55.51} = 0.0177$$

Hence 1 kg water contains 55.51 mole H_2O.

Only in very dilute solutions is the activity of a dissolved substance numerically almost equal to its molar concentration. Naturally, aqueous solutions in which the solutes are mostly polarized compounds do not behave at all like ideal solutions. The ions of the dissociated solids interact with each other and with the water.

The activity of a certain molecule or ion is designated by the symbol a in conjunction with the chemical symbol for the molecule or ion, aCa^{2+}, or the chemical symbol is enclosed in square brackets, $[Ca^{2+}]$.

The activities of dissolved substances and of the solvent can be determined by measurement of the solubility, the dissociation constants, the vapor pressure, the freezing point, and boiling point, as well as by measurement of the electromotive force of galvanic cells (Kortüm 1966b). However, the activities are more easily calculated from the analysis values by multiplying by the activity coefficient f_i; this coefficient is defined as the ratio of the ionic activity to the molar concentration:

$$f_i = \frac{a_i}{m_i} \tag{15}$$

Numerical values for activity coefficients of numerous substances over a wide range of concentrations are given by Haberer (1969a), Harned & Owen (1958), Klotz (1950), and Latimer (1953).

For dilute solutions, the activity coefficients of individual ions can be obtained to sufficient accuracy from equation 16, which is based on the Debye-Hückel theory of interionic interaction, as expressed by the equation

$$\log f_i = \frac{A z_i^2 \sqrt{I}}{1 + \mathring{a}_i B \sqrt{I}} \tag{16}$$

in which A and B are pressure- and temperature-dependent constants specific to the solvent, z_i is the valency of the ion in the solution specified, I is the ionic strength, and $\mathring{a}_i$ is a term dependent on the effective diameter of the ion in the solution. Experimentally determined values for $\mathring{a}_i$ are given in Table 14, and values for A and B as functions of temperature in Table 15.

The values A and B are related to the unit volume of the solution, and the values A' and B' to unit weight of the solvent. The values for $\mathring{a}_i$ are greater than the ionic radii for crystals, possibly as the consequence of the tendency for water dipoles to orient themselves around ions in the form of hydration sheaths. The observed $\mathring{a}_i$ values have not so far led to any definite description of the arrangement of the water molecules in the structure of the layers around the dissolved ions (Robinson & Stokes 1949).

The following equation is an adequate approximation of the negative logarithm of the activity coefficient f_i (Haberer 1969a):

$$-\log f_i = \frac{1.815 \times 10^6}{(DT)^{3/2}} \cdot z_i^2 \sqrt{I} \tag{17}$$

in which z_i = valency of the ion
$\qquad D$ = dielectric constant
$\qquad T$ = absolute temperature of the water

At room temperature the coefficient $(1.815 \times 10^6)/(DT)^{3/2} \cong 0.50$. For strong electrolytes at room temperature the following relationship holds between the mean ionic activity coefficients and the ionic strength (Alekin 1962):

$$-\log f_i = \frac{0.5 z_i^2 \sqrt{I}}{1 + 0.33 \times 10^8 \mathring{a}_i \sqrt{I}} \tag{18}$$

in which z_i = valency of the ion
$\qquad \mathring{a}_i$ = approximate diameter of the ion (between 10^{-8} and 5×10^{-8} cm)

Equation 19 holds for the calculation of the activity coefficients for aqueous solutions for ionic strengths up to 0.2, according to Alekin (1962).

$$-\log f_i = 0.5 z_i^2 \frac{\sqrt{I}}{1 + \sqrt{I}} \tag{19}$$

Finally Alekin (1962) gives another empirical relationship:

$$-\log f_i = 0.298 z_i^2 \sqrt{2I} \tag{20}$$

For weak electrolytes the concentration of each molecule or ion pair must be multiplied by its degree of dissociation, the fraction of the molecules dissoci-

Table 14   Values for $\mathring{a}_i$ for Ions in Aqueous Solution (Klotz 1950)

$\mathring{a}_i \times 10^8$	Ion
2.5	Rb^+, Cs^+, NH_4^+
3.0	K^+, Cl^-, Br^-, I^-, NO_3^-, NO_2^-, CN^-
3.5	OH^-, F^-, HS^-
4.0	Hg^{2+}, SO_4^{2-}, CrO_4^{2-}, HPO_4^{2-}, PO_4^{3-}
4.0–4.5	Na^+, HCO_3^- $H_2PO_4^-$ HSO_3^- $H_2AsO_4^-$
4.5	Pb^{2+}, CO_3^{2-}, SQ_3^{2-}
5.0	Sr^{2+}, Ba^{2+}, Ra^{2+}, Cd^{2+}, S^{2-}
6	Li^+, Ca^{2+}, Cu^{2+}, Zn^{2+}, Sn^{2+}, Mn^{2+}, Fe^{2+}, Ni^{2+}, Co^{2+}
8	Mg^{2+}
9	H^+, Al^{3+}, Cr^{3+}, Fe^{3+}
11	Sn^{4+}

Table 15   Values of Constants for Use in the Debye-Hückel Equation (aqueous solution) (after Manov et al. 1943)

Temperature (°C)	A	A'	B ($\times 10^{-8}$) (cm^{-1})	B' ($\times 10^{-8}$) (cm^{-1})
0	0.4883	0.4883	0.3241	0.3241
5	0.4921	0.4921	0.3249	0.3249
10	0.4961	0.4960	0.3258	0.3258
15	0.5002	0.5000	0.3267	0.3266
20	0.5046	0.5042	0.3276	0.3273
25	0.5092	0.5085	0.3286	0.3281
30	0.5141	0.5130	0.3297	0.3290

ated into ions. Dissociation constants are tabulated in chemistry textbooks. The degree of dissociation α is a measure of the strength of the electrolyte. Weak electrolytes have α-values that vary greatly with the dilution and approach the limiting value zero, whereas strong electrolytes, which have only a small degree of dependence on α, regularly have values approaching unity.

Equations 18–20 lead to quite accurate calculations for ionic concentrations up to 50 meq/l, and sufficiently accurate values for concentrations up to 100 meq/l (Alekin 1962).

The relationships between the activity coefficient of dissolved mono-, di- and trivalent ions and the ionic strength of the solution are shown in Fig. 11. Dissolved molecules such as H_2S and H_2CO_3 have activity coefficients close to unity in solutions up to an ionic strength of approximately 0.5 ($fH_2CO_3 =$ 1.10). The activity coefficients of dissolved gases can be found from the equation (Garrels 1960).

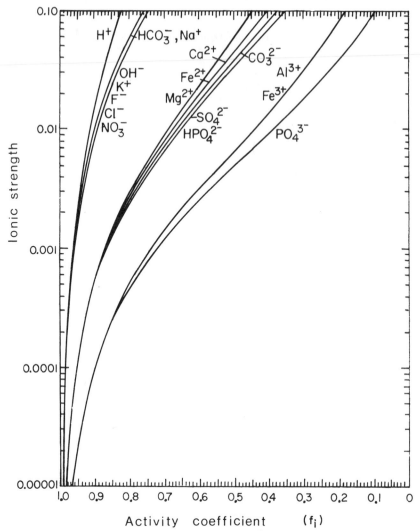

Fig. 11 Relationship between the activity coefficients of dissolved ions and the ionic strength of the solution (after Hem 1961b, 1970).

$$f_{i \text{ gas}} = \frac{C_0}{C_s} \cdot \frac{P_s}{P_0} \tag{21}$$

in which P_s = vapor pressure of the solution at the given ionic strength

P_0 = vapor pressure of pure water

C_0 = absorption coefficient of the gas in pure water (gas volumes at standard temperature and pressure dissolved in unit volume of water at a partial pressure of 1013.24 mbar above the solution)

C_s = absorption coefficient in the solution at the given ionic strength

34

Solubilities may be calculated from the activities of the dissolved forms of a given element (ion and molecules) if all the types containing the element concerned are contained in the result of the chemical analysis, and when the relationships between the concentration of a given type of solute and its activity are known. It is often possible to calculate one useful minimum solubility for the purposes of water chemistry.

The concentration of a given element A in an aqueous solution, as determined by quantitative analysis can be described as:

$$m_A = \sum m_S = \sum \frac{a_s}{f_{is}} \qquad (22)$$

in which m_A = kilo-molarity, obtained as the sum of the kilo-molarities m_S
           of all the species that contain the element A
     a_s = activity
     f_{is} = activity coefficient

For example, a sample of water under a given partial pressure of CO_2 is saturated and the content of the dissolved CO_2 is known; the important dissolved constituents are H_2CO_3, HCO_3^-, and CO_3^{2-}. Hence by neglecting some species present in very small amounts, the limiting value below is found.

$$mA(CO_2) = \sum m_S = mH_2CO_3 + mHCO_3^- + mCO_3^{2-} \qquad (23)$$

The term $mA(CO_2)$ is calculated approximately from the activities of the ion types H_2CO_3, HCO_3^-, and CO_3^{2-}, which can be found from the known partial pressure of CO_2, and from the known activity coefficients fH_2CO_3, $fHCO_3^-$, and fCO_3^{2-}, using the expression

$$mA(CO_2) = \frac{aH_2CO_3}{fH_2CO_3} + \frac{aHCO_3^-}{fHCO_3^-} + \frac{aCO_3^{2-}}{fCO_3^{2-}} \qquad (24)$$

The accuracy of this approximation is also affected by the errors in the determinations of the activities and the activity coefficients. A good approximation to the solubility of a given element generally requires accurate data for all the dissolved molecules and ions that contribute to the solubility, and accurate values of the activity coefficients. Otherwise the calculated value is significantly less than the measured value. In highly concentrated solutions, particularly with concentrations above $1M$, the activity a_i can exceed the molarity m_i, so that the calculated value is higher than the observed value.

Because in the general chemical analyses of natural waters not all the dissolved ions and molecules are determined, the species that enters into the calculations depends on the knowledge and experience of the worker. In dilute inorganic aqueous solutions the dissolved components tend to form individual free ions or molecules. In concentrated aqueous solutions (for ionic strengths above 0.1, but in many cases also for much lower values) the ions, instead of being widely separated in dilute solutions, tend to interfere with one another. Hence new types originate alongside the free ions, and apparently maintain their individuality in the solution. These complex ions can be treated as thermodynamic entities and are in dynamic equilibrium with the free ions. The equilibrium can be expressed by a thermodynamic equilibrium constant, the association constant, or the reciprocal of this, the dissociation constant (Garrels & Christ 1965). Complexes are combinations of cations, which are called the central atoms, with molecules or anions containing free pairs of electrons (bases), which are referred to as ligands. Bases containing more than one ligand atom (e.g., oxalate and citrate) are referred to as multidentate complex formers. Complex formation with multidentate ligands is called chelation, and the complexes are called chelates. Chelates are usually more stable than complexes with monodentate ligands.

Two types of complex species can be distinguished: ion pairs and complexes in the narrow sense. In the ion pairs water molecules are retained by the metal ion or the ligand or both, so that the metal ion and the base are separated by one or more water dipoles. In complexes in the narrow sense, bases and ligands are immediately adjacent (Stumm & Morgan 1970). Garrels & Christ (1965) have pointed out that in natural waters complexing of the major dissolved species H^+, Na^+, K^+, Ca^{2+}, Mg^{2+}, H_2CO_3, HCO_3^-, CO_3^{2-}, Cl^-, SO_4^{2-}, H_4SiO_4, and $H_3SiO_4^-$ is unimportant above the level of the ion pair at low temperatures except in dense brines. For further information concerning complexes, see Davies (1962), Garrels & Christ (1965), and Stumm & Morgan (1970).

The activity of a complex is usually a power function of one or more factors. A few complexes are highly active at low temperatures; others do not have an effect until a critical concentration value of the ions has been reached. The increasing appearance of complexes with increasing concentration prevents the calculation of the ratios of ion activities to the total concentration in highly concentrated solutions or waters with a high ionic content.

Natural waters often contain organic compounds. Some of these organic compounds incorporate metal cations, so that active substances in this category must be taken into account in the calculation. Natural organic substances of microbial origin, such as tartaric acid, citric acid, catechol, and salicylic acid (Scheffer & Schachtschabel 1976) may yield stable soluble complexes.

Water-insoluble complexes (e.g., insoluble humic acids) may irreversibly fix cations, which will occupy two-thirds of the positions of the total binding capacity of about 200–600 meq metal ion per 100 g of humic acid (Förstner & Müller 1974).

The solubility of an element in natural waters can also be determined from the activities of all the dissolved species in equilibrium with the stable solids and from the ratios of the activities of the dissolved species to their concentrations. The activities of all known species can be calculated from the thermodynamic data described above. Errors of calculation originate through incomplete analytic data, especially for concentrated solutions of mixed electrolytes, such as brines. Conversely, the difference between the calculated solubility and the measured value makes it possible to estimate the degree of complex formation.

The activity values in dilute solutions are therefore always rather lower than the actually measured concentrations, on account of the interionic forces, the combination with the molecules of the solvent, and other phenomena. With increasing total concentrations in natural waters, these differences also increase, at least up to a concentration of approximately 5 g/l of dissolved solids. In waters that have concentrations lower than about 100 mg/l dissolved solids, the activity and the measured concentration deviate from each other by generally not more than 10–15%. For concentrations of approximately 1000 mg/l, however, the activity of divalent ions such as Fe^{2+} can amount to only about half the measured concentration values (Hem 1960a, 1961a). For most waters, therefore, activity corrections have to be applied to the measured concentrations to obtain satisfactory accuracy for the calculated relationship. The pH values measured with a glass electrode give the hydrogen ion activity directly ($-\log a_H = pH$) and need no correction. For equilibrium calculations, which provide the accurate determination of the concentrations of individual substances, the activity correction has to be applied in all cases.

Important equilibrium systems include carbonate–carbon dioxide equilibrium, particularly the calcium carbonate–carbon dioxide equilibrium, the buffer systems of natural waters, which keep the pH value almost constant, and the processes dependent on the redox potential and pH value. These systems can be described with sufficient accuracy by means of equilibrium models based on the principle of the law of mass action and the laws of chemical thermodynamics. An introduction to the thermodynamic treatment of equilibrium systems can be obtained from relevant textbooks, for example Garrels & Christ (1965), Glasstone (1947), Krauskopf (1969), and Stumm & Morgan (1970).

For compound systems the thermodynamic state depends, apart from pressure and temperature, on the molar concentration of all the substances present. Chemical reactions are possible in such an equilibrium system only if temperature or pressure, or both, vary simultaneously. Conversely, such changes are directly bound up with a chemical reaction. Thus chemical changes can be observed only as independent variables in a "closed equilibrium system," that is, in a system that is in an apparent state of thermodynamic equilibrium, from which it will depart when the restraints are removed or temperature and/or pressure changes, which will occur only insofar as the heat of reaction is removed or supplied. The law of mass action describes the equilibrium

$$\nu_A A + \nu_B B + \cdots \rightleftharpoons \nu_C C + \nu_D D + \cdots$$

in terms of a constant that is dependent on the given temperature and pressure:

$$K = \frac{[a_C]^{\nu_C} \cdot [a_D]^{\nu_D}}{[a_A]^{\nu_A} \cdot [a_B]^{\nu_B}} \qquad \text{(see equation 8)}$$

This thermodynamic equilibrium constant is accordingly calculated as a product of the activities of the reaction products divided by the product of the reactants raised to the power indicated by their numerical coefficients.

The heat of reaction at constant volume results from the change in the internal energy of a system $\Delta H = \Sigma \nu_i H_i$. The heat of reaction indicates the heat that is exchanged with the surroundings during an irreversible direction of reaction. When ΔH is positive, heat must be supplied from the surroundings so that the temperature will remain constant during the reaction (endothermic reaction); for negative ΔH, on the other hand, heat is given off (exothermic reaction).

To define the heat of formation for all the reactants, it is necessary to employ a specified standard state: for gases this is the state of the ideal gas, for solids and liquids it is usually the state of the pure phase at 1013.24 mbar and 25°C. The important concept of free energy for chemical and geochemical purposes is the difference between the internal energy H and the product of temperature and the amount of the internal disorder, the entropy S. The free energy is also known as the Gibbs free energy or the Gibbs function (Krauskopf 1969). In publications in English the symbol F has often been used instead of G.

$$G = H - TS$$

At constant temperature T (°K),

$$\Delta G = \Delta H - T \Delta S$$

The free energy can be thought of as the energy that appears in an isothermal chemical reaction. Reactions in which the free energy in the reaction products is less than that in the original substances will probably proceed spontaneously unless retarded. The free energy in the original materials at chemical equilibrium is equal to that in the reaction products, and there exists a close relationship between the change of free energy and the equilibrium constants of the law of mass action.

The standard free energy change $\Delta G°$ is the energy of reaction in kilojoules required for the formation of 1 mole of the product from the reactants in the standard state at 25°C and 1013.24 mbar total pressure. Under these standard conditions the standard state of a reactant is stable at the time (e.g., Fe_c, O_{2g}, Hg_l; for explanation of the subscripts, see note a Table 16). The standard free energy change of an element is by definition zero. An activity value of unity is defined for the standard state of a substance dissolved in water. The standard free energy is the change in energy that occurs when the elements form a new dissolved substance under the standard conditions.

Table 16 gives values of the standard free energy change for important substances in water. Additional values can be found in *Enthalpies Libres de Formation Standards, à 25 °C* (1955), Garrels & Christ (1965), Latimer (1953), Perry (1950), Robie et al. (1978), and Rossini et al. (1952).

The standard free energy of a reaction is therefore the sum of the free energy changes of the reaction products in the standard state minus the sum of the free energy changes of the reactants, also in the standard state.

$$\Delta G° = \Sigma \Delta G°_{\text{reaction products}} - \Sigma \Delta G°_{\text{reactants}} \tag{25}$$

The free energy change for a chemical reaction under standard conditions can be calculated from the $\Delta G°$-values of the various substances involved. For example, consider the solution of calcite by the action of H^+ ions: $CaCO_{3c} + H^+ \rightarrow Ca^{2+} + HCO_3^-$, with the following $\Delta G°$-values.

Ion	$\Delta G°$ (kJ/mole)
$CaCO_{3c}$	−1129.51
H^+	0.0
Ca^{2+}	−553.41
HCO_3^-	−587.45

The energy change per mole of reaction products is therefore

$$\Delta G° = (-553.41 - 587.45) - (-1129.51 + 0.0) = -11.35 \text{ kJ/mole}$$

The negative sign shows that energy is released in the reaction and that the reaction will probably proceed spontaneously. It is not possible, however, to predict the speed of the reaction with this method. The standard free energy change is related to the equilibrium constants thus:

$$\Delta G° = -RT \ln K \tag{26}$$

where R is the gas constant 8.315 $J·°C^{-1}·mole^{-1}$ and T is the absolute temperature in K.

For logarithms to the base 10, and at 25°C and 1013.24 mbar, this reduces to:

$$\Delta G° = -1.364 \log K \tag{27}$$

Table 16 **Free Energy Change Values of Important Substances**

Substance	Description	State[a]	$\Delta G^{\circ b}$	Source[c]
Ca	Metal	c	0.0	(3)
Ca^{2+}		aq	−553.41	(4)
CaO		c	−603.487	(5)
CaO_2		c	598.7	(3)
$Ca(OH)_2$		c	−898.408	(5)
CaF_2		c	−1176.920	(5)
$CaCl_2$		aq	−815.92	(4)
CaS		c	−477.7	(4)
$CaCO_3$	Calcite	c	−1128.842	(5)
$CaCO_3$	Aragonite	c	−1127.793	(5)
$CaMg(CO_3)_2$	Dolomite	c	−2161.672	(5)
$Ca(HCO_3)_2$		aq	−1728.31	(4)
$CaSiO_3$	Pseudowollastonite	c	−1544.955	(5)
$CaSiO_3$	Wollastonite	c	−1549.903	(5)
Ca_2SiO_4	β	c	−2146.6	(3)
Ca_2SiO_4	γ	c	−2150.8	(3)
$CaSO_4$	Anhydrite	c	−1321.696	(5)
$CaSO_4$	α	c	−1312.65	(4)
$CaSO_4$	β	c	−1308.21	(4)
$CaSO_4$		aq	−1295.90	(4)
$CaSO_4 \cdot \frac{1}{2}H_2O$	α	c	−1436.16	(4)
$CaSO_4 \cdot \frac{1}{2}H_2O$	β	c	−1435.15	(4)
$CaSO_4 \cdot 2H_2O$		c	−1796.93	(4)
$Ca_3(PO_4)_2$	α	c	−3892.5	(4)
$Ca_3(PO_4)_2$	β	c	−3902.1	(4)
$CaHPO_4$		c	−1681.0	(4)
$CaHPO_4 \cdot 2H_2O$		c	−2154.5	(4)
$Ca(H_2PO_4)_2$	Precipitated	c	−2813.5	(3)
$CaWO_4$		c	−1543.7	(3)
C	Diamond	c	2.8680	(4)
C	Graphite	c	0.0	(4)
CO		g	−137.171	(5)
CO_2		g	−394.375	(5)
CO_2		aq	−386.48	(4)
CH_4		g	−50.708	(5)
C_2H_2		g	−209.3	(4)
H_2CO_3		aq	−623.83	(4)
HCO_3^-		aq	−587.45	(4)
CO_3^{2-}		aq	−528.46	(4)
COS		g	−169.36	(4)
CS_2		g	65.10	(4)
CF_4		g	−635.56	(4)
Cl		aq	−131.256	(4)
Cl_2		g	0.0	(3)
Cl_2		aq	6.91	(3)
HCl		g	−95.299	(5)

Table 16 (*Continued*)

Substance	Description	State[a]	$\Delta G^{\circ b}$	Source[c]
HCl		aq	−131.256	(4)
Fe		c	0.0	(3)
Fe^{2+}		aq	−84.99	(4)
Fe^{3+}		aq	−10.55	(2)
FeO	Wüstite	c	−245.155	(5)
Fe_2O_3	Hematite	c	−742.683	(5)
Fe_3O_4	Magnetite	c	−1012.566	(5)
$Fe(OH)^{2+}$		aq	−234.08	(4)
$Fe(OH)_2$		c	−483.87	(4)
$Fe(OH)_2^+$		aq	−444.6	(4)
$Fe(OH)_3$		c	−695.0	(3)
$FeCl^{2+}$		aq	−150.3	(4)
FeO_2H^-		aq	−379.4	(1)
FeS	α	c	−101.333	(5)
FeS_2	Pyrite	c	−160.229	(5)
$FePO_4$		c	−1138.81	(3)
$FeCO_3$	Siderite	c	−674.33	(4)
FeSe	Precipitated	c	−58.2	(3)
$FeSiO_3$		c	1076.0	(3)
Fe_2SiO_4		c	−1379.375	(5)
$FeMoO_4$		c	−839.1	(3)
$FeWO_4$		c	−1048.4	(3)
H^+		aq	0.0	(3)
H_2		g	0.0	(3)
Mg	Metal	c	0.0	(3)
Mg^{2+}		aq	−456.32	(4)
MgO		c	−569.196	(5)
MgO	Finely divided	c	−566.52	(4)
$Mg(OH)_2$	Brucite	c	−833.506	(5)
$MgCl_2$		aq	−718.83	(4)
MgS		c	−350.0	(3)
$MgSO_4$		aq	−1198.81	(4)
$Mg_3(PO_4)_2$		c	−3784.9	(3)
$Mg_3(AsO_4)_2$		c	−2844.1	(3)
$MgCO_3$		c	−1029.480	(5)
$MgNH_4PO_4$			−1632.9	(3)
Mn	α	c	0.0	(3)
Mn	γ	c	−1.38	(4)
Mn^{2+}		aq	−227.8	(3)
Mn^{3+}		aq	−82.1	(3)
MnO		c	−362.896	(5)
$HMnO_2^-$		aq	−506.2	(3)
MnO_2	Pyrolusite	c	−465.138	(5)
MnO_4^-		aq	−449.7	(3)
MnO_4^{2-}		aq	−504.1	(3)
Mn_2O_3		c	−881.068	(5)

Table 16 (*Continued*)

Substance	Description	State[a]	$\Delta G^{\circ b}$	Source[c]
Mn_3O_4		c	-1282.774	(5)
$Mn(OH)_2$	Precipitated	c	-615.0	(3)
$Mn(OH)_3$		c	-757.8	(3)
MnS	Alabandite	c	-218.155	(5)
MnS	Precipitated	c	-223.2	(3)
$Mn_3(PO_4)_2$	Precipitated	c	-2859.6	(3)
$MnCO_3$	Rhodochrosite	c	-816.047	(5)
$MnCO_3$	Precipitated	c	-813.5	(3)
$MnCO_3$		aq	-1186.1	(4)
$MnSiO_3$		c	-1243.081	(5)
K	Metal	c	0.0	(4)
K^+		aq	-282.44	(2)
K_2O		c	-322.087	(5)
KOH		c	-378.932	(5)
KOH		aq	-439.869	(4)
KCl		aq	-413.723	(4)
K_2S		c	-404.4	(2)
K_2SO_4		c	-1319.662	(5)
K_2CO_3		c	-1069.7	(2)
$KAlSi_3O_8$	Microcline	c	-3742.330	(5)
$KAl_3Si_3O_{10}(OH)_2$	Mica	c	-5600.671	(5)
N_2		g	0.0	(3)
NO		g	86.746	(4)
NO_2		g	51.251	(5)
NO_2^-		aq	-34.54	(3)
NO_3^-		aq	-110.66	(3)
$N_2O_2^{2-}$		aq	-138.2	(2)
N_2O_4		g	-98.352	(4)
NH_3		g	-16.410	(5)
NH_3		aq	-26.63	(3)
NH_4^+		aq	-79.55	(4)
HNO_3		l	-79.970	(4)
HNO_3		aq	-110.57	(2)
NH_4OH		aq	-263.98	(3)
Na	Metal	c	0.0	(4)
Na^+		aq	-262.048	(4)
NaCl		aq	-393.304	(4)
Na_2S		c	-362.6	(3)
Na_2SO_4		c	-1269.985	(5)
Na_2SO_4		aq	-1266.59	(4)
Na_2CO_3		c	-1048.4	(4)
Na_2CO_3		aq	-1052.6	(3)
$NaHCO_3$		c	-852.4	(4)
$NaHCO_3$		aq	-849.5	(3)
Na_2SiO_3		c	-1427.7	(4)

Table 16 (*Continued*)

Substance	Description	State[a]	$\Delta G^{\circ b}$	Source[c]
O_2		g	0.0	(3)
OH^-		aq	-157.403	(4)
H_2O		g	-228.569	(5)
H_2O		l	-237.141	(5)
H_2O_2		aq	-131.759	(3)
O_2^-		aq	54.5	(3)
S	Orthorhombic	c	0.0	(3)
S	Monoclinic	c	0.096	(4)
S		g	182.42	(2)
S^{2-}		aq	92.5	(3)
S_2		g	80.09	(2)
S_2^{2-}		aq	91.3	(3)
S_3^{2-}		aq	88.3	(3)
S_4^{2-}		aq	81.2	(3)
SO_2		g	-300.170	(5)
SO_3		g	-371.046	(5)
SO_3^{2-}		aq	-486.09	(3)
SO_4^{2-}		aq	-742.49	(4)
$S_2O_3^{2-}$		aq	-519.2	(3)
$S_2O_4^{2-}$		aq	-600.4	(3)
$S_2O_5^{2-}$		aq	-791.3	(3)
$S_2O_6^{2-}$		aq	-967.2	(3)
$S_2O_8^{2-}$		aq	-1096.9	(2)
$S_3O_6^{2-}$		aq	-958.8	(3)
$S_4O_6^{2-}$		aq	-1022.8	(3)
$S_5O_6^{2-}$		aq	-956.7	(3)
HS^-		aq	12.6	(4)
H_2S		g	-33.543	(5)
H_2S		aq	-27.38	(4)
HSO_3^-		aq	-527.5	(3)
HSO_4^-		aq	-753.37	(4)
H_2SO_3		aq	-538.38	(3)
H_2SO_4		aq	-742.49	(4)
$HS_2O_4^-$		aq	-592.0	(3)
$H_2S_2O_4$		aq	-586.2	(3)
$H_2S_2O_8$		aq	-1096.9	(2)
SF_6		g	-992.3	(4)
S_2Cl_2		l	-24.7	(3)
SO_2Cl_2		g	-308.1	(3)

[a] Symbols: c, crystalline; aq, aqueous; l, liquid; g, gaseous.

[b] (KJ/mole at 1013.24 mbar and 25°C)

[c] After *Enthalpies Libres de Formation Standards à 25°C* 1955 (1), Garrels & Christ 1965 (2), Latimer 1953 (3), Rossini et al. 1952 (4), and Robie et al. 1978 (5)

These equations are used for the calculation of the equilibrium constants of the reactions that determine the composition of natural waters. This approach is valid only for electrolytic solutions, not for colloids. As the solid phase surface energy is added to the Gibbs free energy, it follows that fine-grained material is less stable and thus more soluble than coarse-grained material. This is independent of the kinetic effects of smaller grain sizes, which result in faster equilibration (Langmuir 1971a).

Thermodynamic calculations of this kind hold only for systems in chemical equilibrium, which may not be completely attained in nature. However, many systems are so close to equilibrium that the thermodynamic treatment yields results only slightly different from reality. Therefore the equilibrium model must always be proved by direct measurements.

1.2.3.2 Solubility in Aqueous Solutions.   Solubility in aqueous solutions is different from that in pure water. It is reduced by the presence of similar ionic compounds in the solution (common ion effect), as will be seen by considering the solubility product, assuming that the activity coefficients are unity. For concentrations of more than a few milligrams per kilogram, the interaction between solutes must be considered. Unlike ionic components can raise the solubility, as will be seen by considering the increase in solubility of $CaSO_4$ with increase in NaCl concentration (Fig. 12).

It is possible to deduce from a chemical analysis of a solution whether precipitation or dissolution of solid species is likely to be occurring. This is ac-

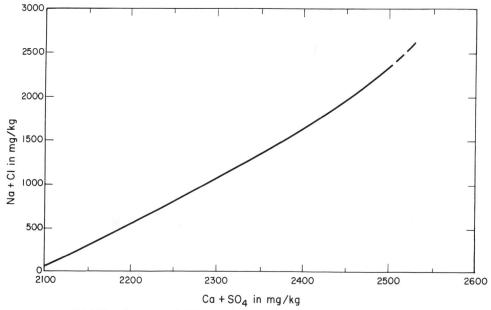

Fig. 12   Solubility of gypsum ($CaSO_4$) in NaCl solution at standard temperature and pressure (after Hem 1970).

complished by comparing the actual ion activity product of the dissolved constituents of a solid species with its solubility product, yielding the saturation index SI,

$$SI = \log \frac{\text{IAP}}{K_{\text{SP}}} \tag{28}$$

where IAP is the ion activity product $\{A\}^x\{B\}^y$ and K_{SP} is the solubility product of the solid species $[A_x\,B_y]$. If SI is less than zero, the groundwater is undersaturated with respect to $[A_x\,B_y]$ and a net dissolution of the solid should occur; when SI is zero, the groundwater is at equilibrium with the solid; when SI is greater than zero, the groundwater is supersaturated with the solid's constituent ions and net precipitation should occur. Kinetic delays may prevent the precipitation from solution of a mineral that is supersaturated in solution (Edmunds 1977).

1.2.3.2.1 *Carbonate–Carbon Dioxide Equilibria.* The carbonate–carbon dioxide equilibria play an important role in natural water chemistry. Particularly important is the calcium–carbonate — carbon dioxide equilibrium system.

Calcium Carbonate–Carbon Dioxide Equilibrium. The solubility of $CaCO_3$ in pure water is low (15 mg $CaCO_3$/l at 20°C); however, in the presence of dissolved carbon dioxide it rises rapidly (Fig. 13). The thermodynamic aspects of the system $CaCO_3$-CO_2-H_2O have been fully treated by Garrels & Christ (1965) as a model for all metal carbonates; at constant pressure and temperature the defining variables in a closed system (i.e., groundwater) are: aH_2CO_3, $aHCO_3^-$, aCO_3^{2-}, aH^+, aOH^-, and aCa^{2+}. These are related by the following expressions:

$$aCa^{2+} \cdot aCO_3^{2-} = K_{CaCO_3} \tag{29}$$

$$\frac{aH^+ \cdot aHCO_3^-}{aH_2CO_3} = K_{H_2CO_3} \tag{30}$$

$$\frac{aH^+ \cdot aCO_3^{2-}}{aHCO_3^-} = K_{HCO_3^-} \tag{31}$$

$$a\,H^+ \cdot aOH^- = K_{H_2O} \tag{32}$$

For the exact solution of these equations, the mass balance equation and the equation for electric neutrality are to be considered (Garrels & Christ 1965).

There are two main questions that refer to two different ways of treating the calcium carbonate–carbon dioxide system. The first, more practical one is the check of equilibrium. This problem is usually solved in Germany by using Tillmans's law and the Langelier index. The second more theoretical one deals with the amount of calcium carbonate that may be dissolved by a

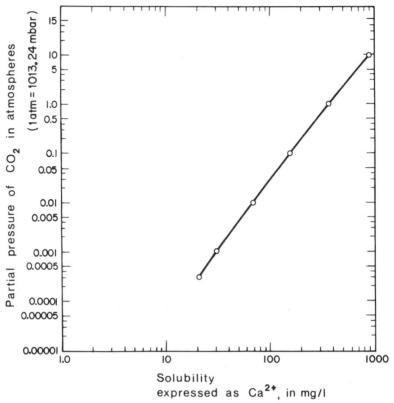

Fig. 13 Solubility of calcite in water in the presence of CO_2 (at 25°C) (After Frear & Johnston 1929).

water in disequilibrium to attain equilibrium. This can be done by solving the nonlinear set of equations mentioned above. Instead of this crucial procedure a graphical approximation may be used for practical purposes.

The equilibrium is defined by the empirical Tillmans's law, established by Tillmans & Heublein (1912):

$$C_{CO_2} = K_T \cdot C_{HCO_3^-}^2 \cdot CCa^{2+} \qquad (33)$$

in which C_{CO_2} = the concentration of the free CO_2 dissolved in water, calculated as CO_2 (molecular weight 44) in meq/l. The free CO_2 may be defined by the equation $K_{CO_2} = P_{CO_2} \cdot a_{H_2CO_3}$, with P_{CO_2} defining C_{CO_2}.

$C_{HCO_3^-}$ = concentration of the "bound" carbon dioxide dissolved in water (i.e., the concentration of hydrogen carbonate ions, by convention calculated as "semibound" carbon dioxide, MW = 22), in meq/l.

CCa^{2+} = the concentration of calcium molecules, calculated in meq/l.

K_T = the temperature-dependent Tillmans's constant

This relationship, which holds only for the pure system, is given in Fig. 14.

The following equation takes into account the effect of other electrolytes present (Hässelbarth 1963), as would be the case with groundwater.

$$-\log C_{H_2CO_3} = \log K_{HCO_3^-} - \log K_{H_2CO_3} - \log K_{CaCO_3} - (2 \cdot \log C_{HCO_3^-}) - \log CCa^{2+}$$

$$+ \frac{3 \sqrt{I} + 1.7I}{1 + 5.3 \sqrt{I} + 5.5I} \tag{34}$$

The external effect on the first ionization constant is based on the equation given by Shedlowsky & MacInnes (1935); for the second ionization constant of carbonic acid the external effect is based on the equation given by Harned & Scholes (1941), and for the solubility product of $CaCO_3$ it is that given by Larson & Buswell (1942).

By using the activity coefficient f, derived from the ionic strength I and the expression

$$\log f = \frac{3 \sqrt{I} + 1.7I}{1 + 5.3 \sqrt{I} + 5.5I}$$

one obtains from equation 34, after eliminating the logarithms and reversing

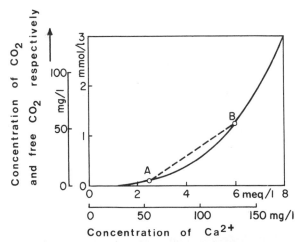

Fig. 14 Graph of the calcium carbonate–carbon dioxide equilibrium $[CO_2] = K_T [HCO_3]^3$ (Tillmans's equation) after Axt (1966). The content of undissociated H_2CO_3 (referred to as milligrams of CO_2 per litre) on the equilibrium curve is called "free associated" carbon dioxide; any excess content above this curve is "free aggressive" carbon dioxide.

the sign and combining the equilibrium constants. Tillmans's constant K_T

$$\frac{K_{HCO_3^-}}{K_{H_2CO_3} \cdot K_{CaCO_3}} = K_T$$

and a modified form of Tillmans's law:

$$C_{CO_2} = K_T / f \, C_{HCO_3^-}^2 \cdot CCa^{2+} \tag{35}$$

Tillmans's law in this form permits the calculation of the concentration of the free carbon dioxide in equilibrium, in the presence not only of the components taking part in the equilibrium, but also other species in the solution. The concentrations of ions involved in the equilibrium are both directly accounted for and additionally in the correction factor in equation 35. The effect of other ions (e.g., Na^+, Mg^{2+}, Cl^-, and SO_4^{2-}) is taken into account only through the correction factor.

Further equilibrium conditions are given by Strohecker (1936), Langelier (1936), and Larson & Buswell (1942) for the range up to pH 9.5.

$$pH_{equilibrium} = pK^* - \log CCa^{2+} - \log CHCO_3^- + \log f_L \tag{36}$$

where quantities in addition to those given in equation 33 are defined as follows:

$pH_{equilibrium}$ = pH of natural water in the calcium carbonate–carbon dioxide
　　　　　　　　equilibrium
pK^* = the Langelier constant.

$$\log f_{Langelier} = \log f_L = \frac{2.5 \sqrt{I}}{1 + 5.3 \sqrt{I} + 5.5 I} \tag{37}$$

The Strohecker-Langelier formula expressed in this way permits calculation of the equilibrium pH not only in the presence of the reactants taking part, but when other components are present in the solution.

Figure 15 shows the relationship between equilibrium pH and the Ca^{2+} and HCO_3^- activities. The values of I and $\log f_{Langelier}$ are given in Table 19, below.

Waters are in calcium carbonate–carbon dioxide equilibrium when the calculated equilibrium concentration for free carbon dioxide is equal to the concentration actually observed (neutral water). If the difference between the two is less than zero (i.e., CO_2 obs $> CO_2$ calc), a net dissolution of $CaCO_3$ is occurring, if greater than zero, a net precipitation of $CaCO_3$ is to be expected. The second criterion for the state of equilibrium is that the calculated equilibrium pH agree with the measured value. If the two criteria do not agree, there has been an error of measurement and/or calculation.

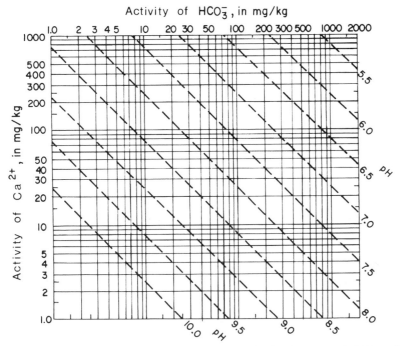

Fig. 15 Equilibrium pH values for Ca^{2+} and HCO_3^- activities for the range 1.0–2000 mg/kg in contact with calcite at standard temperature and pressure (after Hem 1961b).

The following indications serve as a measure of carbon dioxide attack on calcium carbonate:

1 The "excess" carbon dioxide of Tillmans's theory, which represents the difference between the content for free carbon dioxide and that for equilibrium concentration as calculated from equation 35.

2 The Langelier index (calcium carbonate saturation index), after Strohecker et al. (1936) and Langelier (1936):

$$I = \Delta pH = pH_{measured} - pH_{equilibrium} \qquad (38)$$

Negative I-values indicate that carbon dioxide is attacking calcium carbonate, positive values $CaCO_3$ indicate oversaturation, and $I = 0$ indicates that the water is in equilibrium with calcium carbonate.

3 Carbon dioxide attacking calcium carbonate as determined in the marble solution experiment (Heyer test) (Hässelbarth 1963, Höll 1972), where the amount of calcium carbonate in solution removed from solid phase calcium carbonate is compared with the amount of free carbon dioxide involved in the reaction. However, the marble solution experiment requires maintenance of equilibrium, and the temperature of the

reaction must be kept constant because of the strong temperature dependence of Tillmans's constant (Hässelbarth 1963).

Lewis & Randall (1921) have shown that the ionic strength I is found from the concentrations of the individual salts in the solution by using equation 10. The ionic strength can also be calculated by the methods given in Section 1.2.1. For practical calculations a simplified scheme has been given by Hässelbarth (1963), in which the following can be differentiated:

$I_{(1-1)}$ Ionic strength of the 1–1 electrolytes
         ($NaHCO_3$, $NaCl$, KCl)
$I_{(2-1)}$ Ionic strength of the 2–1 or 1–2 electrolytes
         ($Ca(HCO_3)_2$, $CaCl_2$, Na_2SO_4)
$I_{(2-2)}$ Ionic strength of the 2–2 electrolytes ($CaSO_4$, $MgSO_4$)

Trivalent ions occur in natural waters in such low concentrations that their influence on the ionic strength may be neglected. Likewise in practice the influence of 1–1 electrolytes at concentrations less than 0.002 mole/l can be neglected.

The values of the ionic strengths of 1–1, 2–1, and 2–2 electrolytes are tabulated in Table 17, which gives the concentrations in mmole/l.

The ionic strengths of individual salts ($I_{(1-1)}$, $I_{(2-1)}$, and $I_{(2-2)}$) are added to give I_{total}. The procedure is illustrated below, using the water analysis given in Table 13.

Analysis values: $CCa^{2+} = 4.21$ meq/l $CSO_4^{2-} = 0.46$ meq/l
                 $CMg^{2+} = 2.10$ meq/l $CCl^- = 0.36$ meq/l
                 $CHCO_3^- = 5.86$ meq/l $CNO_3^- = 0.04$ meq/1

These values can be expressed as follows:

5.86 meq/l in 2–1 electrolyte $[Ca, Mg(HCO_3)_2] = 2.93$ mmole/l
0.45 meq/l in 2–2 electrolyte ($MgSO_4$) $= 0.22$ mmole/l
0.40 meq/l in 1–1 electrolyte ($NaCl$, KNO_3, NH_4Cl) $= 0.40$ mmole/l

Table 17 shows that

for 2.93 mmole/l $I_{(2-1)} = 8.79 \times 10^{-3}$
for 0.89 mmole/l $I_{(2-2)} = 0.89 \times 10^{-3}$
for 0.40 mmole/l $I_{(1-1)} = \underline{0.40 \times 10^{-3}}$
$I_{total} = 10.08 \times 10^{-3}$

Similarly, correction factors for the Tillmans expression $f_{Tillmans}$ and for the Strohecker-Langelier equation $\log f_L$ over the range of I_{total} required are read from Tables 18 and 19. Hence with the value of I_{total} found above (10.08×10^{-3}), the equilibrium concentration of the free carbon dioxide can be found referring to Table 18 ($f_{Tillmans} = 1.587$). Table 19 gives $\log f_L = 0.16$ for calculating the pH value.

Table 17   Ionic Strengths of Salts (After Hässelbarth 1963)

n (mmole/l)	$I_{(1-1)}$ ($\times$ 10^{-3})	$I_{(2-1)}$ ($\times$ 10^{-3})	$I_{(2-2)}$ ($\times$ 10^{-3})
0.09	0.09	0.27	0.36
0.1	0.1	0.3	0.4
0.2	0.2	0.6	0.8
0.3	0.3	0.9	1.2
0.4	0.4	1.2	1.6
0.5	0.5	1.5	2.0
0.6	0.6	1.8	2.4
0.7	0.7	2.1	2.8
0.8	0.8	2.4	3.2
0.9	0.9	2.7	3.6
1.0	1.0	3.0	4.0
1.1	1.1	3.3	4.4
1.2	1.2	3.6	4.8
1.3	1.3	3.9	5.2
1.4	1.4	4.2	5.6
1.5	1.5	4.5	6.0
1.6	1.6	4.8	6.4
1.7	1.7	5.1	6.8
1.8	1.8	5.4	7.2
1.9	1.9	5.7	7.6
2.0	2.0	6.0	8.0
2.1	2.1	6.3	8.4
2.2	2.2	6.6	8.8
2.3	2.3	6.9	9.2
2.4	2.4	7.2	9.6
2.5	2.5	7.5	10.0
2.6	2.6	7.8	10.4
2.7	2.7	8.1	10.8
2.8	2.8	8.4	11.2
2.9	2.9	8.7	11.6
3.0	3.0	9.0	12.0
3.1	3.1	9.3	12.4
3.2	3.2	9.6	12.8
3.3	3.3	9.9	13.2
3.4	3.4	10.2	13.6
3.5	3.5	10.5	14.0
3.6	3.6	10.8	14.4
3.7	3.7	11.1	14.8
3.8	3.8	11.4	15.2
3.9	3.9	11.7	15.6
4.0	4.0	12.0	16.0
4.1	4.1	12.3	16.4
4.2	4.2	12.6	16.8
4.3	4.3	12.9	17.2
4.4	4.4	13.2	17.6
4.5	4.5	13.5	18.0
4.6	4.6	13.8	18.4
4.7	4.7	14.1	18.8
4.8	4.8	14.4	19.2
4.9	4.9	14.7	19.6

Table 17 (*Continued*)

n (mmole/l)	$I_{(1-1)}$ ($\times 10^{-3}$)	$I_{(2-1)}$ ($\times 10^{-3}$)	$I_{(2-2)}$ ($\times 10^{-3}$)
5.0	5.0	15.0	20.0
5.1	5.1	15.3	20.4
5.2	5.2	15.6	20.8
5.3	5.3	15.9	21.2
5.4	5.4	16.2	21.6
5.5	5.5	16.5	22.0
5.6	5.6	16.8	22.4
5.7	5.7	17.1	22.8
5.8	5.8	17.4	23.2
5.9	5.9	17.7	23.6
6.0	6.0	18.0	24.0
6.1	6.1	18.3	24.4
6.2	6.2	18.6	24.8
6.3	6.3	18.9	25.2
6.4	6.4	19.2	25.6
6.5	6.5	19.5	26.0
6.6	6.6	19.8	26.4
6.7	6.7	20.1	26.8
6.8	6.8	20.4	27.2
6.9	6.9	20.7	27.6
7.0	7.0	21.0	28.0
7.1	7.1	21.3	28.4
7.2	7.2	21.6	28.8
7.3	7.3	21.9	29.2
7.4	7.4	22.2	29.6
7.5	7.5	22.5	30.0
7.6	7.6	22.8	30.4
7.7	7.7	23.1	30.8
7.8	7.8	23.4	31.2
7.9	7.9	23.7	31.6
8.0	8.0	24.0	32.0
8.1	8.1	24.3	32.4
8.2	8.2	24.6	32.8
8.3	8.3	24.9	33.2
8.4	8.4	25.2	33.6
8.5	8.5	25.5	34.0
8.6	8.6	25.8	34.4
8.7	8.7	26.1	34.8
8.8	8.8	26.4	35.2
8.9	8.9	26.7	35.6
9.0	9.0	27.0	36.0
9.1	9.1	27.3	36.4
9.2	9.2	27.6	36.8
9.3	9.3	27.9	37.2
9.4	9.4	28.2	37.6
9.5	9.5	28.5	38.0
9.6	9.6	28.8	38.4
9.7	9.7	29.1	38.8
9.8	9.8	29.4	39.2
9.9	9.9	29.7	39.6
10.0	10.0	30.0	40.0

Table 18 The Tillmans Correction Factor in Terms of Ionic Strength (After Hässelbarth 1963)

I_{total} ($\times 10^{-3}$)	$f_{Tillmans}$	I_{total} ($\times 10^{-3}$)	$f_{Tillmans}$
0.27	1.101	19.2	1.752
0.53	1.124	19.6	1.758
0.8	1.187	19.8	1.761
1.2	1.228	20.0	1.764
1.6	1.260	20.4	1.769
1.8	1.281	20.8	1.774
2.0	1.289	21.0	1.777
2.4	1.314	21.2	1.780
2.8	1.337	21.6	1.785
3.0	1.348	22.0	1.791
3.2	1.358	22.2	1.793
3.6	1.377	22.4	1.795
4.0	1.396	22.8	1.800
4.2	1.404	23.2	1.805
4.4	1.413	23.4	1.808
4.8	1.429	23.6	1.810
5.2	1.444	24.0	1.815
5.4	1.451	24.4	1.820
5.6	1.459	24.6	1.822
6.0	1.472	24.8	1.824
6.4	1.485	25.2	1.829
6.6	1.492	25.6	1.833
6.8	1.498	25.8	1.835
7.2	1.511	26.0	1.838
7.6	1.522	26.4	1.842
7.8	1.528	26.8	1.846
8.0	1.534	27.0	1.849
8.4	1.544	27.2	1.851
8.8	1.555	27.6	1.855
9.0	1.560	28.0	1.859
9.2	1.565	28.2	1.861
9.6	1.575	28.4	1.863
10.0	1.585	28.8	1.867
10.2	1.590	29.2	1.871
10.4	1.594	29.4	1.873
10.8	1.603	29.6	1.875
11.2	1.612	30.0	1.879
11.4	1.616	30.4	1.883
11.6	1.621	30.8	1.887
12.0	1.629	31.2	1.890
12.4	1.637	31.6	1.894
12.6	1.641	32.0	1.898
12.8	1.645	32.4	1.901
13.2	1.653	32.8	1.905
13.6	1.661	33.2	1.909
13.8	1.664	33.6	1.912

Table 18 (*Continued*)

I_{total} ($\times 10^{-3}$)	$f_{Tillmans}$	I_{total} ($\times 10^{-3}$)	$f_{Tillmans}$
14.0	1.668	34.0	1.915
14.4	1.675	34.4	1.919
14.8	1.683	34.8	1.922
15.0	1.686	35.2	1.926
15.2	1.690	35.6	1.929
15.6	1.696	36.0	1.932
16.0	1.703	36.4	1.936
16.2	1.706	36.8	1.939
16.4	1.710	37.2	1.942
16.8	1.716	37.6	1.945
17.2	1.723	38.0	1.948
17.4	1.726	38.4	1.951
17.6	1.729	38.8	1.954
18.0	1.735	39.2	1.957
18.4	1.741	39.6	1.960
18.6	1.744	40.0	1.963
18.8	1.747		

Table 19   The Langelier Correction Factor in Terms of Ionic Strength (After Zehender et al. 1956 and Hässelbarth 1963)

I_{total} ($\times 10^{-3}$)	$\log f_{Langelier}$	I_{total} ($\times 10^{-3}$)	$\log f_{Langelier}$
0.1	0.024	9.0	0.153
0.2	0.033	10.0	0.158
0.4	0.045	12.0	0.166
0.6	0.054	14.0	0.173
0.8	0.061	16.0	0.179
1.0	0.067	18.0	0.185
1.3	0.074	20.0	0.190
1.6	0.081	22.0	0.194
1.9	0.087	24.0	0.198
2.2	0.092	26.0	0.201
2.5	0.097	28.0	0.205
2.8	0.102	32.0	0.210
3.2	0.107	36.0	0.215
3.6	0.112	40.0	0.219
4.0	0.117	45.0	0.224
4.5	0.122	50.0	0.228
5.0	0.126	60.0	0.233
6.0	0.134	70.0	0.238
7.0	0.141	80.0	0.241
8.0	0.147	100.0	0.245

Table 20   Tillmans's Law Constants

t (°C)	K_T (meq/l)	t (°C)	K_T (meq/l)
0	9.371×10^{-3}	26	2.387×10^{-2}
1	9.700	27	2.459
2	1.004×10^{-2}	28	2.539
3	1.039	29	2.617
4	1.076	30	2.703
5	1.114	31	2.804
6	1.156	32	2.903
7	1.199	33	3.012
8	1.244	34	3.117
9	1.291	35	3.235
10	1.339	36	3.356
11	1.389	37	3.482
12	1.441	38	3.613
13	1.492	39	3.748
14	1.548	40	3.889
15	1.606	41	4.025
16	1.667	42	4.176
17	1.729	43	4.323
18	1.790	44	4.486
19	1.857	45	4.643
20	1.926	46	4.806
21	1.999	47	4.986
22	2.074	48	5.162
23	2.147	49	5.356
24	2.227	50	5.544
25	2.311		

The equilibrium concentration for the free carbon dioxide is calculated from equation 35. The values of the constant K in the calculation at the temperature concerned are read off as concentration quantities in meq/l from Table 20. The concentration of the free carbon dioxide in equilibrium in meq/l is then calculated by multiplying by the factor 22 to obtain the result in mg/l.

From Table 20 we can find the value $K = 1.415 \times 10^{-2}$ for a water temperature of 11.5°C. The correct factor f_{Tillmans} is accordingly 1.587. Then from equation 35:

$$C_{CO_2\ \text{equilibrium}} = \frac{K}{f_{\text{Tillmans}}} \, C^2_{HCO_3^-} \cdot CCa^{2+} = \frac{1.415 \times 10^{-2}}{1.587} \times (5.86)^2 \times 4.21$$

$$= 129 \times 10^{-2} = 1.29 \text{ meq } CO_2/l$$

$$= 28.4 \text{ mg } CO_2/l$$

Table 21   Strohecker-Langelier Law Constants for Concentrations

t (°C)	pK^* (meq/l)	t (°C)	pK^* (meq/l)
0	8.901	15	8.515
1	8.878	16	8.492
2	8.851	17	8.468
3	8.825	18	8.445
4	8.798	19	8.422
5	8.771	20	8.400
6	8.745	21	8.376
7	8.718	22	8.354
8	8.692	23	8.333
9	8.665	24	8.311
10	8.639	25	8.288
11	8.614	30	8.196
12	8.589	35	8.100
13	8.565	40	7.996
14	8.540	45	7.923
		50	7.844

Because the water analysis has yielded a value of 77.3 mg CO_2/l, the water is not in equilibrium but contains free carbonic acid, which will attack calcium carbonate.

For the purpose of calculating the equilibrium pH from equation 36, the values of the constant pK^* for the temperatures concerned are given in Table 21 as concentration quantities in meq/1. For a water temperature of 11.5°C the value of pK^* in Table 21 is 8.601, and the correction factor $\log f_L$ = 0.161. Hence from equation 36, using $\log CCa^{2+} = 0.624$ and $\log C_{HCO_3^-} = 0.768$,

$$pH_{equilibrium} = pK^* - \log CCa^{2+} - \log C_{HCO_3^-} + \log f_L$$
$$= 8.601 - 0.624 - 0.768 + 0.161 = 7.4$$

Because the water has a pH value of 6.9, this calculation shows that it is not in equilibrium. This agrees with the result of the calculation of carbon dioxide equilibrium.

It is accordingly necessary to determine the circumstances in which water is able to dissolve calcium carbonate. This can be done by considering the calcium carbonate solubility capacity proposed by Axt (1962). The following holds for the equilibrium condition:

$$\frac{Ca^{2+} \cdot (HCO_3^-)^2}{CO_2} = K_C \cdot \frac{K_1}{K_2} = K \qquad (39)$$

Where there is a capacity for dissolving calcium carbonate, the values are modified by the calcium carbonate solution capacity x in the equilibrium relationship:

$$\frac{(Ca^{2+} + x) \cdot (HCO_3^- + 2x)^2}{CO_2 - x} = K_s \cdot \frac{K_1}{K_2} = K \tag{40}$$

The calcium carbonate solution is conveniently determined from a nomogram. Strict comparison with the conditions for attack on calcium carbonate shows that these and calcium carbonate solution capacity do not coincide, with the exception of equilibrium conditions, because then both values are zero. Moreover water that has a high activity can exhibit only a low calcium carbonate solution capacity, and vice versa. Solution can occur when the values are not too small (calcium carbonate solution capacity not less than 0.1 mmole/l, or calcium carbonate saturation index not less than 1.0) (Axt 1962).

When other electrolytes are present, the concentrations must be multiplied by the appropriate activity coefficients.

Thus the calcium carbonate solution capacity can be represented as an equilibrium relationship based on equation 39 after considering the influence of other ions. Figure 16 is a nomogram for $K = 7 \times 10^{-5}$ in neutral water for the concentrations of Ca^{2+}, HCO_3^-, and CO_2. For water that is not neutral, two points are needed: the first is P_0, which corresponds to the measured concentrations of Ca^{2+} and HCO_3^- (e.g., 2.6 mmole Ca^{2+}/l and 4.0 mmole HCO_3^-/l), and the curve that corresponds to the given CO_2 concentration (e.g., 0.9 mmole/l). For calcium carbonate solution based on the equation

$$CaCO_3 + H_2O + CO_2 \rightleftharpoons Ca^{2+} + 2\ HCO_3^-$$

the point P_0 moves to the right on the diagram at a slope of 45°. Its point of intersection with the CO_2 curve, P_1, is not however reached, because the increase in Ca^{2+} concentration is accompanied by an equal decrease in the CO_2 concentration. The actual point reached, P_x, corresponds to the point that is to be found by experiment. The rise in the Ca^{2+} concentration from P_0 to P_x corresponds to the calcium carbonate solution capacity x (in this example, 0.4 meq/l), which is read off the ordinate (Axt 1962).

Langemuir (1971b) pointed out that the effect of $CaSO_4^0$, $MgSO_4^0$, and $MgHCO_3^+$ ion pairs on the solubility of calcite has to be considered if equilibrium calculations are done.

CaCO₃ Precipitation. $CaCO_3$ can be precipitated if the equilibrium is destroyed by the addition of material, or through pH change. For example $CaCO_3$ would be precipitated if as a consequence of a reduction of free dissolved CO_2, addition of HCO_3^- or Ca^{2+} ions or strong bases that give rise to OH^- ions, the H_2CO_3 or H^+ content fell below the level for equilibrium. Ac-

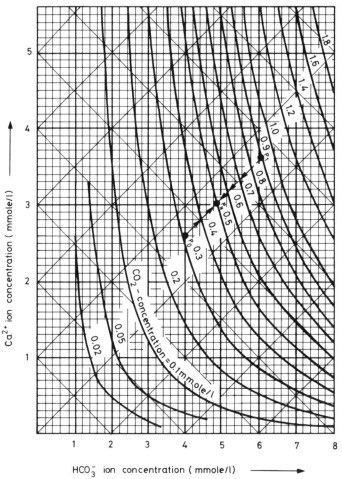

Fig. 16 Calcium carbonate solution capacity as a function of Ca^{2+} and HCO_3^- ion concentration (after Axt 1962).

cording to Orlow (1930, 1931) and Stumper (1934), catalysts are necessary, particularly $CaCO_3$ crystallization nuclei, to start the precipitation of $CaCO_3$ from the oversaturated solution of calcium hydrogen carbonate.

Influence of Pressure on Carbonate Equilibrium.   Pressure does not have an appreciable effect until a pressure of 25 bar is reached, hence at water depths of more than 250 m. The pressure normally occurring in natural waters will have only an insignificant effect on the solubility of mineral material. However an increase in the partial pressure of carbon dioxide in contact with the water raises the solubility of the gas so that indirectly the solution of calcium and magnesium carbonate is in turn raised. By this means there can occur precipitation or solution of carbonates through pressure changes in surface or subsurface water flow.

Solubility of Other Carbonates. Magnesium carbonate (magnesite), $MgCO_3$, nesquehonite, $MgCO_3 \cdot 3H_2O$, and the double carbonate dolomite $CaCO_3 \cdot MgCO_3$, are similarly soluble under the influence of CO_2 gas dissolved in water (Table 22). Ion pairs and complexes affect also the solution of these minerals (Hosteler 1964, Langmuir 1971b).

Table 22 Solubility of Dolomite, Calcite, and Magnesium Carbonate (After Yanat'eva 1954)

Temperature (°C)	pCO_2 (mbar)	Solubility [mmole $(HCO_3^-)_2$ in 1 kg solution]		
		Dolomite	Calcite	Magnesite
0	1013.24	10.74	15.08	22.52
25	1013.24	6.49	9.0	16.50
55	1013.24	6.08	6.09	15.59
70	1013.24	4.58	3.45	15.59
25	1.08	1.42	0.81	1.65

1.2.3.3 Importance of pH with Regard to Solubility. The activity of hydrogen ions, which can be accurately measured by experiment, is of great importance in many reactions between dissolved substances. Hence a_{H+} can often be used as a standard reference activity for a great number of reactions, including those involving oxides, hydroxides, carbonates, silicates, and sulfides. For this purpose, as explained below, the equation of the reaction

$$Fe_2O_{3c} + 3\ H_2O = 2\ Fe^{3+}_{aq} + 6\ OH^-_{aq}$$

can have the dissociation equation of water added to it, thus:

$$\begin{array}{r} Fe_2O_{3c} + 3H_2O = 2Fe^{3+}_{aq} - 6OH^-_{aq} \\ 6OH^-_{aq} + 6H^+_{aq} = \qquad 6H_2O \\ \hline Fe_2O_{3c} + 6H^+_{aq} = 2Fe^{3+}_{aq} + 3H_2O \end{array}$$

In this relationship $[Fe^{3+}_{aq}]$ and $[H^+_{aq}]$ occur as variable activities. Their equilibrium constant is therefore:

$$\log K = \log \frac{[Fe^{3+}]^2}{[H^+]^6} = 2 \log [Fe^{3+}] - 6 \log [H^+]$$

Because pH is equal to $-\log [H^+]$, $\log K = 2 \log [Fe^{3+}] + 6 \text{ pH}$, the pH value determines the solubility of many elements. Only a few ions, such as sodium, calcium, nitrate, and chloride, can be present in solution in normal groundwaters over the whole range of naturally occurring pH values. Most metallic ions are soluble in acid groundwaters as cations, but nevertheless precipitate as hydroxides or basic salts when the pH increases, although for some elements there is a much greater tendency for colloids to form. Iron(III) ions precipitate above pH 3. Iron(II) ions start to be precipitated at pH 5.1 as hydroxide, but they are not fully precipitated even in a neutral solution (pH 7).

The precipitation of aluminum as hydroxide begins at pH 5 and has a maximum at pH 6–7; then aluminum has increased solubility in the alkaline range. Manganese(II) ions tend to be precipitated above pH 8, and magnesium ions above pH 10.5. The formation of sulfides depends on the concentration of S^{2-} and SH^- ions, in addition to the pH value.

Dissolved substances usually alter the pH value, because some ions combine with the H^+ or OH^- ions and thus upset the chemical equilibrium. For example, calcium carbonate alters the equilibrium in the alkaline region. Salts of aluminum and iron have an acid reaction, but seldom reach concentrations high enough to influence the pH values of natural waters.

Water can also contain free acids: most often carbonic acid and organic acids (e.g. humic acids), less often sulfuric acid or hydrochloric acid. The ratio of H^+ ions to these acids will be greater the more dissociated the acids are. This leads to a decrease in the pH value.

During solution of strong bases and removal of acid residues from the solution (e.g., through escape of CO_2), the strong bases give rise to OH^- ions in the solution and raise the pH value.

Waters that contain only weakly dissociated bases or acids tend to have a marked constancy of pH. If H^+ or OH^- ions are introduced into buffered solutions of this kind, they upset the equilibrium so that the acid or base transforms into a salt. Until this happens, the pH of the solution remains relatively constant.

Natural waters usually contain dissolved CO_2 and hydrogen carbonate ions, which form a buffered system with the carbon dioxide. This is the chief reason for the small variations in the pH, generally between 5.0 and 8.0 in natural waters (Schwille & Weisflog 1968, Garrels & Christ 1965).

Very high pH values (> 8.5 pH) are usually associated with water that has a high sodium carbonate content (e.g., pH 11.6 in a Na-Cl-CO_3 water from Mount Shasta, California (see Table 65, analysis 8, below). Quite high pH values are found in hydrogen carbonate–rich waters. Very low values (< 4.0 pH) occur in acid mine waters, the free acids occur as a result of oxidation of sulfides, or in thermal waters, which contain H_2S, HCl, and other volatile constituents of volcanic origin [e.g., pH 0.4 in a thermal water from Yakeyama, Niigata Prefecture, Honshu, Japan (Table 65, analysis 9)]. Relatively low pH values can be attributed to small amounts of sulfuric acid from the oxidation of metal sulfides or to organic acids produced by humification.

In general the soft water coming from silicate rocks has a lower pH than hard water from calcium carbonate-bearing rocks.

The pH values used for the study of solubility require the most accurate measurement possible. These are made almost exclusively with a glass electrode in conjunction with a calomel electrode as the reference electrode. The usual glass electrode permits accurate measurements in the pH range below 11. For measurements above pH 11 special electrodes are used, the corresponding operating instructions are usually enclosed with the apparatus and are not given here (Bates 1954; Gold 1956; Barnes 1964; Kortüm 1966b).

The pH of aqueous solutions is dependent on temperature. This should be allowed for by incorporating temperature compensation and standardizing before and after reading the instrument, especially when there are marked temperature changes. The glass electrode should be rinsed in distilled water after use to remove adsorbed ions. The readings should not be taken until the measured voltage has stopped drifting, indicating that equilibrium between electrodes and solution has been reached. For solutions that in natural circumstances are not in contact with the atmosphere, pH measurements can be upset by the entrance or escape of gases (particularly the pH-controlling CO_2 or H_2S); the same applies to the escape of gases as a result of pressure or temperature changes during pumping processes in wells or on emergence at springs. Hence it is necessary to aim to measure natural waters in situ. During experimental work the measurements can be affected by gas exchange with the atmosphere, and unavoidable temperature changes can cause addition or escape of volatiles, which cause a steady drift in the pH readings. The pH measurements obtained in laboratory experiments that are conducted for prolonged periods can be misleading both because of heating and because of microbial activity inside the flasks. Differences between field and laboratory pH values of 0.5 pH and more can arise in dilute solutions because of the presence of air in the glassware. As little air as possible should enter, and there should be a pH neutral inhibitor in the liquid to stop biological activity. Field measurements are essential for exact measurement of natural conditions.

1.2.3.4 Importance of Redox Potential with Regard to Solubility.

Oxidation and reduction can be broadly defined as loss of electrons and gain of electrons, respectively. The reduction-oxidation (redox) potential serves as a measure of the relative state of the oxidation or reduction in an aqueous system. In a solution that contains various oxidation states of an element, the redox potential is measured as an electrical potential between an inert metal electrode and a standard reference electrode, both immersed in the solution. The redox potential is usually denoted by the symbol E, or, if referred to the standard hydrogen electrode of zero volts, by Eh, E_h, or E_H. The standard reduction potential E^0 holds for a redox pair at activities of unity and standard temperature and pressure. Increasing redox potential is normally indicated by increasing positive values. Every reduced element of a pair can reduce an oxidized element that has a higher Eh potential.

Table 23   Standard Electrode Potentials of Various
Substances (After Schoeller 1962, Latimer 1953,
Stumm & Morgan 1970)

	E^0 (V)
$Mn^{3+} + e = Mn^{2+}$	+1.51
$O_2 + 4H^+ + 4e = 2H_2O$	+1.229
$NO_3^- + 4H^+ + 4e = NO + 2H_2O$	+0.94
$2NO_3^- + 4H^+ + 2e = N_2O_4 + 2H_2O$	+0.80
$Fe^{3+} + e = Fe^{2+}$	+0.77
$H_2SO_3 + 4H^+ + 4e = S + 3H_2O$	+0.45
$O_2 + 2H_2O + 4e = 4OH^-$	+0.401
$SO_4^{2-} + 4H^+ + 2e = H_2SO_3 + H_2O$	+0.17
$Mn(OH)_3 + e = Mn(OH)_2 + OH^-$	+0.1
$NO_3^- + H_2O + 2e = NO_2^- + 2OH^-$	+0.01
$2H^+ + 2e = H_2$	0.00
$S + 2e = S^{2-}$	−0.48
$Fe(OH)_3 + e = Fe(OH)_2 + OH^-$	−0.56

The redox potential is related to the standard free energy change of a reaction, its equilibrium constants, the amounts of the reacting substances present in the equilibrium (see Table 23) and the standard potential, as shown in equations 41–43. Equation 43 is generally known as the Nernst equation.

$$\Delta G^0 = -RT \ln K \tag{41}$$

$$E^0 = -\frac{\Delta G^0}{nF} \tag{42}$$

$$Eh = E^0 + \frac{RT}{nF} \ln \frac{a_{ox}}{a_{red}} \tag{43}$$

in which ΔG° = standard free energy of formation of the chemical reaction (in kJ/mole)

R = universal gas constant (8.315 J · °C^{-1} · mole^{-1})

T = absolute temperature (K)

n = number of electrons, represented by a multiple of e in the redox equation (difference of electrons between the oxidized and reduced substances)

F = Faraday's constant (unit of charge on 1 mole of electrons, 96.564 kJ/V or 96.487 C/mole)

E^0 = standard potential (V) ($a_{ox} = a_{red} = 1$)

Eh = redox potential (V)

a_{ox} and a_{red} = activities of the oxidized and reduced substances, respectively, in the chemical system under consideration

These equations give correct values for equilibrium conditions when the reactions are expressed as oxidations, that is, when e appears on the right-hand side of the chemical equation.

At 25°C the following equations hold, converted to logarithms to base 10:

$$\Delta G^0 = -1.364 \log K \tag{44}$$

$$E^0 = \frac{\Delta G^0}{96.564n} \tag{45}$$

$$Eh = E^0 + \frac{0.0592}{n} \cdot \log \frac{a_{ox}}{a_{red}} \tag{46}$$

The redox potential can be detetermined by voltage measurement. For the voltage between the above-mentioned inert electrode (usually a platinum electrode, less often gold) and the reference electrode (usually a calomel electrode, Glasstone 1949; Zobell 1946a) and the change of the standard free energy the following relationship applies:

$$G_R^0 = E_R^0 nF \tag{47}$$

in which E_R^0 is the voltage of the reaction at an activity of unity for the substances concerned and F is Faraday's constant.

Finally equation 48 gives E_R^0 directly in terms of the equilibrium constant under standard conditions:

$$E_R^0 = -\frac{5.708}{nF} \log K \tag{48}$$

Kortüm (1966a, 1966b), Garrels & Christ (1965), Whitfield (1974), and Marshall & Bergemann (1942) have given further details of measurement methods. Many hydrogeologically important oxidation-reduction reactions are defined by Eh, by Eh and pH, or by Eh, pH, and the activity of the individual elements in the solution, and both Eh and pH can be measured directly.

It is possible to calculate the approximate activity ratios of ion pairs with the help of platinum or gold electrode measurements if: (1) solution equilibrium is approached – dissolved oxygen, sulfide, and hydrogen are slow to attain equilibrium; (2) the ions of each element are present in sufficient concentration ($> 10^5$ M); (3) the ion pair members are electroactive (i.e., they transfer electrons rapidly); and (4) the electrodes are not poisoned by the adsorption of O_2, S^{2-}, Cl^-, or other contaminants. Consequently Eh measurements in nonequilibrium systems, such as natural waters, are technically difficult to achieve (Stumm 1961, Stumm & Morgan 1970). However natural waters with large quantities of oxidizing or reducing agents give measurements of high and low Eh values, respectively (Stumm 1967).

The entrance of atmospheric oxygen into systems of low initial Eh causes a rapid upward shift of Eh values during water pumping from wells or at emergence from springs. The available Eh measurements for natural waters in contact with the atmosphere show a small spread of values, generally between 0.35 and 0.50 V at pH 7 approximately (Hem 1961a). Stratification of Eh values may be observed in groundwater in the higher layers where there is some contact with the air. Besides atmospheric oxygen, effects that tend to give stable conditions can also originate from the solids in contact with the water.

It is thus difficult to measure natural systems without the influence of the atmosphere; to prevent air from entering when the electrodes are inserted involves considerable expense on special equipment. When experiments are made, the measurements are affected by the rapid stabilizing effect of oxygen. However, when atmospheric oxygen can be excluded, very accurate measurements can be achieved, as shown by the in situ measurements of pH and Eh conditions in aerated and water-saturated soils made by Starkey & Wight (1945). Instruments have been specially designed for Eh measurements of groundwater (Hem 1961a).

The solubility of some elements (e.g., iron, manganese, copper, vanadium, and uranium) depends on the oxidation state, which is controlled by the redox potential and the pH of the environment. These controls of solubility can be represented on stability field diagrams, as the following example for iron shows.

In laboratory experiments with simple systems, for which the variables are known and the equilibrium conditions are fulfilled at the electrode surfaces, good agreement can be reached between the solubilities as calculated from pH and Eh measurements and experimentally observed values. In natural waters, however, these criteria are very often not met. Such systems are complex and contain numerous reaction components, for example various soluble iron ions and iron complexes, dissolved oxygen, various organic redox systems, and sulfides (Stumm 1961). Therefore, although in many cases the stability field diagrams with pH and Eh data for natural waters are useful in understanding the observed water quality, great care should be taken in making conclusions.

1.2.3.5 Stability Field Diagrams to Show Dependence of Solubility on pH and Eh. The stability field diagram is a useful model introduced by Pourbaix (1945, 1949), with pH plotted along the abscissa and Eh along the ordinate. Chemical systems so represented usually consist of one or more chemical compounds, usually solids and water at standard pressure and temperature. The limits of the stability field of the ion species or solids at chemical equilibrium can be calculated for given concentrations of dissolved ions with the help of thermodynamic equilibrium relationships. The diagrams are not suitable for predicting reaction rates.

The procedure is now explained by considering an example of iron solubility. Except for those of low pH, most hydrolysis or oxidation reactions in which iron participates proceed quickly enough in natural waters for equilibrium to be reached (Hem & Cropper 1959). The departures from the standard conditions of temperature and pressure in most natural systems hardly affect the position of the stability boundaries, so that the stability field diagram is of considerable practical use in the study of the hydrochemistry of iron.

It can be assumed that under equilibrium the solutions have Eh and pH values inside the stability field of water, beyond which water is decomposed by oxidation to gaseous oxygen and hydrogen ions:

$$2H_2O \rightleftharpoons O_2 + 4H^+ + 4e$$

and below which water is decomposed by reduction to hydroxyl ions and hydrogen:

$$2H_2O + 2e \rightleftharpoons H_2 + 2OH^-$$

The position of the stability field is now calculated from the values for the standard potential E^0 for these reactions, using equations 45 and 46.

The standard potentials are found in published tables, or they can be calculated from the data of the free energy change and equations 44 and 45. The activities of water and gases are by definition unity. The equation for the oxidation reaction is:

$$Eh = 1.23 + \frac{0.059}{4} \log [H^+]^4 = 1.23 - 0.059 \text{ pH} \qquad (49)$$

and that for the reduction reaction is:

$$Eh = -0.83 + 0.059 (14 - \text{pH}) \qquad (50)$$

These equations delimit the field in which at equilibrium the observed Eh and pH values of natural waters lie (Baas Becking et al. 1960). Because of the overpotential (overvoltage) effect in galvanic cells and the long duration of some reactions however, redox potentials can also occur outside the stability limits of the water (Hem 1961a). First consider the stability field diagram in Fig. 17, which is a model of the reactions occurring in an iron solution in distilled water in contact with $Fe(OH)_3$ or $Fe(OH)_2$. The following hydrolysis reactions are important in the determination of the stability limits of iron(III) in distilled water:

$$Fe^{3+} + H_2O \rightleftharpoons FeOH^{2+} + H^+$$
$$FeOH^{2+} + H_2O \rightleftharpoons Fe(OH)_2^+ + H^+$$
$$Fe(OH)_2^+ + H_2O \rightleftharpoons Fe(OH)_{3c} + H^+$$

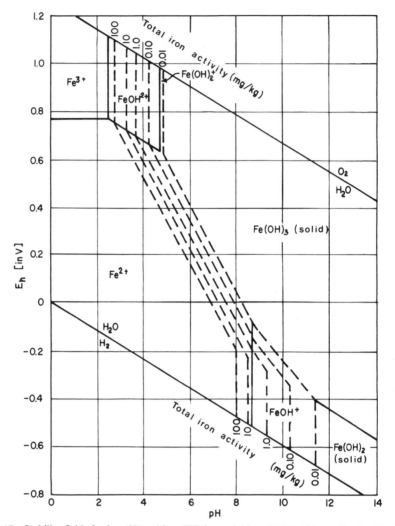

Fig. 17 Stability fields for iron(II) and iron(III) for activities of 0.01–100 mg dissolved in iron/kilogram (after Hem 1961a).

Similar types of iron(II) reactions may be described. The boundaries that separate iron(III) from iron(II) species are determined from the following equilibrium equations:

$$Fe^{3+} + e \rightleftharpoons Fe^{2+}$$
$$FeOH^{2+} + H^+ + e \rightleftharpoons Fe^{2+} + H_2O$$
$$Fe(OH)_2^+ + 2H^+ + e \rightleftharpoons Fe^{2+} + 2H_2O$$
$$Fe(OH)_{3c} + 3H^+ + e \rightleftharpoons Fe^{2+} + 3H_2O$$
$$Fe(OH)_{3c} + 2H^+ + e \rightleftharpoons FeOH^+ + 2H_2O$$
$$Fe(OH)_{3c} + H^+ + e \rightleftharpoons Fe(OH)_{2c} + H_2O$$

The values of E^0 for equilibrium can be calculated from the free energy data given above (Hem & Cropper 1959), and from equation 44, and the Eh values from equation 45. In these calculations the ratios Fe^{3+}/Fe^{2+} are assumed to be constant, when one or more solid phases are present. Under these conditions the activity of the solid phase can be taken as unity and selected values for the activity of the dissolved iron (e.g., 0.01, 0.10, 1.0, 10, or 100 mg/kg) inserted into the calculation. The boundaries between both types of solid phase are independent of the activity of the dissolved iron. For example consider the equilibrium $Fe^{2+} + H_2O \rightleftharpoons Fe(OH)^{2+} + H^+ + e$ with a free energy change $\Delta G^0 = 88.26$ kJ/mole. Using equation 44 the value of E^0 is accordingly $E^0 = 88.26/96.55 = 0.91$ V, and using equation 45 the redox potential is

$$Eh = 0.91 + \frac{0.059}{1} \log \frac{[FeOH^{2+}][H^+]}{[Fe^{2+}]}$$

This simplifies to $Eh = 0.91 - 0.059$ pH if the ratio of the two iron types is unity. The slope of the Fe(II)/Fe(III) boundary varies when the type of iron or the solid changes.

At high or low redox potential the prevailing oxidation level of the ions is dependent on pH alone. The equilibrium (hydrolysis) constants for these reactions can be taken from published values or calculated from free energy data using equation 44.

The vertical boundary lines are found by calculating the pH values, in which the ratio of the activities of both species separated by the boundary has the value unity. In the example there is for the first hydrolysis equilibrium the expression for the hydrolysis constant:

$$K_h = \frac{[FeOH^{2+}][H^+]}{[H_2O][Fe^{3+}]}$$

The activity of water is unity. Because all activity values except $[H^+]$ simplify, the pH value for the $Fe^{3+} - Fe(OH)^{2+}$ boundary is pH $= -\log K_h$. The constant of hydrolysis for this equilibrium is 2.4.

The position of the boundary between the two solid phase ferrous and ferric hydroxides depends on the value that is chosen for the limit of iron solubility. The activities of the solid phases are taken as unity. Five different stability boundaries between solids are shown as dashed lines in Fig. 17. The right-hand boundary line of the stability field is for a minimum dissolved iron content of 0.01 mg/l, determined by the usual 2.2-bipyridin method of water analysis. This constant corresponds to a molar iron activity of 1.8×10^{-7}. The other boundaries have been drawn for dissolved iron activities of 0.1, 1.0, 10, and 100 mg/l. The solid lines that separate the dissolved components or various solid phases, on the other hand, are independent of the iron concentration.

In the presence of CO_2, $FeCO_3$ appears as a solid phase and the solubility

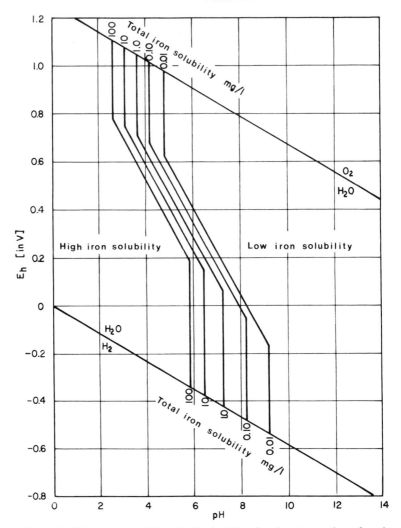

Fig. 18 Iron solubility in terms of *Eh* and pH. Activity of carbonate species referred to 100 mg/kg hydrogen carbonate (after Hem 1961a).

of iron is diminished. Figure 18 shows the solubility of iron at a carbonate activity of 100 mg/kg (as hydrogen carbonate). Carbonate activities of this level or even higher are very common in groundwater. The species include carbonate and hydrogen carbonate ions, undissociated carbonic acid, and dissolved CO_2.

Figure 19 shows the effect of stable sulfur compounds in natural waters on a solution containing iron and carbonate. The required equilibrium constraints for the stability field diagram are often absent in natural waters on account of the low reaction velocity of the sulfur species concerned (Hem

1960b). The diagrams show that when in contact with atmospheric air water has an Eh value of 0.35–0.50 V and usually a pH above 5, and low iron solubility. The stable form under these conditions is $Fe(OH)_3$ or F_2O_3. In reducing conditions, however, solutions with considerable Fe^{2+} ion content can occur. Under sulfate-reducing conditions, iron will be deposited as the sulfide.

Numerous factors, such as Eh, pH, and carbonate and sulfate content, af-

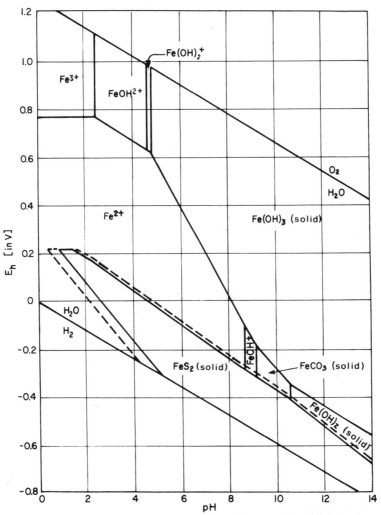

Fig. 19 Stability fields for ferrous and ferric iron species for activities of 0.01 mg/kg dissolved iron and 100 mg/kg carbonate species as hydrogen carbonate. The system contains sulfur referred to 10 mg/kg sulfate content (solid line boundaries) or 100 mg/kg sulfate content shown by dashed lines (after Hem 1961a).

fect the solubility of iron in groundwater; these factors can change very easily during water flow beneath the surface, or when the water table rises or falls. This explains some of the apparently random differences in iron content of water from geologically similar well conditions, and in the iron content of water from a single well after prolonged periods of pumping.

An Eh-pH diagram is a theoretical model, containing only the information that has been used in its construction. It is always an oversimplification because not all the components in the system are known, or all cannot be taken into account. Thus the examples presented here do not include soluble polymers such as $Fe_2(OH)_2^{4+}$ and other soluble complexes, chloride, sulfate, and phosphate complexes, which occur in natural waters. However, these diagrams are useful aids to the interpretation of observations (Stumm 1961).

For application to natural systems, one should finally note that in natural waters in addition to Fe^{2+}-Fe^{3+} and O_2-OH^-, other redox systems are present, such as NO_3^--NO_2^-, NO_2^--NH_4^+, S^{2-}-S_2-SO_3^{2-}-SO_4^{2-}, and Mn^{2+}-MnO_2. These do not necessarily remain in equilibrium with each other because of slow reactions that occur over time. It is seldom possible to determine which of these reactions is influencing the measured Eh. Such "mixed potentials" are usually not amenable to quantitative interpretation (Stumm & Morgan 1970).

1.2.4   Specific Electrical Conductance of Aqueous Solutions

The specific electrical conductance of an aqueous solution of one or more salts is made up of the conductances of the individual cations and anions. Table 24 gives the equivalent conductances or "mobilities" of the most important ions.

Whereas the values given in Table 24 hold for ideal solutions only, with their aid the conductance in microsiemens per centimeter ($\mu S/cm$) can be calculated approximately from the chemical analyses, in which one adds to-

Table 24   Ion Equivalent Conductance (Mobility) at 18 °C (After Fauth 1969)

Ion	Equivalent conductance (S/cm^2)	
	1	1*
Na^+	43.5	1.89
K^+	64.6	1.65
$\frac{1}{2}Ca^{2+}$	51	2.54
$\frac{1}{2}Mg^{2+}$	46	3.78
Cl^-	65.5	1.85
$\frac{1}{2}SO_4^{2-}$	68.3	1.42
HCO_3^-	38.3	0.63

gether the litre values and concentrations (meq/l), or litre values and concentrations (mg/l) of the individual ions. The mobilities increase with rising temperature, decrease with falling temperature, at an average of about 2%/°C.

For an estimate of the total dissolved solids in mg/l in freshwater it is sufficient to multiply the specific conductance of the water in $\mu S/cm$ by the factor 0.65. An exact relationship exists between concentration and conductance. For approximately pure water the conductance divided by 100 gives the concentration of the solution in meq/l to an accuracy of about 5%; for water at 1.0–10.0 meq/l the accuracy of the result is about 15% (Davis & DeWiest 1967). Logan (1961) has proposed the following empirical relationships between conductance C and the total concentration of dissolved solids B in meq/l:

$$B < 1.0 \qquad\qquad\qquad\qquad C = 100B \qquad\qquad (51)$$

$$B \ 1.0\text{–}3.0 \qquad\qquad\qquad\qquad C = 12.27 + 86.38B + 0.835B^2$$

$$(52)$$

$$B \ 3.0\text{–}10.0 \qquad\qquad\qquad\qquad C = B(95.5 - 5.54 \log B) \quad (53)$$

$$B > 10.0 \text{ and } HCO_3^- \text{ as the dominant ion } C = 90.0B \qquad (54)$$

$$B > 10.0 \text{ and } Cl^- \text{ as the dominant ion } \quad C = 123B^{0.9388} \qquad (55)$$

$$B > 10.0 \text{ and } SO_4^{2-} \text{ as the dominant ion } \quad C = 101B^{0.9489} \qquad (56)$$

Because the total concentration in meq/l of the sum of the anions usually deviates slightly from that of the cations, the value B is taken as the average of both sums. The relationships are valid only for values of B below 1000 meq/l.

The reciprocal calculation of the conductance for the water calculations under discussion is based on the fact that the electrical conductance of a salt solution is the sum of the conductances of the component ion species. This calculation is useful in the evaluation of electric borehole logs, for the comparison of calculated and measured conductance in a water to determine coarse errors in the analysis, or for a check on the thoroughness of the analysis.

Guillerd (1941) has multiplied the concentrations of various ions (in mg/kg) by the following factors:

Cl^-	0.27×10^{-5}	Ca^{2+}	0.387×10^{-5}
SO_4^{2-}	0.155×10^{-5}	Mg^{2+}	0.55×10^{-5}
CO_3^{2-}	0.2×10^{-5}	K^+	0.248×10^{-5}
		Na^+	0.27×10^{-5}

The average of the anion and cation sums of the conductances corresponds to the conductance of the water. The specific electrical conductance of rainwater usually ranges between 5 and 30 μS/cm; for fresh groundwater it lies between 30 and 2000 μS/cm, and for seawater between 45,000 and 55,000 μS/cm. The highest values occur in oil field brines with usually more than 100,000 μS/cm (Schoeller 1962, Davis & DeWiest 1967). Waters charged with calcium carbonate and calcium sulfate generally have the lowest, and waters charged with sodium chloride the highest conductance for a given total concentration of the dissolved components.

For further treatment of this topic see Blanquet (1946), Dunlap & Hawthorne (1951), and Martin (1958).

Chapter 2 Geochemical Processes and Water

The geochemical properties of groundwater generally depend on those of the recharge water (atmospheric precipitation, inland surface waters, seawater), and on subsurface geochemical processes. These control the water quality during the course of its underground movement by raising or lowering the amount and kind of the dissolved solids. The scale of these changes is dependent on the chemical and physical properties of the surrounding rocks, the degree of diagenesis in sediments, the water temperature, the salinity of the water and its other chemical content, the volume of water in movement and its velocity, and human influence. Periodic changes of origin and constitution of the recharge water and hydrologic and human factors cause periodic changes in groundwater quality (Nöring 1951a, Chebotarev 1955). Knowledge of the geochemical processes leads to an understanding, under favorable circumstances, of groundwater quality, and on occasion to conclusions about the geochemical processes occurring in the subsurface regions of water movement. The processes, which Degens & Chilingar (1967) refer to as diagenesis or metamorphism of the water (following rock terminology), can have such a powerful effect on quality that statements about the origin of the water on the basis of quality alone are either impossible or fraught with great uncertainty.

Dissolved solids in groundwater can change through diagenetic processes and the properties of the aquifer, and lead to the formation of mineral deposits and enrichment in certain elements (Cadek et al. 1968). Noteworthy in this respect are changes in aquifer permeability by formation of some minerals or removal of others by solution within existing cavities in the rocks.

2.1 DISSOLUTION, HYDROLYSIS, AND PRECIPITATION

Contact between water and adjoining rock causes rock components to go into solution, as well as precipitation of material from groundwater in the rock cavities. Rock-forming minerals become dissolved by chemical weathering. These processes consist chiefly of dissolution, decomposition, and hydration. The attack of the water is raised by the presence of organic and inorganic acids as well as by increase in temperature. Dissolution and precipitation are chiefly controlled by hydrogen ion concentration (pH) and redox potential (*Eh*).

Role of Contact Surface and Contact Time Between Water and Rock, and Temperature. The surface of contact between rock and groundwater is a powerful factor in the mineral solution process because dissolution is a surface phenomenon. Dissolution processes are promoted by increasing the surface area exposed to the solution.

Physical weathering comprises release of stress after the removal of the superincumbent rock material, insolation and frost action as a consequence of diurnal and seasonal temperature changes, frost weathering caused by frozen water expanding in the rock, disintegration through the growth of mineral crystals, expansion of minerals during uptake of water of crystallization, and the breaking apart of rocks by the action of plant roots. Transport by wind, flowing water, and glaciers leads to further decrease in grain size.

Rocks that have a particularly low inherent permeability (e.g., igneous and metamorphic rocks) or have become almost impermeable through diagenesis, contain groundwater mainly in fracture planes. Highly compacted uncleaved clay formations and silica-cemented sandstones can be so impermeable that no additional changes occur under the pressure and temperature conditions near the surface. On the other hand, porous sandstones can be completely permeated by water. The contact surface in pore channels increases with decreasing grain size and pore diameter, in fissure and karst aquifers with increasing joint spacing and decreasing width of fault zones.

Contact time as well as contact area is important in the interaction between rock and water. In general the movement of subsurface water is slower in rock of large contact surface, in porous media, than in those rocks in which water movement is restricted to fissure channels (faulted and karst rocks). Large surface of contact and long periods of contact therefore work together to bring about powerful leaching effects.

The flow velocity tends to decrease with increasing depth, thus the duration of contact with rock increases. Hence it is to be expected that the concentrations of dissolved solids increase downward, as is particularly observable in the water contained in deep groundwater zones, above all those in oil fields (Roger 1917, Schoeller 1956). The influence of temperature on the solution processes should also be noted here. The rate of the solution of a solid is, according to Nernst's law, proportional to the saturation deficit, which increases with the temperature. Because the temperature of the groundwater increases with depth, the solute content (except for alkaline earth carbonates) also increases. Regional temperature distribution over the Earth leads to localized variations in the solvent action of groundwater. The groundwater-rock system tends toward a physical-chemical equilibrium. This equilibrium is reached only after time, and is attained completely, theoretically, after an infinite period.

Typical examples for nonequilibrium waters are described for the carbonate–carbon dioxide equilibrium. Thus groundwaters in carbonate aquifers have been observed supersaturated or undersaturated with respect to the carbonate minerals present. The reasons for these nonequilibria may be different; most of them are caused by kinetic effects (Barnes 1965) connected with the mixing of different water types (Back & Hanshaw 1970, Runnels 1969) or the exsolution of CO_2 (Holland et al. 1964). In other cases the presence of other dissolved substances may be important: gypsum (Back & Hanshaw 1970) humic and fulvic acids, certain aromatic carboxylic acids (mellitic, gallic, tannic), and orthophosphates (Berner et al. 1978). The rate of calcite dissolution may be further influenced by both transport and surface reaction processes (Plummer & Wigley 1976). It seems that very long contact times are necessary.

Back & Hanshaw (1971) calculated for the Tertiary carbonate aquifer system of Florida that the groundwater moving slowly down the gradient (velocity 2–8 m/year) attains equilibrium with respect to calcite in about 4000 ^{14}C years and with respect to dolomite in about 15,000 ^{14}C years.

Dissolution of Salts. The most important soluble salts occurring in relatively large quantities in rocks include $CaCO_3$, $CaCO_3 \cdot MgCO_3$, $MgCO_3$, $NaCl$, $CaSO_4$, and $CaSO_4 \cdot 2H_2O$. The other substances occur only in trace quantities, or are found locally in mineral deposits or in certain types of rock. The most important ions in groundwater are therefore Ca^{2+}, Mg^{2+}, Na^+, Cl^-, SO_4^{2-}, and HCO_3^-.

As noted in Section 1.2 the individual substances show a variation in solubility, which is generally controlled by temperature, pressure, and other factors, in some instances by the pH and redox potential of the solution.

The easily soluble alkalis and alkaline earth metals are dissolved by water and removed. In humid regions this process is generally described as leaching from the soil; it removes those salts from the upper zones of the Earth's crust that are penetrated by groundwater. The less soluble calcium and magnesium carbonates are dissolved by the presence of CO_2 in the water as hydrogen carbonates. The high partial pressure of CO_2 and low temperature of the ground assist this process by increasing the CO_2 solubility. Sulfuric and nitric acids, which are generated under aerobic conditions, particularly by microbiological activity, can also attack carbonates.

Organic acids such as lactic, butyric, and citric acids, which form under anaerobic conditions by microbiological activity (Scheffer & Schachtschabel 1976), can dissolve carbonates to an important extent. These organic acids are not however stable in groundwater (Schoeller 1962).

Hydrolysis. The breakdown of minerals under the influence of H^+ and OH^- ions in the water is known as hydrolysis. The breakdown of silicate minerals, particularly feldspars, is the best example of this process; feldspars make up on average 57.9% of the Earth's crust (Correns 1940), and contain almost 80% silica. Hydrolysis attacks the surface of these minerals, which are otherwise relatively insoluble, and causes the formation of new minerals at the same time as the removal of Ca^{2+}, Mg^{2+}, Na^+, and K^+ ions in solution. The gradual breakdown of silicate minerals, particularly alkali feldspars, has been described by Sticher & Bach (1966).

First, the K^+ ions are freed. This reaction can be considered to be an exchange between H^+ and K^+:

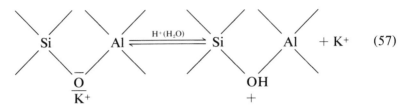

$$+ K^+ \qquad (57)$$

The accumulation of protons in the oxygen of the Si—O—Al bonds leads to a rupture of the Al—O bond and to the formation of a silanol (SiOH) group and an unstable $\equiv Al^+$ ion (58a), which immediately reacts with water (58b).

$$\equiv Si \qquad Al\equiv \qquad \equiv Si$$
$$OH \longrightarrow OH + \equiv Al^+ \qquad (58a)$$
$$+$$

$$\equiv Al^+ + H_2O \rightleftharpoons Al(OH_2) \rightleftharpoons Al{-}OH^- + H^+ \qquad (58b)$$

The same process can bring about acid catalysis and breakdown of the Si—O—Si bonds (siloxan bonds) and cause the formation of a silanol group and an unstable $\equiv Si^+$ ion, which reacts with water and likewise forms a silanol group. However, it is more likely that base-catalytic hydrolysis occurs, wherein the Si—O—Si bond is broken by assimilation of a hydroxyl ion to split off the silicon atom.

$$(59)$$

By this reaction the coordinate number in the intermediate stage is raised above 4. This involves a weakening of the adjacent Si—O bond, which ultimately causes breakdown.

The stability of the Si—O—Si bond also depends on the kind of metallic ions adjacent in the silicate lattice. This stability increases in the order K < Na < Ca < Mg.

The end product of hydrolysis of alkali feldspars can be silicic acid ($HSiO_3^-$) or aluminum hydroxide (60), or silicic acid and clay minerals, for example Kaolinite (61).

$$KAlSi_3O_8 + 8H_2O \rightarrow Al(OH)_3 + 3H_4SiO_4 + K^+ + OH^- \qquad (60)$$

$$2KAlSi_3O_8 + 2H^+ + 9H_2O \rightarrow Al_2Si_2O_5(OH)_4 + 4H_4SiO_4 + 2K^+ \quad (61)$$

The release of potassium ions in this process is important, as shown by equations 57, 60, and 61.

The hydrolysis process illustrated by these equations can also be applied to other silicate minerals, plagioclase feldspar (62, 63), pyroxene (65), amphibole and olivine (64).

$$2NaAlSi_3O_8 + 2H^+ + 9H_2O \rightarrow Al_2Si_2O_5(OH)_4 + 4H_4SiO_4 + 2Na^+ \quad (62)$$
albite   kaolinite

$$CaAl_2Si_2O_8 + 2H^+ + H_2O \rightarrow Al_2Si_2O_5(OH)_4 + Ca^{2+} \qquad (63)$$
anorthite   kaolinite

$$5Mg_2SiO_4 + 8H^+ + 2H_2O \rightarrow Mg_6(OH)_8Si_4O_{10} + H_4SiO_4 + 4Mg^{2+} \quad (64)$$
olivine (forsterite)   serpentine

The stability of iron- and manganese-bearing silicates depends on the redox potential, because these metals can be easily reduced or oxidized. Hydrolysis and oxidation combine to bring about the breakdown of the pyroxene hedenbergite, thus:

$$4CaFeSi_2O_6 + 8H^+ + O_2 + 6H_2O \rightarrow 4Ca^{2+} + 4FeOOH + 8H_2SiO_3 \quad (65)$$
pyroxene (hedenbergite)

The ion Fe^{2+} is released from the lattice and oxidized to ferric oxide, when biotite is broken down to vermiculite and hornblende to chlorite, and iron-rich olivine is transformed to serpentine.

The effect of hydrolysis in the removal of alkali and alkaline earth ions in seepage water and groundwater is to form a residual layer or weathered rind on the surface of the mineral, in which the bulk of the aluminum, silicon, and iron remains, either as clay formed by bonding layers of K^+ ions or as silica, aluminum, and iron oxides and hydroxides (Correns 1940). The rind so formed hinders the process of weathering. Groundwater transports dissolved silica best at pH values above 9, Al^{3+} ions below pH 5 or above pH 7.5, and iron ions only at certain redox potentials (cf. Section 1.2.3), furthermore the aluminum and iron are transported as soluble organic complexes. Hydrogen ions, CO_2, and H_4SiO_4 concentration, as well as the alkali and alkaline earth metal content control the reactions and determine the type of solid weathering products (Kramer 1968).

Metallic ions from the silicates are dissolved as soluble stable complexes, chiefly chelates. The principal chelate-forming agents are organic substances of microbiological origin such as tartaric acid, citric acid, ketoglutaric acid, pyrocatechol, salicylaldehyde, and salicylic acid. Derivatives of o-dihydroxy-benzene can also form stable chelates with cations such as Al, Fe, and Ti, as well as Si (Scheffer & Schachtschabel 1976).

Hydrolysis proceeds most effectively at high temperature, low pH, and low redox potential. Increasing hydrogen ion concentration in the presence of free acids hastens chemical disintegration. Carbon dioxide dissolved in water is the most powerful acid involved. Its effect is based on the supply of H^+ ions for hydrolysis, for neutralization of the hydroxide resulting from the hydrolysis, and on the formation of water-soluble calcium, magnesium, and ferrous hydrogen carbonates that can then be removed from the system (Scheffer & Schachtschabel 1976). Besides carbonic acid there occur locally other free acids in nature, such as sulfuric, nitric, and unneutralized organic acids.

Changes in Water Quality Caused by Dissolution and Precipitation. During the course of its flow groundwater picks up dissolved solids. It is worth noting that a very small amount of soluble substances that can be attacked by the water need be present in the rock for relatively high concentrations to be brought about in the groundwater. A rock of density d, porosity n, and NaCl content a (wt %), contains in 1 dm^3 of rock $a \times d(1 - n)$g NaCl. At equilibrium there is in the pore solution the same concentration per unit volume as in the rock. In a rock of porosity 0.20 and density 2.65 g $\cdot$ cm^{-3} with only 1 part in 1000 NaCl, there is in 2.120 kg of solids 2.12 g NaCl per cubic decimetre, hence in the pore solution in equilibrium with the rock 2.120 g NaCl per litre. For clays the calculation is similar. The latter according to Billings & Williams (1967) contain an average of 1466 mg Cl/kg, corresponding to 2.4% NaCl for $d = 2.2$ g $\cdot$ cm^{-3} and $n = 0.40$, or a concentration of the pore solution of 3.1 g NaCl per litre of water. Even higher concentrations are to be expected in unleached rocks deep below the surface, which can give rise to high concentrations in deep-lying groundwater.

Because of the effect of carbonic acid on carbonates and silicates, the dissolved solids content of groundwater near the surface, except for calcium carbonate-free rocks, consists chiefly of Ca^{2+}, Mg^{2+}, and HCO_3^- ions. These waters can be in carbonate–hydrogen carbonate equilibrium through prolonged residence below the surface, and with constant CO_2 pressure can dissolve more carbonate only on the addition of other ions from outside the system; for example the solution of dispersed gypsum $CaSO_4 \cdot 2H_2O$ causes a rise in the Ca^{2+} ion content and consequently a fall in the number of HCO_3^- and CO_3^{2-} ions by precipitation of calcium carbonate.

Therefore $CaSO_4$-bearing waters often show a below-average carbonate content. The latter can rise with increasing partial pressure of CO_2 or, with decreasing partial pressure or fall in temperature, they can decrease as a result of $CaCO_3$ deposition.

When gypsum is widely distributed throughout the rock the water can ultimately become saturated with $CaSO_4$. With the entrance of geochemically less common $CaCl_2$, or more abundant SO_4^{2-} ions in salts such as Na_2SO_4 or $MgSO_4$, the solubility product is exceeded and $CaSO_4$ is precipitated.

Highly soluble sodium chloride is the commonest salt in highly mineralized groundwater. Long flow channels, very long contact time, and large contact surface serve to increase the NaCl content to equilibrium with the NaCl content in the rock. Values close to saturation are reached in the vicinity of salt deposits and in salt-bearing aquifers.

Schoeller (1962) has derived a general relationship between concentration and dominant ions:

1 For concentrations below 40 meq/l (corresponding to a residue after evaporation of about 950–1090 mg/l) there is an equivalent ratio of

$$(HCO_3^- + CO_3^{2-}) > Cl^- \quad \text{or} \quad SO_4^{2-}$$

2 For concentrations between 40 and 60 meq/l (corresponding to a residue after evaporation of about 1565 mg/l and more) possibly, also

$$(HCO_3^- + CO_3^{2-}) > Cl^- \quad \text{or} \quad SO_4^{2-}$$

3 For concentrations above 60 meq/l

$$SO_4^{2-} \text{ or } Cl^- > (HCO_3^- + CO_3^{2-})$$

4 For concentrations above 180 meq/l (corresponding to approximately 5100 mg/l residue after evaporation)

$$Cl^- > SO_4^{2-} > (HCO_3^- + CO_3^{2-})$$

Higher SO_4^{2-} values can occur in magnesium-bearing waters, but for water with dissolved solids above a concentration of 390 meq/l the following always holds:

$$Cl^- > SO_4^{2-} > (HCO_3^- + CO_3^{2-})$$

Because of the variable solubility of the substances in the water caused by increasing concentration through prolonged circulation or in very long flow paths, the anion ratios alter so much that they bring about a geochemical zonation of the dominant anion type – generally from the water table downward (Chebotarev 1955, Schoeller 1962, Freeze & Cherry 1979):

$$HCO_3^- \rightarrow HCO_3^- + SO_4^{2-} \rightarrow SO_4^{2-} \rightarrow SO_4^{2-} + Cl^- \rightarrow Cl^- + SO_4^{2-} \rightarrow Cl^-$$

For large sedimentary basins Domenico (1972) associated this sequence to three main zones, which correlate in a general way with depth. The upper zone is characterized by active groundwater flushing through relatively well-leached rocks, HCO_3^- as dominant anion and in low concentrations of total dissolved solids. In the intermediate zone groundwater circulates less actively. Here sulfate is the dominant anion. Finally the lower zone with very slow groundwater flow is characterized by high concentrations of Cl^- and total dissolved solids. In this zone highly soluble minerals may be present because the ground flushing is very small.

Travel distance and time tend to increase from the upper zone to the lower zone, but no specific prediction can be made. The time in different basins may range from years to thousands of years for the upper zone and from thousands to millions of years for the lower zone (Freeze & Cherry 1979).

Besides the described general principle of the anion distribution, other localized fluctuations of the anion ratios with increasing depth can be found when anions are removed from groundwater or added by other geochemical processes. Thus Wandt (1960) has noted a relative increase of hydrogen carbonate ions in Pleistocene and Tertiary aquifers in Schleswig-Holstein, especially in confined aquifers.

Among the cations, Mg^{2+} ions tend to increase relatively to Ca^{2+} ions in the course of subsurface flow. During recharge the water will often be saturated with calcium hydrogen carbonate. If this water comes into contact with gypsum- or anhydrite-bearing rocks these soluble sulfates are dissolved. If finally the solubility limit is reached, only the much more soluble $MgSO_4$ and $MgCl_2$ (Table 10) may further contribute to the solutes. Thus the Mg^{2+} content may exceed that of Ca^{2+}. The deposition of $CaSO_4$ from the saturated solution is also induced by a rise in the $Ca(HCO_3)_2$ content in consequence of raised CO_2 pressure, which causes a rise in the content of Ca^{2+} ions, or by falling temperature. Finally Mg^{2+} may rise because of Mg-Ca ion exchange.

Ultimately at high concentrations sodium appears as the dominant ion, so that a geochemical zonation based on the characteristic cation can be defined:

$$Ca^{2+} \rightarrow Ca^{2+} + Mg^{2+} \rightarrow Na^+$$

Precipitation in the sequence $CaCO_3 \rightarrow CaSO_4$ in diagenesis leads to a layered structure of pore filling in sedimentary rocks, in which the $CaSO_4$ filling corresponding to a deep geochemical layering, after tectonic upheavals, can be replaced by $CaCO_3$ as a type of retrograde diagenesis in the alkaline earth hydrogen carbonate field (Meisl 1970).

Dependence of Dissolution and Precipitation on Climate. The climatic elements of precipitation, evapotranspiration, and temperature affect water quality through the different weathering processes, through dilution or concentration of soluble salts by climatically different precipitation and evapotranspiration rates over the climate-dependent vegetation.

Humid regions favor the solution and translocation of salts, particularly $NaCl$ and $CaSO_4$. The dense plant cover typical of this climatic type raises the amount of biogenic carbon dioxide exhaled from the roots and from the biologically degraded organic substances in the ground. Carbon dioxide and humic acids lower the pH, contribute to silicate weathering, and can dissolve iron and manganese. The generally large amounts of circulating water relative to the supply of soluble salts leads to a low mineral content in the water. High groundwater recharge brings about the formation of a powerful lens of fresh water above the saltwater that may be present at depth.

Whereas under moderate to high temperatures the biological and chemical processes proceed rapidly, in cold regions the translocation of materials is slower, so that the dissolved solids in groundwater are mostly very low. The slow breakdown of organic substances leads to a definite enrichment and to the formation of reducing conditions involving solution of iron and manganese together with the occurrence of humus, which discolors the water. The partial freezing of shallow groundwater can seasonally raise the concentration in the remaining fluid water, as has been observed in northwestern Alaska by Feulner & Schupp (1963). Distinct seasonal variation in the dissolved constituents (Fig. 20) is brought about by dilution by snowmelt and rain in the spring and through freezing in the autumn and winter. The Ca^{2+} ion content originates in the aquifer, a partly metamorphosed thin-bedded limestone intruded by granite, and SO_4^{2-} ions from weathered sulfide minerals in the contact zone; the relative high Na^+ and Cl^- content is introduced as cyclic salt in the atmospheric precipitation under the influence of the nearby Bering Sea. The increased concentration in the winter is probably attributable to freezing of the water, whereas that of Ca^{2+} and SO_4^{2-} ions may be caused by reduction of flow because of decrease in the cross section of the aquifer through ice formation from the sides and downward, which in turn increases the rock-water contact time. This decrease in cross section because of inward freezing is presumably the cause of the marked decline in yield during winter.

In arid regions the precipitation is generally insufficient to ensure continuous percolation to the groundwater table. Evaporation occurs in the strata near the surface, causing an increase in soluble salts having their source in deposition, or salt-bearing dust, or weathering from the rocks, especially where the water table is less than 2 m below the surface, so that water rises and can evaporate. The precipitation of $CaSO_4$ and $CaCO_3$ when their solubility limits are exceeded thereby leads to relative decreases in SO_4^{2-} compared with Cl^- ions, and in Ca^{2+} with respect to Mg^{2+} ions because sulphate radicle and magnesium are removed from the solution while Cl^- remains. Crusts of limestone or gypsum (caliche) can form in the soil, or even crusts on the surface. Because $CaCO_3$ is precipitated first, with increasing aridity zones of $CaCO_3$, then $CaSO_4 \cdot 2H_2O$, and finally $NaCl$ can build up. The concentrations of salts are dissolved only by occasional heavy rainfall and flash floods, which give rise to groundwater recharge. The solute content of this kind of percolate will be the greater, the higher the temperature and the more arid the climate. Finally, in arid regions there are salt-bearing strata

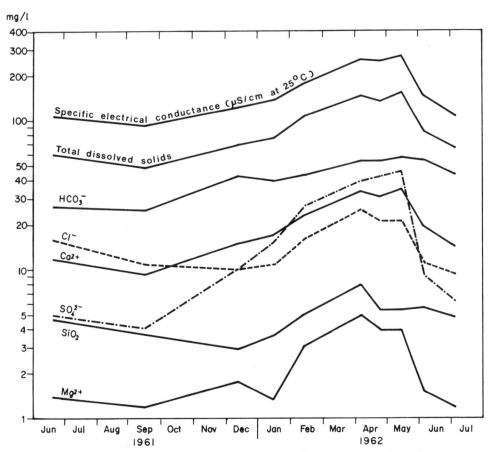

Fig. 20 Seasonal changes in the chemical quality of shallow groundwater in northwestern Alaska due to freezing (after Feulner & Schupp 1963).

and salt domes present near the surface, which can be dissolved away by occasional occurrences of percolating water. The very low groundwater recharge rate hinders or even prevents the formation of a freshwater lens above highly mineralized waters deep underground.

In contrast to humid zones, biogenic CO_2 content in the soil air is reduced in arid climates, and so the HCO_3^- content is relatively low (Schoeller 1941, 1950). In general the dissolved solids concentration in the groundwater is higher than in humid regions. The lowest values are found in regions of rocks that have a low salt content, regions that have metamorphic and igneous rocks, or those that allow the rapid movement of water (karst limestone and volcanic rocks). Aquifers showing a low salt content include sand dunes and porous deposits alongside fast-flowing rivers low in salt, into which the river water percolates because of natural or artificial gradients. Examples are to be found in arid regions, for example the Nile Valley, western South

America, western United States, the area along the Indus River, and the arid parts of the USSR. In many cases however, only groundwater with high concentrations is met and this has to be used for human consumption and irrigation (Section 4.2).

According to Schoeller (1941) the dissolved solids increase from cold climatic regions (high mountains) through temperate zones to totally arid regions like the Sahara; then they decrease again in humid tropical regions. The hydrogen carbonate ion content is quite variable (in 75% of instances between 2 and 8 meq/l), but shows a definite dependence on biogenic CO_2, on which the temperature (cold to warm regions) or the humidity (hot-arid, hot-humid) have a powerful influence.

The SO_4^{2-} and Cl^- ion content increases from very low values in the temperature zones southward toward the Sahara; each of these ion species exceeds the hydrogen carbonate ion content in steppe and desert regions. In humid equatorial regions the percentage contents of SO_4^{2-} and Cl^- ions are again low. The sodium content is generally the same as Cl^-. Calcium and magnesium contents increase in cool and temperature zones, following hydrogen carbonate, and in semiarid and arid regions increase or decrease with the hydrogen carbonate content; in these regions the magnesium content tends to be higher the higher the sulfate concentration.

Garmonov's chemistry–water zone map (Fig. 21) shows the influence of climate on the chemical composition of groundwater, and the climatically controlled zonation of the plant cover and weathering factors. In European Russia Garmonov has differentiated from north to south a zone of hydrogen carbonate–siliceous groundwater in the region of tundra soils, a zone of water-bearing hydrogen carbonate–calcium, in central Russia, and finally a southern zone of dominantly sulfate- and chloride-charged water, which includes a continental margin salt-bearing zone in the depression to the north of the Caspian Sea and in the region to the east of that sea, and a zone of water bearing hydrogen carbonate–calcium in the mountainous part of the Crimea and the Caucasus.

Climate-based zonal belts can be observed across the world. They give rise to variations in the dissolved solids in the water. They are satisfactory as a classification for uniform subsurface conditions only, because rocks of different chemical composition can override and obliterate the effects of climate.

Degree of Concentration and Precipitation. The concentration of dissolved solids in water can be increased, as shown above, by solution and hydrolysis of subsurface material and also by evaporation or freezing, but the effects are generally noticeable only in seepage and shallow groundwater. Deeper groundwater can however be enriched by evaporation under certain circumstances when CO_2 or gaseous hydrogen escapes and carries water vapor along with it (Mills & Wells 1919). With increasing concentration the solubility limit can ultimately be exceeded, but generally only for slightly soluble substances such as $CaSO_4$. Eventually highly soluble sub-

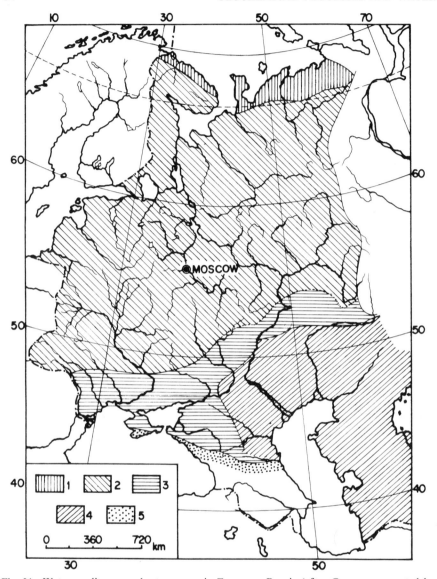

Fig. 21 Water quality groundwater zones in European Russia (after Garmonov, quoted by Alekin 1962). 1, HCO_3–SiO_2 water; 2, HCO_3–Ca water; 3, SO_4–Cl water; 4, continental saltwater zone; 5, HCO_3–Ca water.

stances such as NaCl are deposited under the extreme case of total desiccation during surface evaporation in arid regions; this may also occur at greater depths, as has been demonstrated by Mills & Wells (1919) in oil field aquifers in the Appalachian region.

Dissolution and Precipitation During Mixing Processes. Groundwaters of various qualities can become mixed by flow between different aquifers and multiple-layer aquifers, or at the boundaries between groundwaters of different composition. Mixing processes of this kind can take place during groundwater extraction if different types of groundwater mix in the borehole during pumping or surface water mixes in with the groundwater during recharge and percolation from river courses.

The mixing of different waters can lead to characteristic dissolution and precipitation processes that will impart new properties to the water. The addition of ions of the same kind, especially with only slightly soluble salts, leads to deposition because the solubility product has been exceeded, for example deposition of $CaSO_4$ or $CaCO_3$ during mixture of $CaCl_2$-bearing water with $CaSO_4$- or $Ca(HCO_3)_2$-bearing groundwater. Precipitation of $BaSO_4$ during encounter between Ba^{2+}-rich deep water and sulfate-bearing groundwater has been observed in the Ruhr coal field (Delkeskamp 1908, Patteisky 1952, Fricke 1953b, Buschendorf et al. 1957).

When waters that are in carbonate–hydrogen carbonate equilibrium mix, this condition can continue if the waters are "corresponding," for example if they have the same HCO_3^- content (Axt 1965). In other cases of mixing either calcium carbonate deposition or dissolution may occur. Aggressivity toward calcium carbonate occurs particularly when various kinds of hard water whose free carbon dioxide content lies above the curve of carbonate–hydrogen carbonate equilibrium are mixed together. Line *AB* in Fig. 14 is one example. Mixed waters of this kind can therefore bring about solution of limestone (Tillmans 1932, Zehender et al. 1956), a phenomenon that has been described in karst research as "mixing corrosion" (*"Mischungskorrosion"*) by Bögli (1964, 1969) and in many cases causes cavity formation below the water table. When waters of different temperatures mix together, a displacement of the temperature-dependent carbonate–hydrogen carbonate equilibrium can eventually occur. Raised temperature of the mixed water can lead to partial loss of dissolved gases, particularly CO_2, so that it can become incorporated in deposited $CaCO_3$. Changes in the pH value as a consequence of mixing of strongly acid and basic groundwaters can likewise lead to the deposition of pH-sensitive substances from the solution. Substances can precipitate during mixing of waters of differing redox potentials when their solubility depends on redox potential. For example $Fe(OH)_3$ is deposited when iron-bearing groundwater with negative redox potential mixes with oxygen-containing groundwater with positive redox potential.

Coprecipitation. During precipitation of a particular substance foreign ions are usually brought out of solution as well. They are incorporated by isomorphous substitution with an ion of similar size into the structure of the mineral that is forming (coprecipitation) or into one that has formed (replacement). The replacement of one element in a crystal lattice by another is common in the creation of distinct minerals. The elements that replace one another tend to have similar ionic radii, covalent character, and valence (Krauskopf 1967). The significant feature of replacement and coprecipitation processes is that the new solid phase is more stable in the final solution than the original solid phase. This implies that the new solid phase is more insoluble than the original solid phase, and that the propensity of an ion to be absorbed into a mineral lattice can be expressed by solubility arguments (Jackson & Inch 1980). Stumm & Morgan (1970) proposed for the solid solution of B in A

$$A(s) + B(aq) = B(s) + A(aq)$$

where (s) denotes the solid phase and (aq) the aqueous one, a coprecipitation index that can be calculated from the quotient of the solubility products of the two mineral species:

$$CI = \frac{K_{SP}(A)}{K_{SP}(B)} \qquad (66)$$

Consequently in a solution of radium ions (B(aq)) undersaturated with respect to $RaSO_4$($-\log K_{SP} = 10.4$) but saturated with respect to $BaSO_4$($-\log K_{SP} = 9.5$), it is not unreasonable to expect the coprecipitation of (Ba, Ra)SO_4 (CI $\approx$ 8). The coprecipitation index merely indicates the likelihood of a reaction occurring (CI > 1) but does not guarantee that such will be the case.

Coprecipitation and replacement are of importance in the fixing of trace elements, for example radium in the precipitation of barite, copper in iron hydroxides, and many trace elements during the formation of manganese oxides.

Complexing. The hydrogeological implications of the complexing of metals by inorganic or organic ligands are largely unknown. What is known is that these ligands may complex the transition metal ions (e.g., Fe, Mn, Co, Zn, Pb, Cu, Ni, Cr) and therefore alter their adsorption or precipitation from solution. Duursma (1970) has shown that the adsorption of radioactive cobalt and zinc is significantly decreased in the presence of dissolved organic carbon (DOC) compounds, which are to be found in all natural groundwaters (Leenheer et al. 1974). This decrease is due to the formation of organocobalt and organozinc complexes whose sorption behavior may differ from that of the hydrated metal ions and whose increased solubility in solution reduces the effectiveness of precipitation (Jackson & Inch 1980).

2.2 ADSORPTION AND ION EXCHANGE

Subsurface waters containing solutes interact with organic and inorganic adsorbents. All minerals have on the surface lattice defects that attract

polarized water molecules and H^+ and OH^- ions. In a similar manner dissociated and undissociated organic and inorganic dissolved solids are bound to
the active surfaces. There are all degrees of transition, from adsorption by
Van der Waals forces, which exert only a weak bond between adsorbent and
adsorbate, up to chemical adsorption as a covalent bond (i.e., specific adsorption or chemisorption). With covalent bonding the ions are not only adsorbed onto the surface, but in some minerals become fixed into the crystal
lattice. The exchange of adsorbed ions in stoichiometric proportions by another dissolved ion is termed "ion exchange." Exact distinction between
these processes is hardly possible in practice because almost every ion
exchange process is accompanied by adsorption and desorption. Special analytical techniques may be used to discriminate between exchangeable and
specifically adsorbed species (Suarez & Langmuir 1976).

The subsurface materials that act as adsorbents include clay minerals,
hydrous iron, and manganese and aluminum oxides (Krauskopf 1956b), and
organic substances, particularly humus. Also microbial slimes, plants, and
microorganisms may be important. In addition to these, glauconite, which is
common in marine sands and marls, is an adsorbent (Schoeller 1962), and so
are the rock-forming minerals mica, feldspar, aluminous augite, and hornblende (Renick 1924a, Schwille 1953a), and finely divided quartz. The last
has an adsorptivity lower than that of amorphous silica (Briggs 1905).

The equilibrium between the quantity of substance C_T bound to an adsorbent and the quantity of this substance in solution C_W can be described by
Freundlich's isothermal equation

$$C_T = kC_W^n \qquad (67)$$

where k and n are parameters specific to particular substances. The relation
shows that an increase in the concentration of a solution will raise the adsorbed quantity, and a decrease in the concentration will result in desorption.

The sorptive effective of plants, microorganisms, and microbial slimes
cannot be considered separately from other biological influences; sorption
generally precedes the biochemical reaction and therefore is considered in
Section 2.5.

Adsorption through surface tension effects is naturally dependent on the
dimensions of the external and internal surfaces, hence on density, grain
size, porosity, and extent of the subsurface materials. Clay minerals, hydrous iron and manganese oxides, and organic substances occurring in the
colloidal state or in complexes present especially large surface areas that
consequently possess great powers of adsorption. Fine-grained rocks therefore adsorb better than coarse-grained rocks, porous aquifers better than fissure and karst aquifers.

The surface charge of adsorbents in the saturated and the unsaturated
zones may be due to (1) imperfections or substitutions within the crystal lattice of a particle or (2) chemical reactions of the particle surface involving

the bonding of OH^- or H^+ ions. At pH 7 humic colloids and clays have a net negative charge and thus attract cations. Amphoteric colloids of iron, manganese, and aluminum attract cations or anions depending on the pH of the water (Schoeller 1962) and their crystallinity; basic pH waters and highly crystalline materials tend to favor adsorption of cations.

Ion Exchange. Type, extent, and velocity of an ion exchange process depend on the kinds of dissolved substances and their properties, their ionic composition, or the type and concentration of the dissolved ions and their accompanying "complementary" ions.

The ion exchange process with the competing ions A^+ and B^+ and the cation exchanger R^- (Amphlett 1964):

$$A^+ + B^+R^- \rightleftharpoons A^+R^- + B^+$$

can be described by a selectivity quotient

$$K_B^A = \frac{(A^+R^-)(B^+)}{(A^+)(B^+R^-)} = \frac{(A^+R^-)/(A^+)}{(B^+R^-)/(B^+)} \tag{68}$$

where (A^+) and (B^+) are the concentrations of these ions in solution and (A^+R^-) and (B^+R^-) are their concentrations in the adsorbed phase. The selectivity quotient may be considered as a mass action equilibrium constant if (1) the activities of all four forms may be calculated, (2) the adsorption reaction is reversible, and (3) secondary reactions that are not readily reversible, such as ion fixation due to lattice collapse, do not occur and so prevent the attainment of true equilibrium (Jenne & Wahlberg 1968). The estimation of the activities of the adsorbed phases A^+R^- and B^+R^- has been discussed by Truesdell (1972).

The amount of exchangeable ions in milliequivalents per 100_g solids at pH 7 is commonly known as the ion exchange capacity Q. The various exchange capacities of different subsurface materials are given in Table 25.

Very small particles of all inorganic minerals have low but definite exchange capacities as a consequence of electrically charged lattice defects on the edges of crystals. This capacity increases with decreasing particle size, yet the exchange capacity may be small even for the small grain sizes in the clay fraction grade, as shown in Table 25. The exchange capacity of mineral mixtures, which make up most solid and unconsolidated rocks, depends on the content of exchangeable substances, particularly of the clay and humus content.

Clay minerals formed as autochthonous products of weathering or as allochthonic sediments can influence groundwater quality, especially when they are distributed throughout the aquifer.

Cations are adsorbed on clay minerals in various ways as noted by Grim (1968):

Table 25   Exchange Capacities of Minerals and Rocks (meq/100 g)

| Mineral | For cations | | For anions |
	Grim (1968)	Carroll (1959)	Grim (1968)
Talc	–	0.2	–
Basalt	–	0.5–2.8	–
Pumice	–	1.2	–
Tuff	–	32.0–49.0	–
Quartz	–	0.6–5.3	–
Feldspar	–	1.0–2.0	–
Kaolinite	3–15	–	6.6–13.3
Kaolinite (colloidal)	–	–	20.2
Nontronite	–	–	12.0–20.0
Saponite	–	–	21.0
Beidellite	–	–	21.0
Pyrophyllite	–	4.0	–
Halloysite · 2H$_2$O	5–10	–	–
Illite	10–40	10–40	–
Chlorite	10–40	10–40?	–
Shales	–	10–41.0	–
Glauconite	–	11–20	–
Sepiolite-attapulgite-palygorskite	3–15	20–30	–
Diatomite	–	25–54	–
Halloysite · 4H$_2$O	40–50	–	–
Allophane	25–50	~70	–
Montmorillonite	80–150	70–100	23–31
Silica gel	–	80	–
Vermiculite	100–150	100–150	4
Zeolites	100–300	230–620	–
Organic substances in soil and recent sediments	150–500	–	–
Feldspathoids			
Leucite	–	460	–
Nosean	–	880	–
Sodalite	–	920	–
Cancrinite	–	1090	–

1.   On broken bonds around the edges of silica-alumina units, preferentially on noncleavage surfaces (i.e., fractures). This type of bond occurs particularly with kaolinite and halloysite and with well-crystallized chlorite, and sepiolite-attapulgite-palygorskite; for smectites and vermiculites this makes up some 20% of the total cation exchange capacity. The number of defect localities, hence the exchange capacity, increases as the particle size decreases; for example in kaolinite it increases four-fold from 10–20 μm grain size up to 0.1–0.05 μm; for illite an approximate two-fold increase in

exchange capacity can be observed for grain size less than 0.06 μm compared with grain size in the range 1.0–0.1 μm.

2. On equalization of unbalanced charges that originate by the substitution by trivalent aluminum for quadrivalent silicon in the tetrahedral sheets of the lattice of some clay minerals, and trivalent aluminum by ions of lower valence, particularly magnesium in the octahedral sheet. The exchangeable cations in this case occur mostly on the basal cleavage planes in the larger clay minerals. This type of bond is responsible for about 80% of the exchange capacity of smectite and vermiculite and can have a certain importance in poorly crystalline illite, chlorite, and sepiolite-attapulgite-palygorskite. With the substitution bond the influence of grain size on the exchange capacity is hardly noticeable.

3. By replacement of the hydrogen in exposed hydroxyl groups, which are an integral part of the structure but are found along broken edges of crystals of clay minerals. This process is particularly important in kaolinite and halloysite on account of the presence of an hydroxyl sheet on one side of the basal cleavage planes. Hydroxyl interlayers are important in montmorillonite and vermiculite.

Little is indeed known about the exchange of anions in clay minerals, among other things because research into mineral exchange is made more difficult by the possibility of clay mineral breakdown during the exchange process. The following exchangeable anions occur in clays: SO_4^{2-}, Cl^-, PO_4^{3-}, NO_3^-, and AsO_4^{3-}, in two and possibly three types of anion exchange (Carroll 1959, Grim 1968).

1. Phosphate ions can be adsorbed by kaolinite in place of OH^- ions. In this case the extent of the reaction depends on the accessibility of OH^- ions within the crystal lattice.

2. Anions such as phosphate, arsenate, and borate, which have the same size and geometry approximately as the silica tetrahedron, may be adsorbed by fitting onto the edges of the silica tetrahedral sheets. Other anions, such as sulfate, chloride, and nitrate, which on account of their geometry cannot fit onto the silica tetrahedral sheets, cannot be adsorbed in this way.

In both cases the anion exchange takes place at the edges of clay crystals. Schofield (1940, 1949) has stated however that clay minerals can possess exchange locations on the basal planes in consequence of unequal charges within the lattice, for example through excess aluminum in the octahedral sites, which can exchange with anions. Schofield takes this view because of the observation that clay minerals attract chloride ions when in strong acid solutions, which can be exchanged in their turn for anions.

Zeolites occur frequently in volcanic rocks, especially basalts, and the relatively high exchange capacities of tuffs (Table 25) should be attributed to them. Zeolites are found less often in sedimentary rocks, but they may occasionally be found in larger quantities, for example analcite and apophyllite in oil shales in the Eocene Green River Series, a lacustrine deposit in Utah and Colorado (Bradley 1929). Phillipsite is very widely distributed in deep sea sediments.

Hydroxides and oxyhydroxides of iron, manganese, and aluminum are widely distributed as weathering and precipitation products. Many heavy metal ions such as lead, zinc, mercury, antimony, bismuth, selenium, copper, and complex ions of arsenic and molybdenum are adsorbed on the colloidal particles (Jenne 1968) or coprecipitated and occur in sedimentary iron and manganese ores. Also the bonding of ^{90}Sr in the soil depends essentially according to Tamura (1964), on the effect of hydrated iron and aluminum oxides. Sugarawa et al. (1958) have reported coprecipitation of iodide ions with hydroxides of iron, manganese, and aluminum.

Mattson (1927) has shown that there is sorption of SO_4^{2-} of Cl^- ions in many soils. It appears that the sorptivity of subsurface media for Cl^- and SO_4^{2-} increases with the Fe_2O_3 content and to a smaller extent with the Al_2O_3 content and with the valence of the ions, and with increasing pH. High OH^- concentration tends to drive out anions.

Organic substances, as Table 25 shows, can have high exchange capacities. Besides humus and peat, other organic substances may act as ion exchangers, as Foster (1950) has shown experimentally in groundwaters in the lignite-rich sands of the Atlantic and Gulf coasts, and as Piper et al. (1953) have noted in groundwater in the Pleistocene San Pedro Formation in California.

Exchange Capacity and Ion Species. The bonding affinity with respect to different ions varies between the exchangers. Other things being equal, the affinity for adsorption is greater the higher the valence of the ion and the more difficult it is to remove it from a clay mineral. Hydrogen is an exception because it behaves mostly as a di- or trivalent ion. For ions of the same valence the affinity for adsorption increases with the atomic number and apparently decreases with decreasing ionic size; therefore it decreases with increasing hydrated radius. This explains the low bonding force of strongly hydrated, very small lithium ions. Such ions are an exception: they fit exactly into the spaces of the basal layer of oxygen or, with the formation of new minerals at the water-mineral phase boundary, are incorporated into the lattice, hence are bonded in a nonexchangeable manner or are exchangeable only with difficulty. Potassium, because of its ionic radius of 133 pm, is very firmly bonded in the spaces of the oxygen layers of illite and is therefore very difficult to replace. The same holds for the ammonium ion NH_4^+.

The general series of bonding strengths (Schoeller 1962):

$$H^+ > Rb^+ > Ba^{2+} > Sr^{2+} > Ca^{2+} > Mg^{2+} > K^+ > Na^+ > Li^+$$

varies with different exchangers and changes in the pH value. The sorption capacity of individual sorbents can be differentiated with regard to the sorbate and is partly selective; thus zeolites can selectively take up to 13% by weight of cesium (Schulze & Haberer 1966). Hydrated oxides of iron selectively adsorb zinc, copper, lead, mercury, chromium, molybdenum, tungsten, and vanadium, whereas hydrated oxide of manganese chiefly bonds copper, nickel, cobalt, chromium, molybdenum, and tungsten; clay minerals for example preferentially bond zinc, copper, lead, and mercury (Krauskopf 1956b).

Organic dyes are also very powerfully adsorbed. According to Hendricks (1941) large organic cations are firmly held to the surface by Coulombic and Van der Waals forces on the surface of layer silicates.

Exchange Rates.   Because of the dependence on the special type of bond (i.e., on the species of exchanger), the process of ion exchange may require a considerable time. Equilibrium is more likely to be established by long contact times between solution and ion exchanger, slow flow velocities of water through the rock, and high surface area of water-rock interaction. The cation exchange rate varies with the type of exchanger, the concentration of cations, and the type and concentration of anions. In general the reaction is quickest in minerals in which the exchange takes place predominantly on crystal edges (e.g., kaolinite) and is slower with montmorillonite, vermiculite, and attapulgite, because the adsorption process also requires diffusional transport of the adsorbate into the adsorbent. It is even slower with illite, in which the cations must somehow penetrate between the basal planes, which are firmly bonded together. Chloritic minerals probably have exchange rates similar to those of illite (Schoeller 1962, Grim 1968).

Influence of Temperature.   Temperature influences both the exchange rate and the relative exchange affinities of individual ion species, and the exchange capacity of clay minerals. This exchange capacity decreases slightly with increasing temperature, most noticeably with expanding lattice minerals, in which most of the exchange is on basal planes. Here the exchange capacity is appreciably reduced before loss of swelling characteristic (for montmorillonite 125°C), at which point the capacity drops abruptly (Grim 1968).

Exchange Capacity and pH Value of the Solution.   The bonding strength of various ion species in the exchanger also depends on the pH value of the solution, as the variations in sorption of different radionuclides (Fig. 22) show. The fraction of exchangeable alkali and alkaline earth ions at a pH below 5 amounts to more than 95% of all exchangeable metallic ions.

Ion Selectivity.   Sorption can be influenced by the presence of other dissolved solids and gases. For example strontium is adsorbed better than calcium on clay minerals; nevertheless calcium strongly hinders the sorption of strontium in the soil. The ions are clearly in competition during ion exchange in the soil. The low sorption, hence the high mobility of radiostrontium in a sandy humus in the presence of calcium, is shown in Fig. 23. The effect of

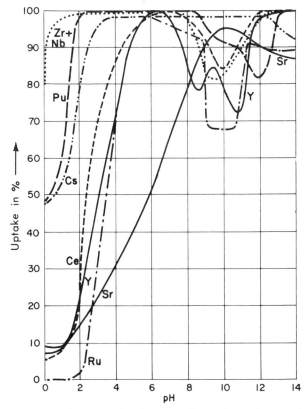

Fig. 22 Influence of pH on the sorption of different radionuclides in the ground (after Brown et al. 1956, Amphlett 1958).

the competing ions on the rate of migration of radionuclides can be described by the distribution coefficient K_d of the subsurface media (in $cm^3 \cdot g^{-1}$) (Higgins 1959), which is a measure of the partitioning of an ion species between the solution and the solid adsorbing phase. This amount is a variable

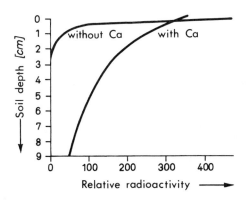

Fig. 23 Influence of Ca^{2+} content in sandy humus-rich soil on ^{90}Sr sorption (Ca^{2+} data in 0.1% aqueous solution) (Herbst 1959, 1961).

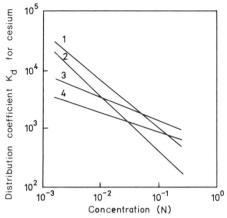

Fig. 24   Distribution coefficient K_d for low concentrations of cesium (in general $< 10^{-8}N = 1.3$ mg/kg) in montmorillonite at different concentrations of sodium, potassium, magnesium, and calcium (after Wahlberg & Fishman 1962).

that describes either the relative affinity of the aquifer matrix for a particular ion or the mobility of the ion in the aquifer or groundwater flow system (Jackson & Inch 1980). The distribution coefficient for cesium as a function of the concentration of the various ionic components in montmorillonite is shown in Fig. 24. Figure 25 gives the same distribution coefficient relative to potassium ion concentration in various clay minerals. The K_d may be affected by reactions other than ion exchange, including precipitation, isomorphous substitution, redox processes, complex ion formation and acid-base buffering (Jackson & Inch 1980, Jackson et al. 1980).

The selectivity quotient is related to the distribution coefficient if the cation exchange capacity Q of the exchanger R^- and the total competing cation concentration in solution C are known and the system under consideration is at equilibrium (Amphlett 1964, Jackson & Inch 1980, Jackson et al. 1980):

$$K_B^A = K_d^A = \frac{[C - (A^+)]}{[Q - (A^+R^-)]} \tag{69}$$

If the contaminant A^+ is present in amounts much less than B^+, (A^+) is much less than C and (A^+R^-) than Q, the relationship can be simplified to

$$K_d^A = \frac{K_B^A \cdot Q}{C} \tag{70}$$

Therefore the distribution coefficient is directly proportional to the cation exchange capacity and the selectivity quotient, and inversely proportional to the total competing cation concentration. If (A^+) and (A^+R^-) are very small, (B^+) is close to C and therefore K_d^A is proportional to $(B^+)^{-1}$. More complex cases, for example involving monovalent-divalent exchange, may be developed using similar reasoning.

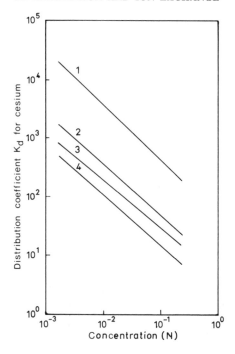

1   Potassium – montmorillonite

2   Potassium – illite

3   Potassium – kaolinite

4   Potassium – halloysite

Fig. 25   Distribution coefficient K_d for low concentrations of cesium (in general $< 10^{-8}N$ $= 1.3$ mg/kg) in various clay minerals as a function of potassium concentration (after Wahlberg & Fishman 1962).

As a consequence of successive ion exchange processes, groundwaters which are normally characterized by specific contents of Na^+, K^+, Ca^{2+}, and Mg^{2+}, can completely alter along a subsurface flow path, the more so the lower the concentration of dissolved solids. The mobility of a sorptive ion below the surface may be described following Higgins (1959) and Mayer & Tompkins (1947):

$$v_A = \frac{v_W}{1 + K_d\,(\rho_b/n)} \qquad (71)$$

in which v_A = the velocity of the ion species (cm · s⁻¹)

v_W = the velocity of the water (cm · s⁻¹)

ρ_b = the bulk density of the solid medium (g · cm⁻³)

n = the porosity

K_d = the distribution coefficient (cm³ · g⁻¹) of the subsurface media in relation to ion species and other ions in the water; K_d is determined in radiochemical analysis as the ratio of dps/g to dps/cm³, where dps stands for disintegrations per second

This expression is the (one-dimensional) retardation equation (Jackson & Inch 1980), since nonzero values of K_d dictate that v_A is smaller than v_W; hence the ion is retarded in its flow through the aquifer. If v_A, v_W, ρ_b, and n are known for a particular aquifer system contaminated by ionic species A^+, then K_d may be determined.

Lai & Jurinak (1972) derived an expression similar to equation 71 based on the convective dispersion equation. Jackson et al. (1980) pointed out that this expression is essentially identical to equation 71 for the case in which K_B^A is constant (i.e., linear exchange isotherm), provided the geochemical reactions that occur (i.e., ion exchange or precipitation) do not affect the ratio ρ_b/n. This proviso may reasonably be assumed to be a necessary condition for equation 71 as well.

The migration in groundwater of contaminants that are sorptively bound onto the subsurface media can likewise be treated like the migration of ions below the surface. In sandy, clay-free, and humus-free aquifers and in groundwater that is saturated with various cations, the distribution coefficient for traces of strontium is $0.5–50$ cm$^3 \cdot$ g^{-1}, that for cesium approximately $1–500$ cm$^3 \cdot$ g^{-1}. Except for fine vermiculitic or montmorillonitic or zeolitic or humic sands, which probably have the highest K_d-values, coarse quartz sand has the lowest, and fine calcite sand the relatively highest distribution coefficient (Davis & De Wiest 1967). With the use of actual values of porosity and density of granular aquifers, it follows from equation 71 that ^{90}Sr and ^{137}Cs migrate significantly more slowly than the groundwater flow and that the velocity of ^{137}Cs in most cases will be lower than that of ^{90}Sr.

Adsorbed contaminants (e.g., radionuclides) can be partially desorbed by other competing ions in the groundwater if these ions occur in sufficiently large concentrations. The higher the ionic concentration, the faster will the contamination move, through successive exchanges.

One consequence of these processes is a separation of the dissolved constituents because of their different sorptive properties. This is particularly important for the behavior of radionuclide mixtures, while they migrate as low-level radioactive waste underground. The distribution of mixed radionuclides of radioactive waste in the percolation zone at Hanford Project (Brown & Raymond 1962) shows (Fig. 26) that the rate of migration of the radioisotopes ^{137}Cs, ^{90}Sr, and ^{106}Ru is variable and decreases in the sequence quoted above. The bulk β^--activity decreases in the direction of propagation. Here it essentially depends on the ^{106}Ru content, which occurs in water almost always as a cationic or anionic complex and is only partly adsorbed (Bovard et al. 1963) — less in anionic form, more in cationic form. Hence radioruthenium is one of the most mobile radionuclides and moves nearly as fast as the groundwater flows (Schulze & Haberer 1966). The radioactive waste at Hanford is disposed into approximately 120 m of Quaternary fluviatile and glaciofluvial deposits, chiefly sands, gravels, and boulder beds, changing to fine-grained sediments (Fig. 27). The underlying lacustrine and fluviatile sediments (more than 100 m thick) of the Ringold Series are 100

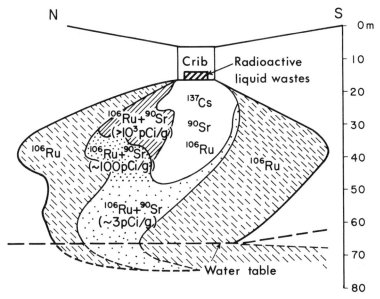

Fig. 26 Distribution of radioisotopes below disposal site (crib) at Hanford Project, Washington State (after Brown & Raymond 1962).

times less permeable than the overlying Quaternary deposits. The water in the basement basalts and interbedded tuffs is confined, so that the radioactive waste flows only in the permeable Quaternary beds and then into the main drainage channel of the river (Fig. 28). According to the measured flow velocities, and by consideration of the half-life of [106]Ru (viz., 1 year) it can be seen that this isotope has decayed to a safe level before it reaches the main river channel.

The K_d-values measured in the laboratory can be compared with those derived by field mapping. This was shown in an intensive evaluation of the migration of ^{90}Sr and ^{137}Cs in the Lower Perch Lake Basin at the Chalk River Nuclear Laboratories, Canada. The granitic gneiss bedrock in the basin is overlain by a sequence of sandy glacial and fluvial sediments. The aquifer system in the valley bottom of the basin sediments is composed predominantly of fine- and medium-grained sands, dispersed in which are thin beds of silts and clays (Fig. 29). The groundwater flows from the recharge area near disposal area A to its discharge in the swamp near Perch Lake and the lake itself. The radioactive waste migration patterns indicate that the silt and clay unit contains permeable windows that are passed by the groundwater flow (Fig. 29) (Jackson & Inch 1980). The mean porosity of the sand is 0.38 and its bulk density value is 1.69 g · cm^{-3}. Approximately 30% of the sands is quartz; 50% is feldspar, the majority of the remainder being mica, sericite, and hornblende. In the clay size fraction ($\sim 0.5\%$) plagioclase and quartz are predominant, whereas minor to moderate amounts of mica and in-

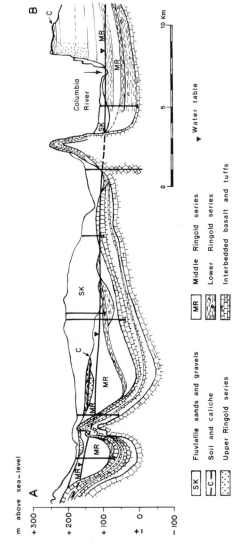

Fig. 27 Geological cross section through the groundwater system at Hanford Project, Washington State (after Brown & Raymond 1962).

98

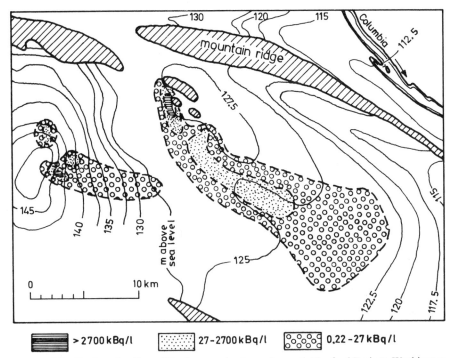

Fig. 28 Distribution of radionuclides in groundwater underneath Hanford Project, Washington State, December 1961 (after Brown & Raymond 1962).

terstratified mica-vermiculite are present. Cation exchange capacities for the sands are of the order of 1 meq/100 g (Jackson & Inch 1980).

In 1954 about 70 m³ of medium-level, liquid radioactive waste containing about 2.2×10^{12} Bq of ^{90}Sr and 2.6×10^{12} Bq of ^{137}Cs were released into a pit at disposal area A lined with lime and dolomite. Another experimental disposal in 1955, containing 11×10^9 Bq of ^{90}Sr and 9.3×10^9 Bq of ^{137}Cs, was not neutralized with lime or dolomite. Since that time these wastes have chromatographically separated into ^{90}Sr and ^{137}Cs fronts (Fig. 30), which are migrating through the sandy aquifer at characteristic velocities much less than the velocity of the transporting groundwater ($\mu w \sim 2 \times 10^{-6}$ m/s). For ^{90}Sr this characteristic velocity is approximately 3% of that of the groundwater; for ^{137}Cs it is about 0.3%. The estimates of K_d for ^{90}Sr and ^{137}Cs are 7 and 80 cm³ · g^{-1}, respectively, by the field mapping method, and 15 and 500 cm³ · g^{-1}, respectively, by the radiochemical analysis method. The field mapping estimates are lower than the radiochemical estimates because only the fronts of the radioactive waste plumes have been mapped. Therefore the former yield an estimate of the fastest radionuclides, whereas the groundwater velocity is a mean value (Jackson & Inch 1980).

Chemical and mineralogical analyses of radioactively contaminated

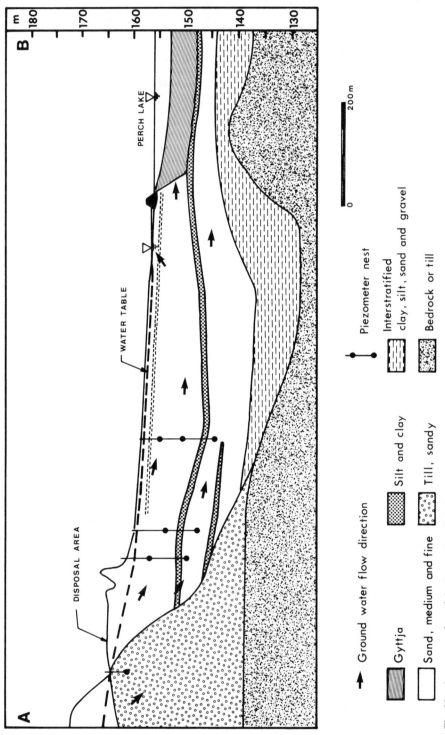

Fig. 29 A cross section of the Lower Perch Lake Basin. (see Fig. 30) (Jackson et al. 1980) (Vertical exaggeration 10×).

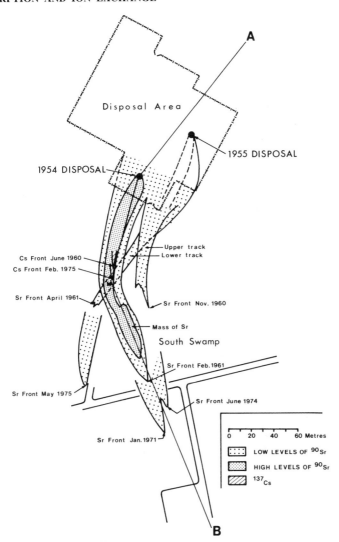

Fig. 30 Plan view of the radioactive waste contaminant migration paths over time. Approximate depth of the front near piezometer nest is 5 m. Parameter v_A in equation (71) is estimated from the field mapping of the path front over time (Jackson et al. 1980).

aquifer sediments showed that most of the ^{90}Sr ($K_d \sim 10$ ml/g) is exchangeably adsorbed, primarily to feldspars and micas; the remainder is either specifically adsorbed to iron, manganese, or aluminum oxides or fixed to unknown sinks. In deeper parts of the aquifer system geochemical retardation of ^{90}Sr may be occurring because of the precipitation of $SrCO_3$. Less than one-half of adsorbed ^{137}Cs is exchangeable with $0.5M$ $CaCl_2$; the high levels of cesium adsorption and fixation ($K_d \sim 10^2$ ml/g) may be ascribed to the

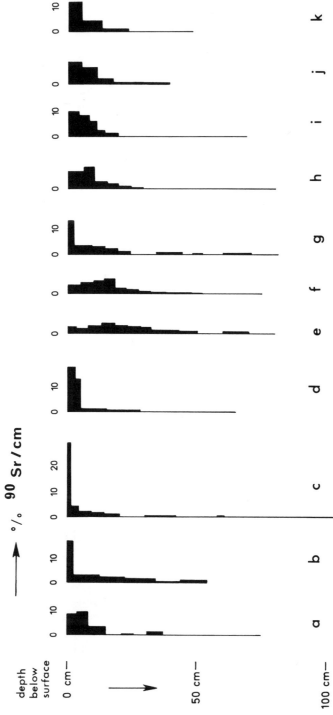

Fig. 31 Distribution of extractable ^{90}Sr in typical central European soils (after Aurand et al. 1972a). (*a*) Rhine Valley chernozem, (*b*) podzol–brown forest soil on windblown sand (dunes), (*c*) podzol–brown forest soil on windblown sand, (*d*) podzol–brown forest soil overlying a groundwater soil with lowered watertable, (*e*) brown–podzolic soil on glacial loam, (*f*) brown–podzolic soil, (*g*) pseudogley, slightly podzolized, (*h*) inceptisol (pelosol–pseudogley). (*i*) brown–alluvial meadow soil on river bank sediments, (*j*) typical gley, (*k*) low moor peat.

presence of the layer silicates vermiculite and biotite. Complexation reactions appear to be unimportant in affecting ^{90}Sr and ^{137}Cs transport and adsorption (Jackson & Inch 1980).

Adsorption Processes in the Unsaturated Zone.   Sorption occurs in the unsaturated zone, particularly in active, humus-rich soil. The soil processes are of very great importance in protecting groundwater from surface pollution. The purifying effect of the sorption processes is first seen in the removal of colored and odorous substances by the soil. Thompson (1850) was the first to investigate the cation exchange process. Even artificial organic substances, such as surfactants (Gottschaldt & Winter 1967), are adsorbed in humus-rich soils.

This phenomenon is particularly noticeable in the bonding of radioactive fallout that affected the whole of the Earth's surface after the first thermonuclear test explosions. By far the greatest proportion of the fission products, including ^{106}Ru and ^{144}Ce in addition to the long-lived ^{137}Cs and ^{90}Sr, which cause the bulk of the radioactive contamination, were detected in the topmost soil horizons (in cultivated soils down to the limit of ploughing), that is, in the top 15–25 cm, and were also partly trapped mechanically (Herbst 1959, Klechkovskii et al. 1958, Walton 1963, Richter 1965, Kloke 1961, Knoop & Schroeder 1958, Gustafson et al. 1958). Figure 31 shows the distribution of ^{90}Sr activity in 11 typical central European soils, which comprise the most important variations in soil chemistry and physical properties. The radionuclides tend to move downward gradually with the percolation water, particularly in sandy soils (Alexander et al. 1960, Herbst 1961, Henzel & Strebel 1967). Hence small amounts of radiostrontium can be detected at depths of 60 cm and more (Fig. 31). This explains the trace occurrence of radiostrontium in groundwater (Wolter & Aurand 1967, Aurand et al. 1971). According to Reissig (1966), however, the leaching process is so slow in most soils that appreciable leaching of ^{90}Sr from the topsoil over a period of 20–30 years must not be expected. The ^{90}Sr rate of movement according to this author lies below 1 cm/700 mm precipitation.

Groundwater can be adversely affected by ion exchange in the soil if potash fertilizers interact with alkaline earth-saturated exchangers in the soil. Whereas K^+ ions may be bound in the soil, Ca^{2+} ions can be desorbed. Hence potassium fertilizers may lead to a rise in the total hardness, especially noncarbonate hardness in groundwater (Matthess 1958).

Exchange Processes in the Groundwater Zone (Saturated Zone).   Exchange processes in the groundwater zone were first demonstrated in this century by Renick (1924a) and Riffenburg (1925) and were subsequently reported by Harrasowitz (1933), Schoeller (1934), Ødum & Christensen (1936), and Schwille (1953a, 1955).

Ion exchange is a very common process in the groundwater zone because of the wide distribution of exchangers and because of the ions involved. One simple proof of this is the occurrence of waters in which the sum of the equivalent weights $(Ca^{2+} + Mg^{2+})$ is less than the HCO_3^- ion concentration. In these cases one must assume that the excess of HCO_3^- ions is related to alkali ions and that the original Ca^{2+} concentration equivalent to the HCO_3^- content was diminished by exchange of Ca^{2+} for bound Na^+ ions. $Na-HCO_3$ waters of this type are just as widely distributed as $Na-SO_4$ waters; for example they are very common in basaltic aquifers (e.g., in the Vogelsberg), as well as in sedimentary rock aquifers in north, south, and central Germany (Löhnert 1970, 1972, Michel 1968, Matthess 1958, Hecht 1964). This process is particularly effective where impregnation with Na-Cl-rich water occurs, particularly with seawater, and there is widespread evidence for Na^+ ion exchange in these circumstances. A similar "regeneration" of exchangers by concentrated Na-Cl water has been noted by Schwille (1953a) in the Mainz Basin region, by Rodis & Schneider (1960) in Cretaceous sandstones of southwest Minnesota, and by Poland et al. (1959), Ødum & Christensen (1936), and Löhnert (1970) in the transition zone between saltwater and fresh water in coastal areas and in areas of inland saline waters. The "regeneration" process is associated with a rise in Ca^{2+} content in water and has been cited as an explanation of Ca-Cl waters.

Several indices are used for the identification of water that is involved in exchange processes:

1. Base exchange index, proposed by Schoeller (1951, 1962). Starting from an initial ion composition $(Na^+ + K^+) = Cl^-$ arising from the solution of alkali chlorides, the equivalent sum $(Na^+ + K^+) < Cl^-$ comes about by exchange of alkalis in the water for alkaline earth in the rock. The "positive" base exchange index is then:

$$I_{BA} = \frac{(Cl^- - (Na^+ + K^+))\ meq/l}{Cl^-\ meq/l} \tag{72}$$

For an initial overall ion composition $Ca^{2+} + Mg^{2+} = SO_4^{2-} + HCO_3^- + NO_3^-$ the equivalent sum $(Na^+ + K^+) < Cl^-$ arises for the case of exchange of alkaline earths in the water for alkalis in the rock. The "negative" base exchange index is then:

$$I_{BA} = \frac{(Cl^- - Na^+ + K^+))\ meq/l}{(SO_4^{2-} + HCO_3^- + NO_3^-)\ meq/l} \tag{73}$$

An imbalance between the equivalent amounts of chloride ions and alkali ions is not always the result of cation exchange. Thus seawater has a "positive" ratio according to equation 72 without this being definitely attributed to cation exchange. On the basis of both equations 72 and 73, groundwater from crystalline rocks and schists almost always possesses a "negative" equivalent ratio, because on the breakdown of silicates more alkali ions than chloride ions are released. Therefore because the causes of the imbalance can be extremely varied, Schoeller (1962) has proposed the neutral index of chloride-alkali imbalance I_D for the ratio given in equation 72.

2. The equivalent ratio $Na^+/(Na^+ + Ca^{2+} + Mg^{2+})$ proposed by Versluys (1915) [cf. "sodium quotient" of Rutten (1949)] does not reflect the following properties: sodium is not only brought in through solution of NaCl in the water, and the ratio of alkalis to alkaline earths is influenced by the solubility of the cation in addition to the cation exchange process, so that the concentration must be also taken into account.

3. The *"titre natronique"* proposed by Delacourt (1942, 1943a, 1943b) $(Na^+ + K^+ - Cl^-)$ in meq/l permits comparisons only if the concentration is also considered.

4. The equivalent ratio $(Na^+ + K^+)/Cl^-$ is also in use.

Sorption and Filter Effect. The filter effect of subsurface media is a complex physical and chemical phenomenon. During filtration the fraction that is too large to pass through the subsurface flow channels in pores and joints is mechanically separated; smaller particles, bacteria, and colloids are retained through the combined effects of sedimentation and sorption. The filtering effect increases with decreasing width of the flow channels and with the duration of the process when the widths of the flow channels are reduced by sedimentation when the material carried by the water is retained on the subsurface media. This process can lead to local clogging of the pore spaces, especially in the infiltration zones at river banks, at percolation basins, and in irrigated fields.

Suspended particles that are very small compared with the pore diameter of the media are adsorbed particularly effectively; these include bacteria (1–2 μm), diatoms (30 μm), SiO_2 particles (20 μm), and flocculated iron hydroxide (10 μm). The filter process proceeds better, the smaller the grain size, the greater the effective surface area, and the thinner the film of water around the suspended particle. Because bonding to the grains in the media depends on small differences in electrical potential, the effective range of which is very small ($< 10^{-2}$ μm), accidental contacts are necessary before the particles become bound to an active surface. The probability of such contacts increases with the complexity of the pore water channels; other causes include dispersion during groundwater flow, Brownian movement of colloidal particles, and the effect of gravity on larger particles (Mintz 1966). Hence the filter effect depends on density, grain size, porosity, and extent of the subsurface filtering agents, percolation and flow velocities, the initial concentration and the properties of the suspended materials, and the temperature, density, and viscosity of the liquid. The filter efficiency can improve with time through the building of a sticky gelatinous coating on the surface of the rock by colonies of bacteria and colloidal material (Huisman & Van Haaren 1966).

Changes in the geometrical structure of the pore spaces through clogging, which can be caused by changes in the dynamic factors of the flow mechanism, can in certain circumstances lead to removal of deposited material and its further migration.

Ion Filtration by Osmosis.   Interlayered beds of clay in aquifers can act as semipermeable osmotic filters in spite of their low permeability. When membranes of this type separate two solutions of different concentrations, then osmotic pressure causes a movement of water molecules to occur in the direction of the more concentrated solution, so that higher hydrostatic pressure builds up in this groundwater body (Hanshaw & Zen 1965, Jones 1968). The following relationship holds for osmotic pressure:

$$P = \frac{RT}{V} \ln \frac{p_0}{p} \tag{74}$$

in which P is the osmotic pressure, R is the gas constant, T the temperature in K, V the mole volume of the water, p_0 the vapor pressure of pure water, and p the vapor pressure of the solution ($= p_0 - p_0 \cdot N$) in which N is the mole fraction of the solution.

Clay beds exhibit differences in electrical potential as borehole logs of self-potential show (Hem 1970).

The regular appearance of a "shale line" shown in electric logs from boreholes in all parts of the world points to the capacity of beds of clay to act as ideal membrane electrodes. Quantitative theoretical treatment of the electrochemical properties of clay beds shows clearly by consideration of a three-phase system of two solutions of electrolytes separated by a bed of clay acting as a membrane that the effect can be described by the equation for a "perfect" electrode (Degens & Chilingar 1967):

$$E = \frac{RT}{F} \ln \frac{c^I}{c^{II}} \tag{75}$$

in which E is the electrochemical potential, R the gas constant, F the Faraday equivalent, and c^I and c^{II} the concentrations of ions in the respective phases. Hence it follows that salts can be retained in clay beds so that the concentration in the pore solutions can rise. The retention of ions depends on the large excess charge that is bound to the clay membrane and hinders the passage of ions bearing the same charge, positive or negative. The separation is therefore brought about through the electrical properties rather than the size of the electrolyte. Water molecules can pass through the membrane layer so that the filtrate shows a lower salt content than that of the original solution (Degens & Chilingar 1967, Von Engelhardt 1960). Filtration by clay membranes must be considered when explanations of raised salt content in groundwater are sought.

2.3 OXIDATION AND REDUCTION

Oxidation and reduction are geochemically important processes, which together with the hydrogen ion activity (pH) determine the solubility of many substances, hence the occurrence of substances in groundwater. In many of these reactions microorganisms are involved, and their participation is noted in detail in Section 2.5.

Above the water table there are pore spaces filled with air which tend to have a higher proportion of carbon dioxide and water vapor, hence less oxygen than normal atmospheric air. This air in the unsaturated zone is defined as ground air, which includes as a subdivision the soil air in the biologically active topsoil. However, zones above the water table are essentially oxidizing. The extent of the oxidizing zone and its ground air composition are dependent on geomorphology and climate. In mountainous regions this zone is deeper than in valleys that carry the main river drainage, and it is deeper (10–100 m) in drier lands than in those with greater rainfall where the water table is usually found at less than 30 m depth. Lindgren (1933) has described a deep-seated water table in the Mapimi lead mine, Durango, Mexico, at 770 m below the surface. The presence or absence of free oxygen in groundwater essentially determines whether oxidizing or reducing conditions will prevail. Percolation water carries dissolved oxygen down with it; besides this, oxygen can be dissolved from the ground air at the water table and can be carried into deeper groundwater levels by diffusion and flow dispersion. In practice, observations show that there is often a supply of oxygen in groundwater, especially in recently recharged ones, ranging from 6 to 12 mg/l. This corresponds approximately to the highest possible concentration at the prevailing groundwater temperature range 0–20 °C (9–14.4 mg/l) in equilibrium, with the oxygen content of the atmosphere compared with ground air (Gerb 1953). In the presence of free oxygen the expression for redox potential holds $Eh = 1.23 + 0.0147 \log O_2 - 0.059$ pH (Schoeller 1962).

Oxidation may also be brought about by the charges on the ions Fe^{3+}, Mn^{3+}, SO_4^{2-}, H^+, and $Fe(OH)_3$ (Schoeller 1962). Dissolved or atmospheric oxygen together with water or water vapor acts as a powerful weathering force, partly by direct oxidation of iron and manganese compounds, and partly by oxidation of carbon compounds in the soil and in biolithic deposits (peat, lignite, bitumen, petroleum, natural gas). Finally carbon dioxide and water are formed and resistant organic compounds can appear as intermediate products, together with compounds of nitrogen, particularly ammonia, nitric acid, and such sulfur compounds as sulfuric acid. The oxidation of the iron sulfides pyrite and marcasite (FeS_2) is given by the equations:

$$2FeS_2 + 2H_2O + 7O_2 \rightarrow 2FeSO_4 + 2H_2SO_4 \qquad (76)$$

or according to Baas Becking (1959):

$$FeS_2 + 8H_2O \rightarrow FeSO_4 + HSO_4^- + 15H^+ + 14e \qquad (77)$$

Because $FeSO_4$ is stable only in anaerobic conditions, it hydrates to melanterite ($FeSO_4 \cdot 7H_2O$) and on occasion to coquimbite [$Fe_2(SO_4) \cdot 9H_2O$] or changes into ferric hydroxide. The sulfates resulting from reactions with sulfuric acid or from the oxidation of sulfides, excluding anglesite ($PbSO_4$) and basic ferric sulfate, dissolve easily and are carried away in the groundwater. The free acids assist in the attack on rock-forming minerals (cf. Section 2.1).

The oxidation processes run faster in warm climates than in cold, and faster in humid and alternating humid and dry climates than in arid climates. The connection between the sulfate content and the effect of oxygen-bearing groundwaters and that of the oxygen in the mine air on sulfide mineral deposits are clearly seen in the groundwater in the Siegerland (Eupel Mine near Niederhövels, Sieg). Down to a depth of 100 m below the surface sulfate contents above 1000 mg SO_4^{2-}/l were found, which gradually decreased downward. Sulfate-free water was present on the newly opened mine bottom at 600 m depth (Heyl 1954, Matthess 1961). In sulfide ore deposits generally the zone above the water table is characterized by a high sulfate ion content in percolation water. Lindgren (1933) quotes a very high value from the central tunnel in the Comstock Lode in Nevada: 209 100 mg SO_4^{2-}/kg. Many elements (e.g., S, Se, As, Cr, V, Mo, W, and U) can be oxidized to easily soluble anion complexes and are therefore especially mobile in the oxidation zone. Other substances, for example iron and manganese, are transported as divalent ions or as colloids, and are deposited as hydroxides under oxidizing conditions, even in relatively acid solutions, and can be concentrated in the boundary region between reducing and oxidizing conditions. Iron then forms ferric hydroxide or hydrated oxide, which inverts to hematite Fe_2O_3 during aging processes; tri- or quadrivalent manganese precipitate as hydroxide or oxide. The importance of this chemical interaction between water and rocks is shown in Fig. 32, in which the characteristic mineral formation processes are clearly seen to be dependent on the Eh and pH conditions occurring in nature. The fence diagram indicates the formation of iron and manganese oxide as typical mineralization in the positive Eh field.

Oxidizing conditions extend to different depths below the surface. In the copper mine at Butte, Montana, the sulfides are oxidizing as far down as 130 m, in the silver-gold veins of Tonopah, Nevada, as far as 230 m, in Tintic, Utah, as far as 530–740 m. At Bisbee, Arizona, oxidized copper ore was found at a depth of 470 m (Lindgren 1933). The oxidation phenomena end chiefly near the fluctuating water table, but they have been detected down to 70 m below this level, when oxygen-bearing groundwater penetrates through fissures in the rock. According to Fairbridge (1967), depth of oxidation can exceed 5000 m, even 8000 m. The depth of the oxidizing zone is also affected by climatically or geologically controlled changes in the position of the water table. In the presence of oxygen-consuming substances, particularly organic matter, but also when Fe^{2+}, Mn^{2+}, S^{2-}, NO_2^-, NH_4^+, H_2, and

$Fe(OH)_2$ are present, the oxygen content of the groundwater is diminished, and when the supply is cut off, possibly because of insufficient circulation, the oxygen is finally consumed (Table 26). The redox values then become more reducing because oxygen content undergoes a gradual decrease from ground level to deeper levels.

Depending on the oxygen demand of the various bacterial species, an oxygen content of about 0.7–0.01 mg O_2/1 at 8°C water temperature has been defined as threshold oxygen concentration for the boundary between oxidizing and reducing conditions. But field observations seem to indicate that the threshold is broader, because the characteristics of reducing conditions appear at considerably higher oxygen contents. This discrepancy probably

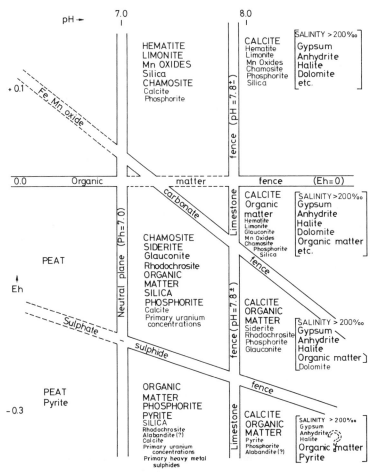

Fig. 32 Fence diagram of sedimentary chemical end-member associations in their relations to environmental limitations imposed by selected *Eh* and pH values. Associations in brackets refer to hypersaline solutions (after Krumbein & Garrels 1952).

exists because for technical reasons groundwater samples are not quasi-point samples but mixtures of waters of different degrees of reduction (Schwille 1976). Reducing conditions can also occur in waterlogged soils, so that generally the depth of the boundary between the reducing and oxidizing zones is controlled by the balance between oxygen supply and oxygen consumption. Below this depth the free sulfuric and nitric acids decrease as well as the oxygen. In the transitional zone from oxidizing to reducing conditions in ore deposits many sulfides and native metals (e.g., Cu and Ag) show secondary enrichment (zone of secondary enrichment).

Oxygen consumption starts with the presence of organic and inorganic compounds, for example sulfur-containing proteins. Incomplete decay leads to enrichment in organic material below the surface. Then if there is an increasing oxygen deficiency, different decay products will be accumulated. Insufficient oxygen supply gives rise to the mouldering process, which yields small amounts of carbonaceous substances in the soil. Increasingly reducing conditions with eventual total absence of oxygen are required for peat and coal formation, and extremely reducing conditions for the putrefaction of petroleum hydrocarbons.

The reducing power of a system increases with decreasing redox potential. The redox potential in an aqueous solution generally falls with rising temperature and pH. As a characteristic of "reduced" water Gerb (1953) and Schwille & Vorreyer (1969) have noted that in addition to the absence of or very much reduced oxygen content there is a noticeable iron or manganese content, the occurrence of H_2S, NO_2^- and NH_4^+, and an absence of nitrate. According to Schwille & Vorreyer (1969) and Golwer et al. (1970) a reduction in sulfate (including absence of sulfate) is also typical (equation 5, Table 26). Ferric hydroxide can be reduced and dissolved in the presence of organic substances (equation 4, Table 26). Baas Becking (1959) has given the following equation for this reaction:

$$FeOOH + CH_4 + 2H_2O \rightarrow Fe(OH)_2 + CO_2 + 7H^+ \qquad (78)$$

$$FeOOH + \tfrac{1}{6} C_6H_{12}O_6 + H_2O \rightarrow Fe(OH)_2 + CO + 3H^+ \qquad (79)$$

The reduction of manganese can be described in a similar way (equation 3, Table 26). The occurrence of hydrogen sulfide is explained by the consumption of oxygen chemically bound in sulfates during microbial breakdown of organic substances, for example, the reaction given by Thimann (1964):

$$2CH_3CHOH \cdot COOH + H_2SO_4 \rightarrow 2CH_3COOH + 2CO_2 + H_2S + 2H_2O \qquad (80)$$

The hydrogen sulfide that occurs when sulfur-containing proteins (e.g., the aminoacid cystine) break down can react with iron and other heavy metals to form sulfides (Baas Becking et al. 1960, Garrels 1960). Besides hydrogen sulfide and other sulfides, elemental sulfur appears in a reducing environment following the breakdown of sulfur proteins, or from microbial reduction of soluble sulfate to hydrogen sulfide. In this way considerable sulfur deposits have been formed (Feely & Kulp 1957). Under reducing conditions therefore there is precipitation of heavy metal sulfides, a concentration of organic substances in coal and oil deposits, and enrichment in several rare elements due to adsorption by organic materials and coprecipitation in metal sulfides.

The sulfate content is a measure of the "bound" oxygen available on the breakdown of the organic substances in reducing, anaerobic, environments. For example Golwer et al. (1970) have measured up to 1600 mg/l sulfate ions (on average about 600 mg/l) in a strongly polluted "reduced" groundwater. When the complete amount of oxygen bound in the sulfate is used, an average of 400 mg O_2/l is available for the oxidation of organic substances. For the reduction of sulfates the following conditions are necessary: a completely anaerobic environment, the presence of hydrogen ions, and an energy source – in general an organic substance (e.g., petroleum, natural gas, peat, coal, organic contaminants) – and a biological transfer mechanism for the hydrogen.

On the other hand microbial nitrate reduction starts a low oxygen content (Van Haaren 1966) and proceeds via nitrite to elemental nitrogen or to ammonia. The importance of nitrate-bound oxygen is also to be noted when considering the breakdown of organic substances. The absence of nitrate has been known as a typical factor of gasoline-contaminated groundwater since this was first reported by Müller (1952).

Champ et al. (1979) proposed a thermodynamic model of the sequential utilization of oxidized species (e.g., O_2, NO_3^-) in aquifers that have an excess of reduced dissolved organic carbon (DOC) rather than dissolved oxygen. Table 26 lists the individual redox reactions of this sequence in the order of their decreasingly negative values of the free energy change at pH 7, $\Delta G°(W)$. Consequently they postulated that in a confined aquifer containing excess DOC and solid phase Mn(IV) and Fe(III) minerals, the oxidized species are reduced in the following order: dissolved oxygen, nitrate, (solid) manganese oxide, (solid) ferric hydroxide, sulfate, dissolved CO_2 (i.e., HCO_3^-), and dissolved nitrogen. Occurrences of such a sequence have been observed in groundwater flow systems by Jackson & Inch (1980), Edmunds (1977), and Golwer et al. (1970, 1976).

Gerb (1953) has assumed that geological time scales would be necessary for the growth and decay of reduction zones. Research into massive groundwater pollution has however shown the surprisingly fast growth of a reducing environment following a definite pattern of events. Starting from flat-shaped pollution sources there appears in the groundwater a zone of reduction, a transitional zone, and an oxidation zone. The zone of reduction contains "reduced" water with a high content of ferrous iron (maximum 700 mg/l), manganese (maximum 155 mg/l), ammonia (up to 1460 mg/l), and free dissolved carbon dioxide (maximum 2046 mg/l) as a breakdown product.

Table 26 Redox Processes in a Closed System with Water Containing Excess Organic Material (DOC) and Initially O_2, NO_3^-, SO_4^{2-}, and HCO_3^- (Champ et al. 1979, modified after Stumm & Morgan 1970)[a]

Reaction	Equation[b]	$\Delta G°$ (W)[c] (kJ at pH = 7.)[c]
1. Aerobic respiration	$CH_2O + O_2(g) = CO_2(g) + H_2O$	−502.4
2. Denitrification	$CH_2O + \frac{4}{5} NO_3^- + \frac{4}{5} H^+ = CO_2(g) + \frac{2}{5} N_2(g) + \frac{7}{5} H_2O$	−476.9*
3. Mn(IV) reduction	$CH_2O + 2MnO_2(s) + 4H^+ = 2Mn^{2+} + 3H_2O + CO_2(g)$	−340.4
4. Fe(III) reduction	$CH_2O + 8H^+ + 4Fe(OH)_3 = 4Fe^{2+} + 11H_2O + CO_2(g)$	−116.0
5. Sulfate reduction	$CH_2O + \frac{1}{2} SO_4^{2-} + \frac{1}{2} H^+ = \frac{1}{2} HS^- + H_2O + CO_2(g)$	−104.7
6. Methane fermentation	$CH_2O + \frac{1}{2} CO_2(g) = \frac{1}{2} CH_4 + CO_2(g)$	− 92.9
7. Nitrogen fixation	$CH_2O + H_2O + \frac{2}{3} N_2(g) + \frac{4}{3} H^+ = \frac{4}{3} NH_4^+ + CO_2(g)$	− 80.4

[a] Model: Recharge of natural waters through a reactive soil zone into a confined aquifer system containing MnO_2, $Fe(OH)_3$, and excess DOC.

[b] (g) = gas, (s) = solid.

[c] $\Delta G°(W) = \Delta G° + (RT/nF) \ln [H^+] p$, $[H^+] = 1.0 \times 10^{-7} M$, where p = stoichiometric coefficient for H^+; for other definition see equation 43.

112

The adjacent transitional zone exhibits free oxygen at times, whereas in the zone of oxidation a permanent oxygen supply is present. With the renewed appearance of free oxygen there is a precipitation of iron and manganese and renewed oxidation of hydrogen sulfide to sulfur and sulfate (Golwer et al. 1970, 1976). The following reactions are typical of the reducing zone: reduction of nitrates, nitrites, sulfates, and iron and manganese oxides during breakdown of organic substances. The characteristic products of reactions of this kind are H_2, H_2S, CH_4, and possibly other hydrocarbons, also S^{2-}, NO_2^-, NH_4^+, Fe^{2+}, and Mn^{2+} ions, and CO_2, Figure 32 shows the typical formation of organic substances and pyrite in the zone of reduction.

Edmunds (1973, 1977) pointed out that in a natural confined aquifer system the oxidizing conditions of the water can change along the flow path into reducing conditions with the typical characteristics of the respective zones. Such oxidation-reduction barriers affect the behavior of minor and trace constituents (e.g., iron and other heavy metals).

In the zone of artificially caused variations of water table level, oxidizing and reducing conditions can alternate from time to time. The reducing power of the groundwater in the permanently saturated region of generally low oxygen content is greater than in the zone of water table fluctuations, where however the generally high oxygen content is relatively greater in acid than in weakly acid to neutral groundwaters (Buchholz 1961). The redox potential, which in soils may reflect the presence of iron(II) ions and H_2S, is displaced at low oxygen content from the aerobic into the anaerobic zone: in podzols with 5% O_2, and in chernozems with 2.5% O_2. Less than 5% O_2 could be observed in podzols with Fe(II) ions extractable with H_2SO_4, HCl, and H_2O, and large numbers of anaerobic bacteria, up to a few hundred million per gram of soil (Grecin & Jun'-schen 1960). Visual indications, such as red or bleached horizons are characteristic of redox conditions in which ferric iron is reduced to the ferrous state. However, transport of ferrous iron in the groundwater flow also plays an important part and explains many local differences. Iron impoverishment and bleaching in the zone of water table fluctuation requires the removal of the ferrous iron, and this also applies to weakly reducing conditions. Red horizons need a supply of iron coming in either with the groundwater or from the weathering of dark minerals (i.e., biotite) and an enrichment through sorption, precipitation after the solubility product has been exceeded (Buchholz 1961).

Short-term fluctuations in redox potential in the zone of water table fluctuation can lead to sudden changes in groundwater quality. Well-known examples of this are very rapid and irregular increases in sulfate content following a drop in water table if sulfides are oxidized to sulfates in the zone of ground air; when the water rises again, the sulfates will be dissolved. An example of this was seen in the Breslau manganese incident in 1906 (Prinz 1923). A waterworks was located in an area of sulfide-bearing sediments in a river floodplain above Breslau. In the cone of depression of the waterworks the oxygen in the ground air oxidized iron sulfide, it released ferrous iron, sulfate, and hydrogen ions, which at the low prevailing pH react with manganese oxide to release manganese ions. During a flood, when percolating waters entered the groundwater, the iron content of the well water rose to 400 mg/l, and manganese to 200 mg/l. Possibly also the organic components in the water percolating into the aquifer had brought about reducing conditions, which raised the solubility of iron and manganese.

Similar phenomena were observed in a 91.5 m deep municipal well at Selma, Johnston County, North Carolina, which was located in a pyrite-bearing, sericite schist. The well was pumped with a steady drawdown of 36.5 m over 13 months. During this time the sulfate content did not deviate significantly from the initial content of 13 mg SO_4^{2-}/kg. During pumping the sulfide was evidently oxidized in the cone of depression, because after a rest period of 4 months the water pumped on the first day contained 1330 mg SO_4^{2-}/kg and 365 mg/kg iron, and gave a very acid reaction (pH 2.5). Afterward a decrease in iron and sulfate content was noted, when water from outside of the cone of depression was drawn in (LeGrand 1958).

The cycle of diagenesis — deposition, then solidification, followed by weathering after uplift to form land — is oxidizing in the initial and final phases: the sediments are in general deposited under oxidizing conditions; euxinic conditions are exceptional, as in the Black Sea (Murray & Irvine 1895), where reducing conditions have become established under a distinct thermocline. During diagenesis the redox potential generally moves in the negative direction and does not reverse except under the influence of oxygenated and carbon dioxide-containing groundwater near the surface by weathering in the oxidizing zone (Fairbridge 1967).

In summary it can be stated that the redox conditions in groundwater are controlled by oxygen-consuming reactions, involving the breakdown of organic substances and the oxidation of reduced substances. Microorganisms play an important part in the reactions; this topic is treated in more detail in Section 2.5.

2.4 GAS EXCHANGE BETWEEN GROUNDWATER AND ATMOSPHERE

Gas exchange between groundwater and the atmosphere takes place across two boundary layers, groundwater–ground air and ground air–atmosphere. Gases dissolve in groundwater according to the laws stated in Section 1.2.1. At a given temperature the prevailing pressure determines whether the gas is totally dissolved (in this case the so-called limiting pressure is exceeded) or whether bubbles of the gas also occur. A gas-water mixture of this kind occurs in nature, chiefly involving CO_2, but mixtures with nitrogen, methane, or hydrogen sulfide are also observed. Aquifers containing gas-water mixtures have hydraulic properties different from those containing purely liquid groundwater. In porous media the groundwater flow is hindered by the presence of bubbles, particularly in the region of pressure relief in the immediate vicinity of wells. The gas bubbles can gradually rise and finally escape into the ground air. In addition to this form of gas mobility, which is generally in an upward direction only, gases in the groundwater can be transported by dispersion and very slow diffusion processes. Dissolved gases (e.g., CO_2), diffuse in a direction determined by the concentration gradients. In zones with high carbon dioxide content in deeper parts of the groundwater body, CO_2 moves upward to the water table (Golwer & Matthess 1972). High concentrations can occur at depth as a consequence of biochemical processes or through postvolcanic degassing of magmas (Golwer & Matthess 1972). At the water table equilibrium with the ground air sets in and most of the carbon dioxide diffuses out of the water. The diffusion of oxygen on the other hand proceeds from above, downward. Oxygen is consumed during oxidation and breakdown processes in the groundwater, and at the water table level more oxygen is continuously dissolved at a rate controlled by the partial pressure of oxygen in the ground air above.

The rate of diffusion of a gas in groundwater can be found from the equation:

$$\frac{ds}{dt} = -KF \frac{dc}{dx} \cdot f \cdot P \tag{81}$$

in which $\dfrac{ds}{dt}=$ the quantity (mole $\cdot$ d^{-1}) of the substance diffusing through unit area (1 cm^2) per unit time.

$K =$ coefficient of diffusion for 1 cm^2 cross section, stated as (cm^2 $\cdot$ d^{-1}). The diffusion coefficients of different substances as a function of temperature can be found in published tables (e.g., Landolt-Börnstein 1923, 1931, Lerman, 1979).

$F =$ cross section normal to direction of diffusion

$\dfrac{dc}{dx} =$ concentration gradient (mole $\cdot$ l^{-1} $\cdot$ cm^{-1} = mole $\cdot$ 1000 cm^{-3} $\cdot$ cm^{-1})

$P =$ effective porosity for gas

$f =$ correction factor to take into account irregularities in the flow paths (tortuosity)

This procedure yields a minimum limiting value because the concentration gradient as generally measured is too low. Furthermore, molecular diffusion is a very slow form of material transport in groundwater, but this can be assisted by hydrodynamic dispersion. The rate of transport of carbon dioxide through the boundary layer between groundwater and soil air can be calculated from an equation given by Münnich (1963b), here modified to suit subsurface conditions:

$$\frac{ds}{dt} = w \cdot \Delta c \cdot F \cdot P \tag{82}$$

in which $\frac{ds}{dt}$ = the amount of the substance passing through unit area per unit time $(cm^3 \cdot d^{-1})$

w = transfer velocity $(cm \cdot d^{-1})$

Δc = effective concentration difference

F = cross section (cm^2)

P = effective porosity for gas

An upper limiting value is obtained with this calculation because the transfer velocity of a carbon dioxide gas front given by Münnich ($w = 7.2$ cm $\cdot$ d^{-1}) may have been too high because of incomplete exclusion of convection currents during the experiment.

Above the water table freely mobile air is found in the voids of the active soil layer, and on occasion arises in the other surface layers, which following von Pettenkofer (1871) is called ground air. A part of the voids in this zone is filled with water, so that according to Russel & Appleyard (1915) the ground air mostly occupies only 10–20 vol% of the subsurface media even in porous material. Besides the more or less evenly distributed pore spaces there also occur in the soil layer cracks, fissures, and animal burrows, which raise the overall porosity. In the deeper parts of the unsaturated zone there are, insofar as hard rocks are concerned, other voids caused by faults, parting planes between beds, shrinkage cracks caused by cooling of igneous rocks, and in karst regions solution channels, all of which can form interconnected passages for water to flow along.

The ground air deviates from the composition of the free atmosphere. In particular it contains, as a result of biological activity, more carbon dioxide and less oxygen. Above the water table gases can accumulate after they have escaped from the groundwater. Enrichment in CO_2, H_2S, and CH_4 in well shafts can cause asphyxiation, and methane brings a danger of fire or explosion. Hydrogen sulfide can have a corrosive effect on wells and pipes. Gas exchange between ground air and the atmosphere depends essentially on the following factors:

1. Expansion and contraction of the ground air as a result of temperature changes. Occasionally rapid temperature changes can intensify the gas exchange (Russel & Appleyard 1915). According to Romell (1922) the diurnal temperature changes in the soil cause less than 1/800 of the gas exchange.

2. Effects of density variations as a consequence of temperature differences between soil and atmosphere. Colder air can penetrate into the soil and displace the relatively warmer air there. This process, according to

Romell (1922), brings about only 1/240–1/480 of the gas exchange in the soil. With fractured surface layers and in karst, an air exchange of this type can be accomplished quite extensively. Ice caves can originate in this way.

3. Gas exchange caused by changes in atmospheric pressure. This process depends on the outflow of part of the ground air during low pressure periods as a consequence of rise in volume, and penetration of atmospheric air into subsurface voids as a result of rise in pressure due to reduction in volume. The aeration effect of barometric changes depends on the degree of the pressure changes, their frequency, and the thickness of the gas-filled subsurface (Bouyoucos & McCool 1924). The pressure adjustment of the ground air to that of the free atmosphere follows in a surprisingly short time. Bouyoucos & McCool (1924) have shown that down to 3 m depth the same pressure variations can be measured in the ground air as in the atmosphere above the ground. The diurnal variation in pressure depends on the time of the year and the geographical location. It is greatest in autumn, winter, and spring, and least in the summer (Bouyoucos & McCool 1924, Schmidt & Lehmann 1929). The influence of air pressure variations on gas exchange is however considered to be quite small (Buckingham 1904, Giesecke 1930). Romell (1922) has estimated that only 1% of the average gas exchange in the soil is caused by atmospheric pressure changes.

4. Pressure and suction effect of wind. According to Romell (1922) wind can cause a considerable amount of ground aeration in areas devoid of vegetation; however, in his opinion the effect should be responsible for less than 1/1000 of normal gas exchange in the soil.

5. Removal of ground air by percolation water. Percolating rain, melted snow, and irrigation water remove the air from the voids. The percolate itself transports dissolved gases. Probably an appreciable part of the O_2 input is from recharging seepage water. Romell (1922) has estimated the portion of the influence of rainwater on the gas exchange in the subsurface as 1/16 to 1/12 of normal aeration of the soil.

6. Changes in water table level. A ventilation effect can result from changes in the water table, which on rising pushes out ground air, and draws in atmospheric air when falling.

7. Diffusion. Gas exchange is effected in the active soil layer and in not too thick surface layers according to Buckingham (1904), Romell (1922), Lundegårdh (1924, 1927), and Penman (1940a, 1940b), essentially by a diffusion process.

While diffusion represents a continuously active factor in gas exchange (Penman 1940a), the other factors are effective only when certain meteorological or hydrological changes occur (e.g., changes in air temperature, precipitation, air pressure, and groundwater level). These factors accelerate gas exchange at irregular intervals. This holds especially for highly permeable surface layers, in which the high proportion of fissures and solution pipes relative to the total void space favors the effectiveness of meteorologically controlled factors. When equation 77 is used to calculate the diffusion rate, the diffusion coefficient of CO_2 or O_2 through air with 79 vol % nitrogen and 21 vol % $O_2 + CO_2$ at 1013.24 mbar and 25°C, $K = 13.9$ cm$^2 \cdot$ d$^{-1} \cdot$ mbar^{-1} can be used (Buckingham 1904, Penman 1940a, 1940b), if the concentration gradient dc/dx is given in mbar $\cdot$ cm^{-1}.

Penman (1940a, 1940b) gives the value 0.66 for the tortuosity correction factor f; according to Van Bavel (1952) it is 0.6. Albertsen (1977) showed that the tortuosity correction factor varies from 0.56 for coarse sand to 0.70 for fine sand. This factor takes into account the net cross section area that is available for gas molecule movement, and the longer pathways in the intricate pore system. Porosity is the most important subsurface factor that controls diffusion; texture, structure, moisture content, soil use, and cultivation are of less importance (Buckingham 1904, Blake & Page 1948, Hannen 1892, Raney 1950). The effective porosity to gas is determined by the water content and its periodic changes. The effective porosity to gas in the unsaturated surface layers can range between 0 and 50%. When the effective pore space is below 20–22%, diffusion rates fall off rapidly with decreasing porosity according to Lemon & Erickson (1952). Field measurements in undisturbed soils show that diffusion rates approach the value zero before zero gas effective porosity is actually reached (from 10–12%) (Blake & Page 1948). This phenomenon probably depends on local interruption of the passages between the pores by soil moisture. There exist empirical relations between tortuosity and water content for clayey (De Jong & Schappert 1972) and sandy materials (Albertsen 1978, Albertsen & Matthess 1977, 1978).

For exchange of the ground air with the atmosphere, the permeability of the subsurface is also of importance. In general a subsurface with low permeability permits no quick diffusion. Between the effective porosity, which determines the diffusion, and the permeability, which influences the gas movement, no simple correlation exists. The problem has been discussed recently by Lerman (1979).

In situ permeability measurements have yielded values between 0.65 and 41 darcys depending on plant cover and moisture (Evans & Kirkham 1950). Disturbances in permeability are caused by stony ground, cracks, fissures, and animal burrows.

Gas exchange between groundwater, ground air, and atmosphere is of great importance for the supply of oxygen to groundwater and to the removal of volatile impurities and decay products from groundwater (Golwer & Matthess 1972).

To summarize, one can state that the composition of the ground air is influenced by the properties that determine the air capacity and permeability of the subsurface: texture, structure, water content, and amount of organic material. Materials of coarse texture or grainy structure, which usually remain free from water, and hence giving rise to high air permeability, favor the overturning of the ground air.

2.5 BIOLOGICAL PROCESSES

Groundwater quality is influenced directly and indirectly by biological processes. Single- and multicelled organisms living in groundwater have adapted themselves to use the dissolved solids and suspended material in their metabolism and in turn release the products of their metabolism into the groundwater. Furthermore, higher plants can extract nutrients from it. In ponds fed only by groundwater and in surface waters generally, planktonic and sessile organisms can change the quality of the water that emerges temporarily at the surface or percolates to the groundwater. In such cases the possibility of the effect of photosynthetic organisms, among which algae are the most important, is one peculiarity to be noted.

Indirect influences include changes in soluble salts of the soil through the microbial breakdown of insoluble substances and the temporary withdrawal of nutrients by higher plants, and the increase in the CO_2 content of the ground air through root respiration and microbial activity. Finally human and animal waste products contribute to the supply of soluble salts. There seems to be virtually no environment anywhere at or near the Earth's surface where the pH and *Eh* conditions will not support some form of organic life (Baas Becking et al. 1960). Fairbridge (1967) concluded there is no near-surface environment (other than volcanic) that is not in some way modified by organic metabolic processes. Biological transformation of material usually hastens geochemical processes. The most effective transformations are therefore to be expected in biologically favorable environments, especially in warm humid climates.

Significant changes in water quality caused by animals and man are to be expected only when human settlements and large-scale animal husbandry (factory farming) are involved. Detailed effects are considered in Section 2.6.

Microbial Metabolism.   Microbial metabolism is of great importance to the quality of groundwater. Microorganisms do not affect the direction of any reaction that results from the thermodynamic constraints of the system and can be deduced from the stability field diagrams given in Section 1.2.3.5, but they do affect the rate of reaction (Hem 1961a). For example sulfides, elemental sulfur, and thiosulfate in the soil can oxidize without any microbial aid. However oxidation by microbial processes proceeds much more quickly. Thus under optimum moisture and temperature conditions for microbial activity, the physical and chemical factors are relatively unimportant compared with the biological (Alexander 1961).

Autotrophic microorganisms, which use CO_2 as the sole source of carbon for metabolism, obtain the necessary energy from sunlight or from the oxidation of inorganic substances, perhaps ferrous iron, and manganese. Most microorganisms can however use organic substances as carbon and energy sources, transforming these from energy-rich complex compounds into

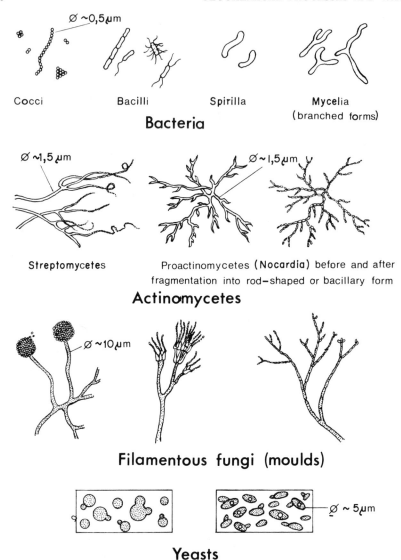

Fig. 33   Important microorganisms in subsurface metabolism (after Davis 1967).

simpler ones and finally into CO_2 and water. Bacteria, Actinomycetes, and moulds take part in this breakdown of organic materials (Fig. 33) (Davis 1967, Fuhs 1961a, 1961b, Kuznetsov 1962, Kuznetsov et al. 1963, Wallhäuser 1967).

Microorganisms occur in very large numbers everywhere in the soil, but also in the water. In 1 g of soil there are quite often as many as 25 billion microbes with a dry weight of about 1 mg. Their geochemical importance is based on their smallness and adaptability, their rapid breeding rate, and their high metabolic rate. Thus an individual bacterium can divide approximately every half-hour, which means that theoretically in 24 hours under optimum conditions it could give rise to 2^{48} individual cells. In practice this number is never reached because the necessary nutrients are absent and the inhibiting metabolic waste products cannot be removed fast enough. The metabolic rate of microorganisms is considerable: within 1 hour they can metabolize about 1000 times their own weight.

Three main groups of microorganisms are present in soil: bacteria, Actinomycetes, and fungi. The bacteria types are mostly cocci or rods. The cell cross section is about 1 μm, the length 2–5 μm. The Actinomycetes are chiefly represented by the genera *Streptomyces, Nocardia,* and *Micromonospora.* The fungi include moulds including Phycomycetes, Ascomycetes, Basidiomycetes, and the asexual *Fungi imperfecti,* and yeasts, which possess the capabilities of single-celled increase and transforming into mycelial-like filaments (Davis 1967). In cm^3 of soil the filaments can total 30 m (moulds and Actinomycetes), corresponding to a dry weight of about 0.2 mg.

The microbial populations of groundwater show considerable variations with respect to near-surface soil horizons and surface waters rich in nutrients. Gram-negative rod bacteria particularly occur in groundwater, especially the genera *Achromobacter* and *Flavobacterium.* These in contrast to Gram-positive bacteria do not retain the blue crystal violet dye. In smaller numbers Gram-positive, nonspore-forming rods, *Micrococcus* and (in some branches) *Nocardia* and *Cytophaga* types occur. *Micrococcus* is an autochthonous microflora adapted to low nutrient concentrations and the aquifer concerned (Wolters & Schwarz 1956).

The composition of the microflora and the bacterial count are influenced by changes in the environmental factors, such as percolation of nutrient-rich surface waters, or by pollution of groundwater. Thus a massive reproduction of microbes in the nearby groundwater extraction work was observed in connection with the impounding of River Danube water near the town of Ybbs/Donau. Under the influence of reducing conditions brought about by this event, there occurred in fluctuating numbers *Chromobacter* and *Achromobacter* types, iron- and manganese-storing *Chlamydobacteria,* sulfur bacteria, *Desulfovibrio, fusarium* and *monosporium* species, and occasionally protozoa and their cysts (Weber 1961).

One of the most powerful environmental factors is the presence or absence of oxygen. Microorganisms can be divided into anaerobic and aerobic types. Oxygen is a deadly poison to the obligate anaerobic types. Other simple organisms, particularly bacteria, use oxygen bound in complex organic molecules or in inorganic molecules (especially nitrates and sulfates) instead of free oxygen. Others can live in the presence of oxygen or without it (facultative anaerobes). Another group (aerobes) requires free oxygen for its metabolism. With few exceptions the Actinomycetes and fungi belong to this group.

The great majority (at least 80%) of microbes grow on solid subsurface media, forming concentrations of cells (Stundl 1967) and coating the soil particles with slimes; but passive transport is possible through loosening of cells and propagation in the groundwater. Sorption processes however hold the microbes, and local differences in the water quality can influence their growth (Wolters & Schwartz 1956). With active colonization the occurrence of microorganisms can be compared with the conditions in an active trickling filter or in biocenoses in slow sand filters as described by Schmidt (1963).

The population density depends on the supply of nutrients and the removal of harmful metabolic products. For the same solution content, fast-flowing water supplies more nutrients and removes more poisonous excreta and therefore permits a greater density of colonization than slow-flowing or stagnant water (Schmidt 1963). Intensive growth is also made possible in the groundwater zone through sorption of nutrients by underground humus and minerals and by gelatinous or colloidal substances secreted by microorganisms, which offer an ideal surface for sorption processes (Husmann 1963).

Subsurface microbe colonization is as a general rule greatest in the nutrient-rich humic upper soil layers and decreases generally with falling nutrient supply at greater depths. This leads to the false interpretation that no microorganisms can live at depth. In the presence of energy sources, particularly organic substances, which permit life, microbe metabolism can also take place below the humus layer (Stundl 1965, 1968). Whereas aerobic microorganisms are tied to free oxygen availability for their distribution, which generally decreases downward, anaerobes and facultative anaerobes are found at various depths.

Sulfate-reducing bacteria have been found in a rock sequence containing limestone, anhydrite, and sulfur at approximately 470 m depth, in petroleum deposits at 1000 m (Vorob'eva 1957), and in marine sediments even at 2700 m under a pressure of 300 bar (Davis 1967). Possibly the microbes have survived in this environment, still favorable for them ever since deposition millions of years ago. Living bacteria have been isolated from anthracitic coals at 540 m depth and from Permian salt deposits in Kansas and in Germany by using suitable processes to avoid any possible contamination by recent organisms (Davis 1967, Gurevich 1962). Evidence for long periods of activity of microorganisms deep below the surface is given by the occurrence of hydrogen sulfide and the low sulfate content in newly drilled groundwater bodies and in oil field brines. Freshly isolated sulfate reducers develop best in media that correspond to the salt content of their former surroundings (Davis 1967). On the other hand, microorganisms can penetrate the subsurface with the recharge of meteoric waters or water introduced during drilling or afterward.

Besides the most important carbon compounds, nutrients necessary for microbe colonization are chiefly nitrogen compounds, such as ammonia or organic compounds of nitrogen, and phosphates (Fuhs 1961b), and in the anaerobic zone oxygen compounds like nitrate and sulfate. Correspondingly sulfate-poor, strongly mineralized oil field brines in the Paleozoic strata of the Caucasus (Dagestan) show bacterial counts of only 10–100/ml, whereas in sulfate- and hydrocarbon-bearing waters of the Tertiary strata in that region 100–1000 bacteria per millilitre maintain a state of sulfate reduction and denitrification (Mekhtieva 1962, Davis 1967). Under suitable conditions high microbe counts can occur, for example 10,000 to 1.72×10^6 per millilitre sulfate-reducing and hydrocarbon-digesting bacteria in the oil field brines of Borislav and Skhodniza, USSR (Osnizkaya 1958).

Microorganisms do not always behave in the laboratory in the same way that is observed under natural conditions. A strain that was isolated from a mineral deposit with a temperature of 120–140°C could tolerate only 70°C in the laboratory, and died after 15 minutes at 80°C (Wallhäuser 1967).

Microorganisms are so adapted to the subsurface physical and chemical conditions that a spatial ecological and geographical distribution of them in groundwater (with respect to depth and region) may be recognized (Gurevich 1962).

The depth to which microbial activity is possible is determined by the nutrient supply, and in addition pH, Eh, salt content, groundwater temperature, and permeability of the aquifer. At temperatures above 100°C most organisms will probably die. Impermeable rocks will impede the circulation of nutrient-bearing water for the support of microorganisms. Aquifers that in the geological past have been cut off from groundwater circulation or show temperatures above 100°C can be recolonized by bacteria that are brought in with groundwater. Some bacteria can endure pressures above 1700 bar (Davis 1967).

The production of surfactants by bacteria should be noted (*Bacillus subtilis, Aspergillus niger, Pseudomonas aeruginosa, Candida lipolytica, Desulfovibrio desulfuricans, Aerobacter aerogenes, Mycobacterium phlei, Acetobacter aceti*). These natural surfactants are the degradation products of organic matter, which result from autolysis of cells or in part probably from the excreta of living cells. These products may change the free pore space, which may explain the influence of bacteria in the flow behavior of oil and water below the surface (La Rivière 1955).

Sulfur and Sulfur Compounds. One characteristic reaction in the groundwater zone is microbial catalysis of the reduction of sulfate by members of the obligate anaerobic genus *Desulfovibrio,* which extract from organic substances the energy necessary for life and the hydrogen for the reaction given in equation 83. In the example given by Sarles et al. (1956) sodium lactate (equation 84) provides the carbon and hydrogen source, and the necessary oxygen comes from calcium sulfate.

$$SO_4^{2-} + 8H^+ + 8e \rightarrow 4H_2O + S^{2-} \qquad (83)$$

$$3CaSO_4 + 2(C_3H_5O_3)Na \rightarrow 3CaCO_3 + Na_2CO_3 + 2H_2O + 2CO_2 + 3H_2S \qquad (84)$$

The *Desulfovibrio* group, which are chiefly represented by *D. desulfuricans*, are adapted to the specific salt content and temperature of the aquifer. *D. desulfuricans* var. *aestuarii* can tolerate an NaCl content up to 200 g/l, and *D. desulfuricans* var. *thermodesulfuricans* can withstand temperatures of 80°C (Baranik-Pikowsky 1927, Elion 1927, Gahl & Anderson 1928, Baars 1930, Wallhäuser 1965).

Microorganisms make possible at normal temperatures sulfate reduction that would otherwise require much higher temperatures. Nonbiological reduction of sulfate with the organic substance presupposes temperatures of 700–1000°C according to Ginter (1930). The experiments made by K. von Karitschoff (1913, in Schoeller 1962), produced a reduction of Na_2SO_4 and $MgSO_4$ solutions using petroleum or petrol partly with, partly without $CdCl_2$ as a catalyst only at temperatures above 96°C. Likewise at temperatures near the boiling point reduction of alkali sulfates and alkali–iron sulfates took place without any contribution from the bacteria and catalysts (Behre & Summerbell 1934, Schoeller 1956).

It has been known for a long time that groundwaters that are in contact with petroleum and natural gas deposits have a deficiency or even absence of sulfate (Potilitzin 1882, Engler & Höfer 1909, Rogers 1917, Renick 1924b); this phenomenon was attributed very early to microbial activity (Meyer 1864, Plauchud 1877, 1882, Etard & Olivier 1882, Beyerinck 1895). The microbial origin of sulfur deposits connected with petroleum and natural gas occurrences could be proved by sulfur isotope determination (Feely & Kulp 1957).

Hydrogen sulfide and other sulfides in groundwater can also originate from the breakdown of sulfur-containing organic substances, for example sulfur-bearing proteins, involving aerobic or anaerobic microorganisms (Fig. 34).

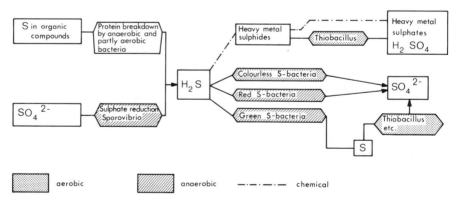

Fig. 34 The chief microbial transformations of sulphur and sulphur compounds (after Schwartz 1958).

Subsurface hydrogen sulfide, sulfur, sulfides, and other sulfur compounds such as thiosulfate, tetrathionate, and sulfites can be oxidized by colorless bacteria of the genera *Thiobacillus, Beggiatoa, Crenothrix, Thiotrix,* and *Thioploca* (equations 85–88; cf. also equation 76; cf. Section 2.3), which as autotrophic organisms obtain the energy for carbon assimilation by chemosynthesis. Thus H_2S is first of all converted to free sulfur, which can be temporarily stored inside or outside the cell before it oxidizes to sulfuric acid. Many types oxidize only the hydrogen sulfide, to elemental sulfur, which is then oxidized by other species (Schwartz 1958, Sarles et al. 1956, Porter 1950, Kuznetsov et al. 1963, Thimann 1964, Zajic 1969).

$$2H_2S + O_2 \rightarrow 2H_2O + 2S \qquad (85)$$

$$2S + 3O_2 + 2H_2O \rightarrow 2H_2SO_4 \qquad (86)$$

$$2FeS_2 + 2H_2O + 3O_2 \rightarrow 2FeSO_4 + 2H_2S \qquad (87)$$

$$Na_2S_2O_3 + H_2O + 2O_2 \rightarrow Na_2SO_4 + H_2SO_4 \qquad (88)$$

These reactions are effected throughout by aerobes, except for *Thiobacillus denitrificans*, which lives in alkaline waters, obtaining the oxygen needed for the oxidation of sulfur by reduction of nitrates:

$$6NO_3^- + 5S + 2CaCO_3 \rightarrow 3SO_4^{2-} + 2CaSO_4 + 2CO_2 + 3N_2 \qquad (89)$$

The anaerobic red and green sulfur bacteria only occur in surface waters (Gmelin 1953), hence can affect groundwater quality only in exposed groundwater occurrences and in bank-filtered river water.

Nitrogen Compounds.   In the nitrogen cycle not only are few specialized types involved, for example nitrifying and denitrifying bacteria, but some of its processes are quite critically affected by other types of bacteria, *Streptomyces,* and fungi.

The reduction of nitrate (denitrification) begins in the presence of organic substances having oxygen contents of approximately 0.5 mg/l (Huisman & Van Haaren 1966), through the action of nitrate-reducing and denitrifying microorganisms. Nitrite is formed as the first product of reduction, through both physiological groups, and is partly broken down by denitrifying organisms to elemental nitrogen, which escapes into the ground air. Other types of bacteria form ammonia from the nitrite (Farkasdi et al. 1969, Schmidt & Kampf 1961). *Desulfovibrio desulfuricans* can reduce nitrate directly to ammonia (Schmidt 1963). Ammonia can also be formed by microbial breakdown of organic compounds containing nitrogen.

Under aerobic conditions ammonia is oxidized to nitrite and nitrate by nitrogen bacteria, such as *Nitrosomonas, Nitrococcus,* and *Nitrobacter* (Van Haaren 1966, Schoeller 1962):

$$NH_3 + 2O_2 \rightarrow HNO_3 + H_2O \tag{90}$$

$$4NH_3 + 3O_2 \rightarrow 2N_2 + 6H_2O \tag{91}$$

The oxidation of ammonia can therefore give rise to an increase in nitrate content. Huisman & Van Haaren (1966) have reported an increase in nitrate content from 1–3 mg/l up to 12 mg/l when the infiltration in a groundwater recharge installation was stopped for some time.

Iron and Manganese. The occurrence or absence of iron and manganese in groundwater depends on their oxidation state and the pH conditions. Certain bacteria play a part in the reduction or oxidation of iron and manganese, also some fungi, protozoa, and algae (obligate or facultative) as catalysts (Oborn 1960, Oborn & Hem 1960, Baas Becking & Moore 1960, Schweisfurth 1966, 1968).

Iron and manganese occur below the surface as widespread insoluble iron(III) and manganese(III) or -(IV) compounds. Their transformation into the divalent state is aided by microbial catalysis (Stumm & Morgan 1970), which requires the presence of organic substances serving as nutrients and sources of energy. The oxygen consumed during their breakdown produces the suitable reducing environment for the solution of iron and manganese (Karbach 1961, Schweisfurth 1966). Under suitable nutrient conditions for a large microbe population (up to about 50×10^6 germs per gram of subsurface media), and favorable Eh and pH values, high iron concentration occurs locally (Oborn & Hem 1960). Microbial oxidation of iron is primarily brought about by members of the genera *Ferrobacillus (F. ferrooxidans), Crenothrix,* and *Gallionella (G. ferruginea),* which as chemoautotrophs produce the energy required for their synthesis of organic matter from the oxidation of ferrous to ferric iron (Schweisfurth 1966). Microbial breakdown of organic complexes of iron can lead to the deposition of iron.

It should be noted that during the oxidation of iron sulfides through formation of iron sulfate and sulfuric acid, sulfur-oxidizing bacteria (*Thiobacillus thiooxidans, T. ferrooxydans*) contribute to the solution of iron, so that microbial reduction and oxidation can lead to solution of iron.

Oxidation of divalent manganese to tri- and quadrivalent states can occur inorganically with atmospheric oxygen at a pH above 7.5, when catalytic sorption supports this reaction. Diatoms, flagellates, fungi, and Schizomycetes participate in biological oxidation (Schweisfurth 1968).

Massive growth of bacteria that process iron and manganese can cause ochreous coating on well screens. This effect is particularly noticeable at redox potentials above -10 ± 20 m V for iron(II) contents above $0.2-0.5$ mg/l in the zone of high flow velocity close to the well filter pack (Hässelbarth & Lüdemann 1967).

Manganese is occasionally also concentrated by nonspecialized microorganisms. In bank-filtered river waters charged with wastewater, Brantner

(1966) observed saprophytic Gram-negative rods that were absorbing manganese as Mn(III) chelate. With the microbial breakdown of the chelate formers, in this case citrates, the manganese was being liberated and oxidized from the trivalent to the quadrivalent state. Because in these organisms no excretion mechanism exists, the manganese accumulates in vacuoles in the cells, which are thereby inflated. After death of the bacteria the quadrivalent manganese is liberated.

Organic Substances. The heterotrophic of chemoautotrophic microorganisms break down organic substances that occur in soils, peat, lignite, coal, bitumen, and natural oil and gas deposits, as well as dissolved in groundwater, to make their own cell structures and obtain energy for their life processes. Under ideal conditions all organic compounds are ultimately converted to CO_2, H_2O, NH_4^+, N_2, NO_2^-, NO_3^-, SO_4^{2-}, phosphate, and other inorganic compounds. Fats, and partly proteins also, occur as intermediate products on the breakdown of carbohydrates, particularly organic acids of simpler structure. The most important connecting links are C_2 particles, the "activated" acetic acid (Karlson 1970). During breakdown of hydrocarbons, organic acids, alcohols, and other compounds can also be found as intermediate products (Zobell 1946b). Microbial oxidation of hydrocarbons differs from this only in the first stage in the reaction, the breakdown of the molecule under the combined effect of two enzymes and a catalytic metal compound (Foster 1962). The remaining oxidation processes are not specific for hydrocarbons.

In aerobic conditions molecular oxygen is used in the transformation. In anaerobic conditions microbial breakdown can proceed either as a methane fermentation or by reduction of sulfates and nitrates (Table 26).

Equation 92 gives the decomposition of a carbohydrate:

$$2C_6H_{12}O_6 \rightarrow 2CH_3(CH_2)_2COOH + CH_4 + 3CO_2 + 2H_2O \qquad (92)$$

Here the decomposition of butyric acid to acetic acid is accomplished by *Methanobacterium suboxydans*, and that of acetic acid to CO_2 and H_2O by *Methanosarcina methanica*. Other organic acids and proteins are broken down in a similar way (Häusler 1969, Ginsburg-Karagitscheva 1933, Jeris & McCarty 1965).

The breakdown of hydrocarbons by methane fermentation following equation 93 yields, as the negative value of the free energy shows, a surplus of energy.

$$4C_nH_{2n+2} + (2n - 2)H_2O \rightarrow (3n + 1)CH_4 + (n - 1)CO_2 \qquad (93)$$

By using the thermodynamic values given by Rossini et al. (1953) the $\Delta G°$ values per mole work out as: ethane -34.0 kJ, propane -59.45 kJ, hexane -141.85 kJ, and *n*-undecane(hendecane) -254.56 kJ (Fuhs 1961a).

An energy surplus also results from the breakdown of paraffins under the sulfate reduction given in equation 94:

$$(3n + 1)Fe^{2+} + (3n + 1)SO_4^{2-} + 4C_nH_{2n+2} \rightarrow (3n + 1)FeS + 4nCO_2 + (4n + 4)H_2O$$

$$\tag{94}$$

$$\Delta G° = -18.26n + 13.11 - \Delta G° \; C_nH_{2n+2}$$

The free energy values $\Delta G°$ are: methane -80.51 kJ, ethane -174.88 kJ, hexane -513.80 kJ, and n-undecane -938.89 kJ (Fuhs 1961a).

Evidence for these processes has been obtained from field observations and by the use of enriched sulfate-reducing cultures. The use of ^{14}C-labeled hydrocarbons has enabled some experimental difficulties to be overcome, for example obtaining pure cultures of denitrifying, sulfate-reducing, and methane-producing organisms, as well as the use and analysis of hydrocarbons. The unequivocal evidence for the oxidation of pure hydrocarbons by *Desulfovibrio desulfuricans* was first obtained with the help of ^{14}C-labeled methane, ethane, and n-octadecane (Davis 1967). The reduction of sulfates labeled with ^{35}S in groundwater samples from petroleum- and bitumen-bearing rocks by active sulfate-reducing bacteria can be regarded as indirect proof of anaerobic microbial hydrocarbon breakdown (Ivanov 1966).

Numerous microorganisms widely distributed in soil and water play a part in the breakdown of subsurface organic substances. Thus hydrocarbons can be decomposed by almost 100 species of bacteria, fungi, and yeasts, particularly those belonging to the genera *Pseudomonas, Achromobacter, Flavobacterium, Nocardia (Proactinomyces)*, and *Mycobacterium*. The microorganisms are more or less strongly specialized for definite organic nutrients. For many groups the best substrate varies from species to species, and for some well-known species like *Pseudomonas aeruginosa* from strain to strain (Fuhs 1961a). Substrate-specific breakdown can for instance be used to isolate naphthas from mixtures containing aliphates and to identify paraffins in trace amounts (Tausz & Peter 1919). Abnormal enrichment of specific hydrocarbon-decomposing bacteria can be used to locate petroleum deposits (Smirnova 1961, Foster 1962, Völz & Schwartz 1962), a procedure that has been used with considerable success in the USSR. This specialization for certain substrates cannot however be generalized. There is an important breakdown process designated as cooxidation by Foster (1962), which is the degradation of certain substances that can be decomposed together only with the substrate that is obligate for the microorganisms. For example *Methanomonas methanica* in the presence of obligate methane or methanol also metabolizes other substances such as ethane, propane, butane, amino acids, and glucose (Fuhs 1961a). The effect of microbial metabolism is also increased for a mixed culture in which a microbe species can make further use of the metabolic products of its neighbors of other species.

Microbial decomposition involves aliphatic, olefinic, naphthenic, and aromatic compounds (Chambers et al. 1963, Haccius & Helfrich 1958, Fuhs 1961a, 1961b), phenols (Meissner 1953, Stundl 1956, Zobell 1946b), surfactants (Husmann 1963, Frank 1965, Heyman & Molof 1967), natural and synthetic rubber (Fuhs 1961a, 1961b, Zobell 1946b), pesticides (chlorinated hydrocarbons: Hill & McCarty 1967, aromatic nitrogen compounds: Germanier & Fuhrmann 1963), benzene, substitute benzene (Zobell 1946b, Alexander & Lustigman 1966), natural substances such as strychnine and atropine (Bucherer 1965), antibiotics (Wallhäusser 1951a, 1951b), and humus generally (Tepper 1963). The rate of decomposition of organic substances varies, partly on account of their varied composition (specifically, their different solubilities, branching hydrocarbons and aromatics being slower than single chains to decay) and partly on account of the constituents of the microflora (Fuhs 1961b).

The decomposition process occurs only in the fraction actually dissolved in water, therefore in hydrocarbons that are not really soluble in water it proceeds only at the phase boundary in contact with water. The process is accelerated by dispersion of the substrate, emulsification, or adsorption to porous media. Adsorption and decomposition processes in biological breakdown cannot really be separated (Husmann 1963). Thus the purifying effect is itself helped in the subsurface by sorption of various substances such as benzene, phenol, detergents, and various industrial wastes onto the soil humus, the microbial slimes, and the microorganisms (Stundl 1956, Husmann 1963). In the sorption of detergents by microorganisms there appears to be a species-related pH dependence (Hartmann & Mosebach 1966). The rate of decomposition increases with the content of free oxygen as well as with temperature (up to about 55°C). Thus the velocity of microbial oxidation of oil under aerobic conditions amounts to as much as 0.5 g per day per square metre of oil surface. Under anaerobic conditions the oxidation is much slower and depends on the presence of nitrates, sulfates, and phosphates as sources of oxygen. Thus approximately 4 mg of nitrate has been found necessary for the complete oxidation of 1 mg of a typical mineral oil (Pilpel 1968). Söhngen (1913) has reported an average microbial oxidation performance at 28°C of 15 mg of petroleum and approximately 8 mg of paraffin in 24 hours per 2 cm^2 surface area of culture liquid.

Even if the decomposition generally proceeds faster under aerobic conditions, some substances are decomposed only partly or not at all in aerobic conditions and are broken down more rapidly under anaerobic conditions. This happens for example with hydroquinone and pyrogallol (Meissner 1953, Stundl 1956), and to some extent this holds for chlorinated hydrocarbon pesticides (Hill & McCarty 1967). These circumstances are particularly effective when a phase of anaerobic decomposition precedes an aerobic one. for example the anaerobic-aerobic method in sewage treatment (Kollatsch 1968).

The decomposition of organic substances expressed in terms of 5-day biochemical oxygen demand (BOD_5) can be described by the following function due to Kramer (1958):

$$A_T = \frac{A_0}{10^{T/C}} \qquad\qquad (95)$$

in which T = soil depth (cm)

$$C = \text{constant} = \frac{T}{\log (A_1/A_2)}$$

A_0 = 5-day BOD of sewage used in the test (mg/l)
A_T = residual pollution of the percolated sewage at soil depth T (mg/l)
A_1, A_2 = residual pollution at soil depths 1, 2 (mg/l)

For wastewater from a sugar factory Kramer (1958) has determined that in the top 30 cm of the soil there is a maximum loading of 1300 g BOD_5 per cubic metre of soil; this corresponds to the daily performance of fully charged treatment plants, or almost double the value for highly loaded droplets. The subsurface oxygen demand for this considerable waste output is met in the aerobic zone from oxygen dissolved in water, and in the anaerobic zone from chemically bound oxygen, particularly nitrate and sulfate. With complete absence of free oxygen, oxygen-bearing ions, or organic compounds from groundwater, no (or only very slow) breakdown of pollutants occurs (Thews 1971). In a comprehensive appraisal of breakdown processes one should also note that the total disappearance of a substance does not necessarily indicate purity because of intermediate products of higher toxicity and resistance to decomposition may have been produced (Schwarz 1967). Complete breakdown of organic substances in water is not reached, and they are omnipresent in groundwater (Huisman & Van Haaren 1966).

Control of Microbiological Activity by Poisons.   Microbiological metabolism can be checked by various inorganic and organic substances, which inhibit the transformation of materials or kill off the microorganisms that are responsible. Hence the same inhibitor can have effects of different strengths on the intermediate processes of biological metabolism. Substances that can be decomposed in low concentrations can have a damaging effect on the microorganisms when present in higher concentrations. Thus for example naphthenic acid at high concentrations exhibits selective effects (Fuhs 1961a, 1961b). Poisonous waste products of metabolism act as inhibitors, which if insufficiently removed can concentrate in the percolate and in the groundwater. This has been noted by Stundl (1965) as the cause of "fatigue" phenomena in soils irrigated or sprayed with sewage.

    Thus it is known that cell multiplication in *Pseudomonas aeruginosa* is inhibited by the production of peroxide as the first intermediate product of the breakdown of cyclohexane, when it temporarily reaches high concentrations (Fuhs 1961a). Ammonia as a metabolic product can also inhibit microbial activity (Albertson 1961). The various antibiotics formed by animals and plants, which can be detected in natural soils and have been widely used for medicinal purposes, can act locally as temporary inhibitors when they build up in high concentrations. At lower concentrations they generally favor the growth of numerous microorganisms and are then decomposed by them. The breakdown processes are aided by sorption onto soil particles (Wallhäuser 1951a, 1951b).

Toxic heavy metal cations can inhibit microbial metabolism when they act as fermentation poisons and block the enzymatic breakdown chain and respiration. As limiting values for the inhibition of microbial metabolism, the following have been obtained: 0.5 mg Cu/l, 5.0 mg Zn/l, and 8.0 mc CR(VI)/l. When these metals occur together the individual toxic effects can be added, but no multiplicative effect has been noted (Kraft & Knoll 1970). A microbe population can adapt itself to more or less constant heavy metal cation concentration over long periods, presumably by changes in its enzymes (Reimann 1969).

Populations of microorganisms can adapt to cyanide so that the cyanide is decomposed, most advantageously at concentrations of about 4 mg CN^-/l. With increasing concentration, especially above the critical level of 10 mg/l, the species composition changes and microbial activity, particularly nitrification, decreases until biological growth ceases completely (Böhnke 1966). Adapted biocenoses can decompose up to 50 mg CN^-/l at uniform concentration and at temperatures of at least 10°C (Knie 1966).

Many hydrocarbons, such as benzene, have a strongly toxic effect toward microorganisms. Thus *Paramecium* is killed in a short time in benzene concentrations above 1:100. Bacterial protein breakdown is distinctly inhibited by benzene mixtures of 1–2 cm^3/l (Stundl 1956).

Above certain concentrations hydrocarbons in mineral oils and related substances can severely impair animal life or make it impossible, but they act only selectively on native bacteria and fungal populations below the surface. Disinfectants such as salicylic and benzoic acids are tolerated up to tenths of a percent and are microbially attacked if they occur as intermediate products from the breakdown from aromatics (Fuhs 1961b, Zobell 1946b).

The toxic effect of phenols is very variable according to whether mono-, di- or polyvalent phenol is involved. For example, *Paramecium* is killed by smaller doses of hydroquinone than of resorcin and phloroglucin (Stundl 1956). Aerobic phenol breakdown is still possible at 500 mg/l of monovalent phenol concentration (Stundl 1956).

Surfactants can also harm subsurface microorganisms (Hartmann 1966). The effect of pesticides varies. Most chlorinated hydrocarbons are only slightly soluble, and in general no appreciable inhibition prevents the breakdown of a standard peptide solution (exceptions are Chlordan 16% and Endosulfan 6%). However two of the chlorinated carboxylic acids 2,4-D (dichlorophenoxyacetic acid) and 2,4,5-T (trichlorophenoxyacetic acid) distinctly caused inhibition (96 and 97% respectively), and adaptative microorganisms adjusted themselves only relatively slowly. Of the organic phosphor pesticides the most water-soluble (Dimethoate and Mevinphos) had an inhibiting effect (8 and 27%, respectively, Philipp & Quentin 1969).

It is possible that bacterial adaptation is positively or negatively affected by temperature-dependent structure changes of the water (Davey & Miller 1966).

Microorganisms can adapt themselves to some extent to unfavorable environmental conditions that are even hostile to life. This comes about through change in their specific composition, through dying off of injured species and multiplication of resistant species and strains, and furthermore through enzymatic adaptation, namely the activation of the system of the various inherent enzymes of the bacteria that is required for metabolism of the nutrients present in the culture medium.

High NaCl content has only a small adverse effect on microbial activity, because the respective population is adapted to the biotope concerned, and in any event minor changes lead to the operation of selection mechanisms or to formation of adapted strains (Fuhs 1961a, 1961b). *Pseudomonas* bacteria tolerate NaCl concentrations up to 15% (Vorob'eva 1957), *Microspira* up to 18% (Ginsburg-Karagitscheva 1933), *Desulfovibrio desulfuricans* var. *aesturarii* up to 20% (Baranik-Pikowsky 1927, Baars 1930), and hydrocarbon-oxidizing bacteria up to 25% (Fuhs 1961a, 1961b).

Inhibition of microbial activity is important for the elimination of subsurface bacterial pollution. The microbes carried by water are largely eliminated as they pass through the subsurface media. For example the count of 30,000–70,000 microbes/cm^3 in the inflow diminished to 100–1000 microbes/cm^3 in the outflow from a soil layer 80 cm thick (Stundl 1964). This effect shows a certain dependence on the depth of the soil column through which the water flows, as Stundl (1965, 1968) has shown by experiments with soil columns of various lengths on which stable manure was spread (Table 27).

The coliform count (*E. coli* index) also decreases as a definite function of

Table 27   Microbe Count in Percolate as a Function of Filter Thickness (After Stundl 1968)

Thickness of soil column (m)	Microbe count on agar base nutrient		
	Maximum	Minimum	Average
0.75	6300	290	2800
1.00	1650	50	320
1.25	840	20	230
1.50	520	20	120
1.75	960	10	130
2.00	750	10	160

Microbe count

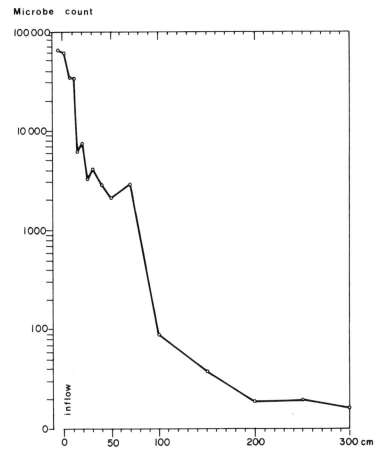

Fig. 35 Elimination of aerobic saprophytes (incubation temperature 22°C) during passage through a sand filter (after Schmidt 1963).

filter thickness. The coliform count of the percolate from soil columns 1.50–2.00 m long was in the majority of cases negative in 100 cm³ (Stundl 1965).

Schmidt (1963) has proved however that the elimination of entrained microorganisms from a continuously irrigated sand filter is several orders of magnitude better in the first few centimetres than in the deeper layers (Fig. 35). The effect of the subsurface on foreign microorganisms can be viewed as a combination of various biotic and abiotic mechanisms whose power varies as the case may be.

Sorption onto subsurface media is of particular importance with fine grain size. Thus it has been found that coliform bacteria have moved through coarser materials deeper than through finer materials, after they have passed the uppermost biologically effective soil layers (Butler et al. 1954). Microbes can also be removed from groundwater by sorption onto precipitating flocs of iron hydroxide (Meyer 1960). Except for spore-forming types, bacteria that do not find enough nutrients for their propagation below the surface can remain capable of multiplying for a surprisingly long time (often several months). The antagonistic effects toward foreign microorganisms are formed by secretion of antibiotic substances (algae, actinomyces) or by direct assimilation as nutrients (protozoa, lower metazoa).

The various survival times of coliform bacteria and pathogenic intestinal bacteria (e.g., *Salmonella*) found by experiment in different loams and sandy soils depend at least partly on the antagonistic microorganisms described above (after 60 days all microbes were still alive; *E. coli* survived up to 120 days). The survival time was lowest in clay beds with a high microbe content of autochthonous bacteria, greatest in a podzol with low microbe content. Loess loam and degraded chernozem showed microbe survival times somewhere between these limits (Glathe et al. 1963a, 1963b).

Finally, many observations support the view that bacteriophages are very important partners in the control of microbes (Schmidt 1963).

With horizontal motion through sandy gravelly sediments a flow path of 50 m at the most is sufficient to eliminate coliform bacteria, provided the flow velocity is not too high (Kanz 1960). In a study in floodplain gravels in the Munich plain with an undisturbed flow of 22 m $\cdot$ d^{-1} liquid manure and a suspension of *Serratia marcescens* with about 1.3×10^9 coliform bacteria, about 50×10^9 *Serratia marcescens* were injected directly into the groundwater. In a well 70 m away, which was in constant use, after 15 hours (corresponding to 112 m $\cdot$ d^{-1}) none of the introduced microbes was found in the well water, although the passage of the polluting media was clearly proved by the detection of tracers. The favorable result may have been due to the efficient elimination by the natural aerobic microbe population in this well-aerated stratum (Kanz 1960), but it also may have been the result of a chromatographic effect that brought about different retardation and transport velocities of the substances and particles moving with the groundwater.

In other cases coliform bacteria traveled distances of 3–77 m in sandy and gravelly aquifers (Butler et al. 1954). *E. coli* also traveled in gravel and sand deposits as far as 135 m (Schinzel 1968), and intestinal bacteria as much as 850 m in medium-grained sands with gravelly lenses and having a daily velocity of 40 m (Weisman 1964).

Organisms Occurring in Groundwater. The organisms living in water-filled pores of fissured and karst rocks and in the interstitial spaces within sands and gravels feed on the waterborne organic waste and so contribute to the purifying effect of the subsurface zones. There is a fauna adapted to porous sands, using for nutrients the debris of plants and colonies of bacteria living on the surface of the grains. The fauna includes turbellaria, nematodes, oligochetes, copepods, and halacarids. The population density depends on the amount of organic debris available and on the suitability of environmental factors, and indirectly on the effect of incoming dissolved organic substances that are nutrients of the established bacteria colonies. The composition of the species and their numbers of individuals are continually adjusting to the arrival of organic matter (Husmann 1966a, 1966b, 1968).

The nutritional supply is generally greatest in the boundary zone between surface water and groundwater, for instance in gravelly-sandy river beds. Here also there is a development of the faunal elements that have passed from the surface water into the interstitial water in the gravel bed, which are in the position of facultatively building up types that can live in the groundwater (stygophile species). By this means the species spectrum can expand and the number of individuals can increase because of the rich supply of nutrients. With increasing distance from the river bed the number of species and individuals of the fauna within the rock pores decreases until finally only stygobiotic (exclusively living in groundwater) and high grade stygophile groundwater organisms remain. If the population is to be maintained, no prolonged breaks in the supply of organic matter can occur (Husmann 1968).

Influence of Higher Plants on Groundwater Quality.   Higher plants absorb mineral substances through their roots and these are stored in their cells during growth. Besides carbon, hydrogen, and oxygen, large amounts of calcium, silicon, potassium, sulfur, nitrogen, magnesium, sodium, and trace elements such as strontium, molybdenum, copper, boron, and zinc are taken up. By this process many soluble soil components are retained in the plant tissues that otherwise would be leached by the percolation water and added to the groundwater. The retention of soluble substances by higher plants can be used for the purpose of improving the water quality. Höll (1963) recommends in place of coniferous woods, which transmit considerable amounts of nitrate to the groundwater of catchment areas, planting stands of rosebay willowherb (*Epilobium angustifolium*), which keeps the soil free of nitrate; hence the percolate as well. Harth (1965) recommends planting encircling screens of deeply rooted forest trees, which in shallower groundwater diminish the solute content derived from the unused fraction of fertilizers. Finally one should mention the possibility of removing phenols by plaited rushes (*Scirpus lacustris*) from the water of recharge basins (Czerwenka & Seidel 1965). The precipitation water can however dissolve out plant nutrients from surface parts of plants, essentially through discontinuities in the cuticle. The losses are variable, depending on the part of the plant involved, the nutrients, environmental factors such as day or night, temperature, dew formation, plant species. Thus beans, maize, and gourds lost almost 4% of the total ^{45}C from their leaves against 1% from tomato leaves.

In a similar manner beans and gourds lost 1.6 and 2.4% ^{42}K from the leaves compared with 0.3 and 1%, respectively, from leaves of tomato and sugar beet. Leaching losses are also known from conifers and other woodland trees as well as from cereals. The leaching effect depends on the physiological age of the plant at the time. With increasing maturity there is a tendency to increasing loss of material. The cause of this is the increasing permeability of the cells, the accumulation of salts in the mature leaves, and the initial hydrophobia of the young leaves. Leaching by dew is stronger than during periods of heavy rainfall. Besides leaves, leaching is also effected by fruits, flowers, stems, and branches (Tukey & Tukey 1959). Table 28 shows the range of leaching ability of inorganic materials.

Table 28   Relative Leaching Rates of Isotopes from Young Gourd and Bean Leaves (After Tukey & Tukey 1959)

Low	Medium	High
^{22}Na	^{45}Ca	^{55}Fe–^{59}Fe
^{54}Mn	^{28}Mg	^{65}Zn
	^{35}S	^{32}P
	^{42}K	^{36}Cl
	^{90}Sr–^{90}Y	

When plant material dies and decomposes, the mineral residues are liberated and can be leached out if they occur in soluble form, and passed on to the groundwater. This seasonal and biologically determined rhythm in the supply of soluble materials in the soil zone causes for example seasonal changes in nitrate, carbon dioxide, and sulfate content in groundwater (Röhrer 1933, Becksmann 1955a, Matthess & Hamann 1966).

The CO_2 content in the ground air is generally 1–2 orders of magnitude higher than that in the atmosphere, which amounts to 0.03 vol % on average. The CO_2 content in the ground air is determined by root respiration of higher plants (Clements 1921, Möller 1879, Sachs 1860, 1865, Lundegårdh 1924, 1927), and by microbial breakdown of organic substances (Stocklasa & Ernest 1905, Bristol 1920, Wollny 1880b, Russel & Appleyard 1915, Waksman & Tenney 1927) in the strongly active upper part of the soil profile. Corresponding to the dependence of biological activity on temperature and moisture, there are seasonal (Pettenkofer 1871, 1873, Schloesing 1889, Boynton & Reuther 1939, Wollny 1880a, Ebermayer 1890, Russel & Appleyard 1915) and daily (Russel & Appleyard 1915) changes in the CO_2 content as well as dependence on the type of plant cover (Wollny 1880a, Möller 1879, Ebermayer 1890, Russel & Appleyard 1915, Lundegårdh 1924, 1927).

Because the microorganisms (bacteria, actinomyces, fungi) produce the major part of the carbon dioxide in the ground air, the periodic drying out of the top zones, which is prejudicial to biological activity, affects the CO_2 content of the ground air (Russel & Appleyard 1915, Lundegårdh 1927). Russel

& Appleyard (1915) attributed the reduction of the CO_2 content of the ground air in winter to lower microbiological activity in the soil. Furthermore, there is a definite tendency toward increased CO_2 content with depth, even if not in all cases (Pettenkofer 1871, 1873, Fleck 1876, Möller 1879, Schloesing 1889, Wollny 1886), and a connection between the CO_2 content and soil type (structure, texture) (Möller 1879, Wollny 1881, 1886, Ebermayer 1890, Lundegårdh 1924, Fleck 1876).

The importance of the microorganisms, especially bacteria, moulds, and algae, is clearly demonstrated by the cessation of CO_2 production on the death of the soil organisms through heating or through antiseptics (Stocklasa & Ernest 1905). Distinctly raised CO_2 content of the ground air has been determined in regions of contaminated soils and groundwater (Smolenski 1877, Pettenkofer 1873, Golwer et al. 1970, Golwer & Matthess 1972). The organic material content (humus) and the supply of fertilizers generally increase CO_2 production in the soil (Boussingault & Levy 1853, Lundegårdh 1924, 1927, Wollny 1889, 1880b, Appleman 1927), likewise cultivation of the soil (Möller 1879, Appleman 1927). The supply of soluble sulfates undergoes seasonal variations. The higher plants take up SO_4^{2-} ions in seasonally variable amounts to meet their sulfur demand. The total sulfur content in beech leaves rises steadily from June to October, so that the fraction of water-soluble sulfate in the leaves increases sharply during the yellowing phase in autumn. In raw beech humus on the other hand the soluble sulfate content in summer is low and increases sharply with leaf fall in October and November. With conifers the seasonal changes are not so marked, but a rise in the percentage content of soluble sulfate can be seen in raw spruce humus (Blanck & Evlia 1932). These variations in supply of soluble sulfates depend on two overlapping processes: microbial sulfur mineralization, which occurs in warm seasons more strongly than in cold, and the uptake of sulfate by higher plants. With the end of the vegetative period the uptake of leachable sulfates by higher plants ceases, while microbial production of sulfate continues (Stremme 1950). In addition soluble sulfates are added to the soil by leaf fall. Two springs in the Bunter Sandstone of Spessart, Germany, show seasonal changes in the CO_2 and sulfate content where the catchment areas are chiefly covered with deciduous and mixed conifer forest (Matthess & Hamann 1966).

Figure 36 shows the seasonal changes in the free carbonic acid content, rising values reflecting the rise of the CO_2 content of the ground air during the vegetative period (Becksmann 1955). The sulfate content of the water of both springs shows on the other hand a decrease during the vegetative period. In the late autumn of 1961 and winter of 1961–1962 the graph rises, which can be attributed to leaching of the increased sulfate supply in the forest soil toward the end of the vegetative period. In the extremely early and cold winter of 1962–1963 the frozen ground impeded groundwater percolation. The leachable sulfate ions were retained in the soil and were not leached until after the thaw, and arrived in the subsurface with the percolate.

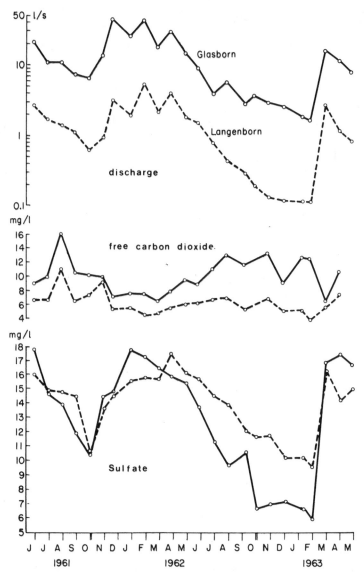

Fig. 36 Yield and content of free carbon dioxide and of sulfate in two springs in the Bunter Sandstone (Glasborn and Langenborn) (after Matthess & Hamann 1966).

The impeding of percolation in the winter of 1962–1963 is also reflected in the spring yield (Fig. 36), the variations in which follow those of sulfate.

The evidence for the biological origin of the variation in the CO_2 and sulfate content, observed in the two springs, is derived from the opposite trends in the behavior of the two solutes. The supply of CO_2 in the ground air is greatest during the vegetative period and leads to high values in the percolate, in contrast to the supply of sulfate ions in the soil, which is greatest after the vegetative period. If on the other hand the changes of the CO_2 and SO_4 content were attributable to the chemical quality of the precipitation, parallel changes of CO_2 and SO_4 content would be expected, because the CO_2 and SO_3 supply in the atmosphere is greatest in winter because that is the period of maximum burning of fossil fuel for heating of buildings.

It should be noted that according to Höll (1963) high sulfate values in winter runoff were indicated by the Bollsee-Nord lysimeters, located on a light soil in an oak-birch forest. Bayly & Williams (1966) attribute high sulfate content in Lake Edward, South Australia, to metabolic processes in the pinewoods of its catchment area.

2.6 MAN-MADE FACTORS

Man-Made Pollution of Groundwater.   Man alters the quality of groundwater in many ways, both directly and indirectly. Direct effects are caused by natural and synthetic substances, which enter the groundwater as a result of human activities. Indirect effects leading to changes in groundwater quality are caused by human interference with hydrological, physical, chemical, and biological processes without any addition of substances. For example the CO_2 content in the ground air is raised by the cultivation of the soil through stimulation of microbial activity, so that the solvent ability of the percolating water for calcium and magnesium carbonates increases, and consequently in many cases the alkaline earth content of groundwater increases. In addition, changes in groundwater temperature caused by human activity (Section 3.2) indirectly affect the chemical quality of the groundwater through temperature-dependent solvent action of the water. There are transitions between direct and indirect effects, for example extraction of bank-filtered river water from polluted rivers. The infiltration of the river water is induced by artificially reversed groundwater gradients, which bring about the introduction of pollutants into the groundwater.

Man-made changes lead to groundwater pollution, defined by Buchan & Key (1956) and Milde & Mollweide (1970) as the partial or total loss of usability of the water for human consumption or processes. This definition does not include all the numerous changes that are caused by man but do not lead to the impairment of the hygienic quality of groundwater. For example water with traces of urochrome from human excreta is accordingly not considered as polluted, because the urochrome content below 1 mg/l can be regarded as harmless from the standpoint of health (Wolter 1967).

For a definition to be of practical use it should be based on the limiting values of water quality. Thus Matthess (1972a) has proposed the following definition: anthropogenically polluted groundwater is groundwater in which the total dissolved and suspended solids caused directly or indirectly by man is higher than the maximum permissible concentrations relative to the limiting values that are laid down in national and international guidelines for potable or industrial water. Because natural groundwater unaffected by man may contain constituents that exceed the limits as defined, the pollution in these cases can be defined by the values that exceed the natural variations of the constituents concerned in a specified water.

It follows from this definition that the natural quality of the groundwater in the region must be known beforehand, and so it can occasionally happen that supposed human factors are in fact the result of natural conditions. Semmler (1961) mentions for example a yellowing of some well water over a long period. This effect was assumed to have been caused by mining, but on more detailed examination it proved to be due to humic acid picked up during groundwater flow through a peat deposit.

For the identification of pollution the most unequivocal criteria possible are required. In this respect the commonly determined inorganic dissolved solids are only suitable in exceptional cases: the potassium content as an indicator of groundwater contamination by sanitary landfills proposed by MacLean (1969) and Davison (1969) is almost always useless, because potassium is strongly adsorbed below the surface and does not migrate with the polluted groundwater. Raised HCO_3^-, Cl^-, NH_4^+, and NO_2^- content, and a high value in the permanganate test can serve as indicators of pollution; however it has to be proved at any given time and place that these indications are not attributable to geological origins (Lüning & Heinsen 1934, Keller 1942).

The identification of intestinal bacteria points to recent pollution, but their absence is not a valid argument against the presence of pollution, because these microorganisms tend to be adsorbed or filtered during flow through porous media and also die off after some time underground. More suited to be chosen as criteria are substances like the organic components of oil field brines or waste from refineries, which can be identified in the groundwater by suitable methods (infrared spectroscopy, gas chromatography) without any doubt (Maehler & Greenberg 1962). The identification of artificial constituents in polluted groundwater is the most convincing proof of human causes.

As a result of the disposal of gaseous, liquid, and solid wastes, groundwater can be poisoned by chemical substances and infected by pathogens, and its potability may also be impaired by troublesome odors, taste, and coloring. Also substances that can be used in agriculture and forestry as fertilizers, insecticides, and herbicides may pollute the groundwater. Their harmful effect on groundwater quality is to some extent caused by inappropriate application, such as overdosing or by the chemical nature (resistance to decomposition) of the substances used.

Indirect groundwater pollution can be caused by organic substances such as petroleum products when they give rise to reducing conditions involving iron and manganese dissolution (Kölle & Sontheimer 1968, 1969, Back & Langmuir 1974).

The pumping of groundwater near a coastline may cause seawater encroachment, especially when the freshwater body is overpumped. Examples of this phenomenon are known from many coasts in the United States (Todd 1960a) and in Europe (Matthess 1961).

Lowering of the water table in agricultural land reclamation or in mining activities increases the supply of oxygen, so that sulfides can be oxidized to sulfates and free sulfuric acid and the pH is shifted into the acid range. Such conditions have been described from reclaimed coastal marshlands (Calvert & Ford 1973) and from mine waters (Biesecker & George 1966, Lindgren 1933, Matthess 1974). The generally acid mine waters (pH 3–4) may contain considerable amounts of dissolved heavy metals from the dissolution of metal sulfides. The cessation of mining does not stop the pollution, which may continue for many decades.

Artificial radionuclides that originate when elements are bombarded with nuclear particles are an important group of man-made waste materials — nuclear waste. If the process yields elements of lower atomic number than that of the original elements, it is described as fission; if it yields elements of higher atomic number it is called fusion. Neutrons formed as fission products can bring about further fission reactions (chain reaction). The most important radionuclides originating through fission and fusion are ^{89}Sr, ^{90}Sr, ^{106}Ru, ^{131}I, ^{137}Cs, ^{144}Ce, ^{3}H, and ^{14}C, whose behavior in the groundwater is the same as that of the natural isotopes of the elements concerned, they are therefore described with the respective elements.

Groundwater pollution by artificial radionuclides can occur as a result of the processing of uranium ore deposits, the production, use, and reprocessing of nuclear fuels and explosives, the disposal of radioactively contaminated cooling water and the leakage of volatile radionuclides in reactors, and through the numerous technical and medicinal uses of radiochemicals. Besides these sources of radioactive contaminants there are also atmospheric and underground nuclear tests, which release considerable amounts of radionuclides. The spread of radionuclides after atmospheric explosions and their influence on groundwater is considered below for individual radionuclides. The same also holds for the radionuclides ^{90}Sr, ^{103}Ru, ^{106}Ru, and tritium, which can be expected to occur as a result of underground nuclear explosions in or near groundwater. The resulting effects depend on the type and power of the nuclear explosion, the rock type, the water quality, and the regional hydrogeology. According to Higgins (1959) and Batzel (1960) groundwater occurrences in cavernous limestones, coarse gravels, highly permeable volcanic rocks, and strongly fissured rocks generally are exposed to the danger of contamination because they do not possess sufficient sorption capacity to retard effectively the spread of radionuclides. Under favorable sorption conditions and low groundwater velocities the long-lasting dangers to regional aquifers can be neglected. Pollution can be caused by excessive amounts of tritium of the order of 10^{16} Bq, which can be released from large nuclear fusion explosions near aquifers and so enter the groundwater circulation (Stead 1963).

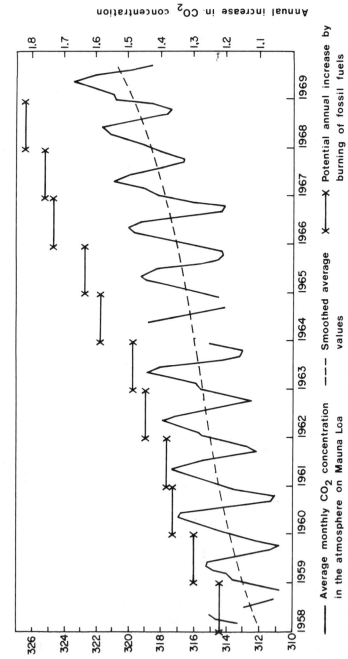

Fig. 37 Change in CO_2 concentration caused by burning of fossil fuels (from Man's Impact on the Global Environment 1971).

142

Pollution by Gases. Industrially or domestically generated gases chiefly consist of oxides of carbon, sulfur, and nitrogen, and to a smaller extent chlorine. Radioactive gases (e.g., ^{131}I) can escape from atomic energy plants or thermonuclear test explosions. All these gases can dissolve in rain and seepage water and so enter groundwater.

The CO_2 content of the atmosphere increases in the course of time because of the burning of fossil carbon compounds (coal, natural gas, petroleum). From 1850 to 1950 10% of the total amount of CO_2 (2.3×10^{18} g) present in the atmosphere in 1950 has been produced in this way (Man's Impact 1971). Since 1958 the atmospheric CO_2 content has risen annually by about 0.7 mg/kg (i.e., approximately 0.2% of the total stock in the atmosphere). This amount lies about 50% below the annual CO_2 output from the burning of fossil fuels (see Fig. 37), probably because of the dissolution of CO_2 by the ocean and its assimilation in the biosphere (Eriksson 1970, Man's Impact 1971). A portion of the carbon dioxide content is present initially as carbon monoxide, which comes chiefly from man-made sources, especially exhaust gases from motor vehicles (Heicklen 1976). Carbon monoxide eventually oxidizes to carbon dioxide in the stratosphere (Eriksson 1970).

The geochemically noteworthy increase in CO_2 content of the atmosphere can bring about only a minor change in the CO_2 content of the groundwater because the major source of its CO_2 content is to be found in the active soil zone. On the other hand the rise in the SO_2 content of the air can cause an important change in the groundwater sulfate content.

The sulfur contained in fuels [lignite 0.2–1% (5%), coals 0.5–1.5%, and heating oils 1.0–4.2%] reappears mostly as volatile SO_2 in flue gases (in lignite 56.5–79.5% of the original sulfur content, in coals 75.1–100%, in heating oils 55.4–70%) (Spengler 1964). Part of the sulfur, especially any original sulfate present, remains in the ash and other by-products (Muhlert 1930). A small fraction of the SO_2 is oxidized to SO_3 in stoves and furnaces, and this in part precipitates in the colder parts of the flue. Thus soot can contain as much as 75% sulfuric acid. The exhaust gases of all types of internal combustion engine, the smelting of copper, lead, and zinc ores, and the manufacture of sulfuric acid also contribute to the SO_2 content of the atmosphere. The worldwide man-made emission of SO_2 amounted to 146×10^9 kg/year for the years 1963–1966 (Heicklen 1976).

The SO_2, which reaches the atmosphere with the flue gases, is partly absorbed by soil or by plants (Ericksson 1952), and partly after further oxidation in the atmosphere is formed into sulfuric acid and precipitated onto the ground.

The rise in the SO_2 content of the atmosphere in the downwind side of a source of emission (coal furnaces, cellulose factories, sulfide ore smelters, etc.) is dependent on the amount of sulfur dioxide discharged, the chimney height, the wind velocity and direction, the distance from the chimney, and the magnitude of exchange flux of air masses in the free atmosphere. The width of the polluted zone ranges from 3 to 6 km depending on chimney height and weather conditions (Pittner 1952, Stratmann 1955, Lahmann 1970).

In winter the sulfate content in atmospheric precipitation in heavily populated and industrialized regions is due to the greater use of fuel during cold weather than in summer (König 1893, Riffenburg 1925, Strell 1955, Gericke & Kurmies 1957). The SO_4^{2-} content of the precipitation is therefore influenced by the SO_2 amounts in the industrial and domestic flue gases, the type and amount of the precipitation, the wind direction and velocity, and the seasonal changes in flue gas discharge. Besides flue gases Gorham (1955) cites soot, ash, dung, and dust from cultivated and uncultivated estates as possible sources of sulfate-releasing air pollutants. The SO_4^{2-} content of the precipitation is according to Eriksson (1952) a sensitive indicator of atmospheric pollution.

Accordingly industrial regions with considerable SO_2 emission have high SO_4^{2-} content in the precipitation (30–450 mg/l or even more), clearly contrasted with nonindustrialized cities (with characteristic SO_4^{2-} content 15–30 mg/l), and rural areas (< 15 mg/l) (Drischel 1940, Junge 1954, Matthess 1961).

Of the oxides of nitrogen, the only important man-made pollutants are NO and NO_2, which account for another 53×10^9 kg/year (computed as NO_2) of the 500×10^9 kg/year emitted worldwide from natural sources. About 93% of the worldwide man-made urban emissions originates from the combustion of coal and petroleum products (Heicklen 1976).

Man-made emissions into the atmosphere are considerable (Table 29). Besides gases there are finely dispersed materials, forming soluble and insoluble aerosols that are primarily removed by rain or snow. Very small particles combine with water droplets in the clouds and are later removed when rain falls. Larger particles can either precipitate in a similar way or be dragged down directly with falling raindrops (Man's Impact 1971, Heicklen 1976).

The relatively high bromine content of atmospheric particles is explained by man-made pollution, and is attributable to bromine in automobile exhaust gases (Lininger et al. 1966, Man's Impact 1971). Chlorine occurs in cement works flue gases (Riehm 1961) and in flue gases from furnaces that burn chloride-bearing coal (Junge 1963, Meetham 1950); it also is produced when

Table 29 Estimated Amounts of Industrial Emission on the Earth in 1968 (from Man's Impact 1971)

Industry	Amount ($\times 10^9$ kg)
Iron and steel	915
Cement	513
Cereal handling, mills, warehousing, storage	112
Paper	90

waste containing polyvinyl chloride (PVC) is burnt. Fluorine comes from burning of waste that includes spray containers of various domestic products (Reimer & Rossi 1970). The occurrence of lead in the air and rainwater (Lininger et al. 1966, Golwer 1973, Golwer & Schneider 1973), and of pesticides as well (Breidenbach 1965, Weibel et al. 1966, Tarrant & Tatton 1968), is caused by man. For example on January 26, 1955, in rainwater in Cincinnati there were detected among other things 0.5 mg chlordan/kg dry residue, 0.6 mg DDT/kg, and 0.003 mg dieldrin/kg (Weibel et al. 1966).

The occurrence of radionuclides in rain has already been mentioned in connection with the tritium cycle (Section 1.1.2). The most important radionuclides, which have entered the atmosphere from nuclear explosions and have been detected in groundwater, include ^{3}H, ^{14}C, ^{36}Cl, ^{90}Sr, and ^{103}Ru. The reported concentrations of these radionuclides in groundwater lie far below the permissible limits.

Pollution by gases can also occur in the area of underground gas storage by leakage from the injection shafts and through the confining strata (Milde & Mollweide 1969).

Pollution by Liquids. Groundwater pollution can arise from organic liquids that come into direct contact with groundwater or are dissolved in infiltrating rain and surface water. Waters in which dissolved or suspended pollutants are contained, for example sewage and polluted surface water, often act as liquid pollutants.

Among organic liquids the most important are crude oils and their derivatives that are liquid under normal environmental conditions such as gasoline, lubricating oils, light and heavy heating oils, electrical transformer oil, milling lubricants, and other emulsified products, such as drilling fluids, tar, and its derivatives produced by coke ovens during low temperature carbonization of coal and oil shales. Most petroleum products (gasoline, petroleum, kerosene, diesel fuel, light and heavy heating oil and motor lubricants) make water unpalatable by affecting its odor and taste, when the concentrations exceed the threshold of odors (Table 30). The ranges of the concentrations for the threshold of odors are caused by the different sensitivities of the persons tested.

Table 30   Threshold Limits for Odors from Petroleum Products (Committee "Mineralöl und Wasser" 1968)

Type of petroleum product	Threshold concentration for odors (mg/l)
Benzol	1–10
Gasoline	0.001–0.01
Diesel fuel	0.001–0.01
Heating oil, extra light	0.001–0.01
Kerosene	0.01–0.1
Petroleum	0.01–0.1

Table 30 shows that the odor from 1 mg of gasoline, diesel fuel, or heating oil can make 1000 litres of water unpotable. This statement when applied to groundwater is seldom of practical importance, because below the surface a thorough mixing of oil and groundwater in the requisite proportion $1:10^9$ does not happen. By field experiments with extralight heating oil a mixing ratio of oil to water of $1:10^4$ was observed (Käss 1969). Furthermore, highly diluted mineral oil mixtures are more rapidly decomposed by microorganisms (Section 2.5), and this holds for the sorption effects of sediment and rock materials (Section 2.2). Numerous cases of pollution, some of them lasting several decades, have been described (Schwille 1964, 1966, 1976, Parizek et al. 1971).

Pollution can also arise through the use of additives in petroleum products, such as antiknock additives or corrosion inhibitors. These are partly pure organic compounds, organometallic compounds (e.g., compounds of heavy metal with fatty acids), phosphorus-, chlorine-, nitrogen- and sulfur-containing compounds, or pure inorganic additives. The diversity of the substances involved always makes it difficult to predict their polluting effect. Some of these additives are compounds that can react with water or are relatively easily soluble. For example lead compounds in solution originate from lead-containing antiknock additives (Committee "Mineralöl und Wasser" (1969).

Pollution by Solids.   Solids in solution or suspension are the most frequent and widespread cause of groundwater pollution. Here the following are noteworthy:

1.   Solid pollutants that occur in natural waters or in industrial or domestic sewage are suspended or dissolved and enter the groundwater by way of well shafts, recharge lagoons, storage basins, the sewerage system, or surface runoff.

2.   Solid pollutants that are deposited on or below the Earth's surface and are leached by rain and percolation water. To this group belong the majority of soluble solids and suspensions found in solid industrial and domestic waste, but also solid constituents of emissions (e.g., fly ash) and of the fallout from nuclear test explosions, which eventually arrive on the Earth's surface and then may be gradually transported into the groundwater.

3.   Bacteria and viruses, which can reach the groundwater with infiltrating polluted water (e.g., with sewage or polluted surface waters).

Effect of Sewage and Wastewater.   Urban wastewater consists of domestic sewage, rain and melt water, water from street washing, and effluent from municipal and industrial premises (Sierp 1939a). Industrial waste includes effluent from canteens and sanitary installations, the condensation, cooling, and washing water of industry, and true manufacturing wastewater (Sierp 1939b).

In the Federal Republic of Germany in 1968 about 13.3 million m³ of wastewater from domestic and business sources was discharged daily into public sewerage systems; about 74% of this amount had been treated, whereas 5.3 million m³ of treated industrial wastewater and 19.8 million m³ of untreated cooling water daily had been released directly into the rivers (Bundesminister des Innern 1971).

Table 31   Normal Mineral Composition of Domestic Sewage (Feth 1966)

Constituent	Amount (mg/l)
Total dissolved solids	100–300
B	0.1–0.4
Na^+	40–70
K^+	7–15
Mg^{2+}	3–6
Ca^{2+}	6–16
Total N	20–40
PO_4^{3-}	20–40
SO_4^{2-}	15–30
Cl^-	20–50
HCO_3^-	60–90

Besides the usual constituents (Table 31), domestic sewage contains surfactants and smaller amounts of other organic substances. Industrial effluents exhibit extraordinarily variable composition (cf. Bucksteeg 1969a).

The different disposal methods for effluents have varied effects on groundwater quality.

Discharge to Surface Runoff. The discharge of effluent into the main drainage channel of a river system affects the groundwater quality in the neighborhood of the surface runoff. River waters seep underground because of natural or artificially induced gradients or when effluent-loaded river water sinks and displaces less dense groundwater (Martini 1952). Furthermore, river waters can penetrate below the surface by artificial groundwater recharge, or via sewage-irrigated fields, and when flood water in the area of inundation percolates downward. Increasing pollution of surface water makes itself particularly noticeable in bank-filtered river waters through an increased load of odorous substances, a disturbing content of iron, manganese, ammonia, and other indicators of pollution (Holluta et al. 1968, Holluta 1960). Phenol is very often reported in bank-filtered river water (Stundl 1956).

Oil field brines, which have a high content of salt and organic substances, can cause pollution of the groundwater through leaky well shaft casings and standpipes, or after coming to the surface can percolate into the groundwater. The case of direct pollution is less probable (Nöring 1958) and should be less common than indirect pollution (Pettijohn 1972). For example the water of Turman Creek, Sullivan County, Indiana, contains up to 10 g Cl$^-$/ kg coming from oil field brines, and at low levels it sinks completely into the sand and gravel deposits of the Wabash Valley (Jordan 1962).

In the River Leine valley, West Germany, an urban waterworks with wells lying several hundred metres from the river, which was charged with the effluent from potash mines, had to be abandoned on account of pollution by salt (Lang & Bruns 1940).

Infiltration from Septic Tanks and Drains. Liquid pollution in ground-water can enter from leaky septic tanks (Miller 1973) and sewerage systems. A part of the pollutants is retained and broken down through filtering, oxidation, and the activity of microorganisms in the unsaturated zone above the water table.

After emerging from two sewers and an oxidation installation, water that was polluted with surfactants ($>$ 0.2–6 mg/l) migrated through sandy-gravelly beds within 1–3 years over some 300–500 m distance (Bahr & Zimmermann 1965).

In two Hungarian municipalities the groundwater was polluted by polluted wastewater from metal plating works, which was discharged into soakaway drains in unconsolidated sediments. In one case the strongest pollution (24 mg Cr(VI)/l) occurred in a well 19 m distant from the soakaway. The chromate pollution could be detected up to a distance of 50 m. In the second case in two wells 56 and 80 m away from the soakaway 0.14 and 1.6 mg CN$^-$/1, 2.1 and 0.04 mg Cr(VI/l, and 1.0 and 0.8 mg Ni/l levels of pollution were detected, respectively. The pollution moved with the groundwater flow over a distance of 100–120 m within 1–1½ years. The transport velocity was affected by the infiltration of 6–10 m^3 of liquid waste per day and the extraction of groundwater from nearby wells. Whereas zinc and Cr(III) were effectively retained below the surface, nickel, cadmium, and copper were retained in insignificant amounts; cadmium and copper apparently migrated as cyanide complexes. The free cyanide and the chromate were hardly adsorbed at all (Csanády 1968).

There are well-known cases of pollution through disposal into septic tanks or disposal pits of urban or industrial liquid waste including oil, solvents, acids, paint and grease-trap fluids (Leggat et al. 1972), phenols (Stundl 1956), chromium and cadmium (Back & Langmuir 1974), arsenic (Balke et al. 1973), and bacteria (Langmuir & Jacobson 1973).

Irrigation and Spray Irrigation. Sewage and the solids dissolved in it can be put to use as fertilizer on forests and croplands (Back 1973, Sopper & Kardos 1973, Parizek et al. 1967, Parizek & Myers 1968, Overman 1973). This kind of disposal leads to a purification of the waste by the soil in agricultural and forestry areas. The purification process depends on filtering, adsorption, coprecipitation, and the breakdown of biodegradable substances in the soil. It is particularly influenced by the soil structure, the temperature, and the soil use. As a rule the purifying effect rises with increasing concentration of the wastewater. The major part of the purification takes place in the top-most soil horizon. The underground passage decreases considerably the contents of nitrogen, phosphate, boron, and heavy metals (Bouwer 1968, Cherry et al. 1973, Sopper & Kardos 1973). Soil treatment of wastewater can lead to a rise in the inorganic salt content in the groundwater lying below. The local use of phenol-bearing wastewater from gasworks (e.g., for agricultural fertilizers) gives rise to high phenol content – up to 216 mg/l has been observed – and unpleasant odors appear in the groundwater (Lingel-bach et al. 1962).

Ground Disposal into Deep Strata. Discharge into suitable storage rocks with the help of injection wells can be a successful method of wastewater disposal (Keller 1958, Finkenwirth 1968, Van Everdingen & Freeze 1971, Aust & Kreysing 1978); however cases have been known of such waste-water rising again and entering groundwater circulation and even surface waters. One example of wastewater percolation is to be seen in the salt-bearing waste from the potash works in the Werra area (West and East Germany). Here, according to Finkenworth (1968) 500 million m^3 of potash wastewater was discharged underground between 1928 and 1967. At present the injection rate of wastewater amounts to about $1 \ m^3 \cdot s^{-1}$ with an average chloride content of 150–170 g/l. The 27 injection wells have depths of 200–700 m.

In the United States also increasing amounts of wastewater are dis-charged underground, for example 1.9 million m^3 of low radioactive waste-water in the year 1958 alone in a well in the National Reactor Testing Station in Idaho (Schmalz 1961).

Disposal of Sewage (Drying and Burning). With sewage disposal by burn-ing there is the chance of an adverse affect on groundwater when dangerous gases such as hydrochloric acid vapor or sulfur dioxide are generated during burning, because these can enter the ground with the precipitation.

Besides all these possibilities, the overflow of wastewater into the ground has to be mentioned. As a result of corrosion, war, earthquakes, and ground settlement damage, the sewerage system in an industrial or a municipal area can become leaky, allowing liquid to seep into the ground and mix with the groundwater there. This will result in the consumption of oxygen and even-tually the development of reducing groundwater zones with high concentra-tions of reduced iron and manganese (Motts & Saines 1969, Langmuir 1969, Leggat et al. 1972).

Disposal of Radioactive Wastewater. Safe disposal of wastewater is one of the greatest problems in the use of atomic energy (Belter 1963, De Laguna 1962, Glueckauf 1961). The problem is clearly demonstrated by the estimate that in the United States alone in the year 2000 there will be amounts in the order of 10^{18} Bq of ^{90}SR and ^{137}Cs produced monthly, and that 37 GBq of ^{90}Sr in 10^8 m^3 water can make it unusable for drinking if the potable water safety limit as laid down by the U.S. Public Health Service is applied.

The disposal process used depends on the radioactivity of the waste, its chemical quality, and the local conditions. Wastewater of low radioactivity $(< 2.7 \times 10^7$ Bq/l), which is produced in large amounts, in a few cases several thousand cubic metres per day, has very often been disposed into the ground. For this purpose underground cavities have been prepared in suitable chosen formations (e.g., clays: Sobolev et al. 1967), or the waste has been allowed to percolate through infiltration basins (Marter 1967) or through boxlike wooden reservoirs (Linderoth & Pearce 1961), or has been stored in concrete containers (Merritt & Mawson 1967). Also deep disposal wells have been used, into which the waste is injected at great depths (some hundreds of metres to over 1000 m) (Spitsin et al. 1967).

When estimating the mobility of radionuclides below the surface, the following points must be considered particularly carefully: the concentration of radionuclides, the concentrations of the stable isotopes of the radionuclides concerned, the pH of the groundwater, and the kind and concentration of other dissolved solids (including the total salt content, density, viscosity, and surface tension), as well as the hydrogeological properties and the condition of the sediment or rock (porosity, clay content, adsorption and ion exchange capacity, capillarity, permeability), and finally the temperature (cf. Section 2.2). Local mineralogical and geochemical differences, and deviations in the composition of the waste, cause variations in the mobility of the contained substances. According to Brown et al. (1956) radionuclide mobility decreases generally in the following order: nitrate > ruthenium > cesium > strontium > rare earths > plutonium.

In the Chalk River area of Canada ^{137}Cs is however retained particularly strongly in the aeolian and fluvial sand aquifer, so that according to Evans (1958) the mobility series was: ^{106}Ru > ^{90}Sr > ^{144}Ce > ^{137}Cs.

At the Savannah River atomic energy plant in Aiken, South Carolina, the waste has been discharged into large ponds or settling basins located on the sands, silts, and clays of the Tuscaloosa Formation. The clays are mostly kaolinitic. Here the most mobile radionuclide is tritium and the next ^{90}Sr, while ^{141}Ce, ^{144}Ce, ^{137}Cs, ^{103}Ru, ^{106}Ru, and Zr-^{95}Nb did not reach the deep groundwater, evidently because of sorption onto subsurface material and radioactive decay. ^{239}Pu was not observed to leave the seepage basin areas (Reichert 1962).

The underground disposal of low activity fission products is based on the fact that most radionuclides are retained below the surface and only a part move with the percolate. Through installation of an observation well network (at Hanford more than 600 wells), the spread of the radioactive waste and the changes in direction of groundwater flow and velocity caused by the introduced water can be monitored (Bierschenk 1961, Brown & Raymond

1962). The emergence of mobile dissolved constituents (nitrate, ruthenium, tritium) and the water temperature indicate the approach of the radionuclides dissolved in the groundwater (Amphlett 1958). By significant shifting of the infiltration points and limitation of the amounts of waste discharged, it can be firmly predicted that the radioactive waste discharged moves so slowly that the radioactivity falls below the maximum permissible limits before the material reaches the main river system (cf. Section 2.2).

Most of the strongly radioactive waste (with more than 10^{13} Bq/l) originates with the reprocessing of nuclear power plant fuels. Most of them have dangerous chemical properties as well as being radioactive. One of the commonest types of waste is an acid water with a high NO_3^- and Al content. Wastewater of this type is mostly stored in tanks on or near the surface or in artificial rock cavities, either as encapsulated solids as in the salt mine Asse II, the collecting point for radioactive waste in Germany (Holtzem & Schwibach 1967), or directly underground in suitable storage rocks. Thus the very low permeability of crystalline rocks at the 500 m depth at the Savannah River plant at Aiken, South Carolina, and the overlying shales in conjunction with the exchange capacities of the sediments above them, led to the expectation that in the vaults in the crystalline rocks the introduced radionuclides will be retained far beyond the 600 years required for their radioactive decay to safe levels (Proctor & Marine 1965).

The storage on land of radioactive substances requires in all cases careful hydrogeological tests, to avoid contamination of any groundwater in use or of potential use.

In the testing of reactor safety, possible contamination of the local groundwater also needs consideration. For the worst possible case arising from the reactor itself or from natural causes, complete protection of groundwater in the area is desirable. Normally radioactivity can be disregarded when the substances are removed from the reactor installation itself (e.g., when the spent nuclear fuel is taken away to a reprocessing plant for regeneration).

Consequence of Solid Waste Disposal on Groundwater Quality. Solid waste, includes ash, fly ash, garbage, building waste (including demolition rubble), slag from metal works, tips associated with coal, salt, and potash mines, and pyritic waste from slate quarries.

The three basic processes for the processing and disposal of solid waste are composting, incineration, and disposal in sanitary landfills. Composting permits biological oxidation of the organic components in refuse to a hygienic, unobjectionable, agriculturally useful compost (Strauch 1964). The materials that cannot be turned into compost, for example much chemical waste, and synthetics, plastics, metals, and glass, must be dealt with in other ways. Other disadvantages are the widespread distribution of the refuse compost and the leaching of its soluble inorganic components by rainwater.

Incineration leads chiefly to a reduction in volume. After burning, if the residues cannot be otherwise harmlessly made use of, they must be dumped. They contain leachable salts (Fresenius & Schneider 1972, Wolfskehl & Boye 1966) so that dumping requires care. Considerable pollution can arise from ash. In the groundwater below a gravel pit used for ash dumping Haupt (1935) observed increases in the residues after evaporation to 29 times that of the unaffected groundwater value (5200 mg/l vs. 179 mg/l), and of the sulfate content to 32 times (1372 mg SO_4^{2-}/l vs. 43.2 mg/l). Similar pollution originated from a slag-containing railway embankment (Lang & Bruns 1940). At high temperatures of combustion in the presence of oxygen some salts can be transformed into a state that is insoluble, or soluble only with difficulty, or they may form insoluble glasslike substances with silica (Wolfskehl & Boye 1966).

Table 32   Substances Dissolved from Solid Waste (Wasmer 1969)

Ion	First 10 months (kg · m^{-3} solid waste)	In 18 months (kg · m^{-3} solid waste)
Na$^+$	0.50	0.70
K$^+$	0.475	0.67
Ca^{2+}	0.672	0.76
Mg^{2+}	0.097	0.136
Cl$^-$	0.66	0.97
SO$_4^{2-}$	0.166	0.189
HCO$_3^-$	1.72	2.31

Sanitary landfills chiefly consist of partly treated or untreated solid waste including reject material from refuse composting plants and residues after burning. Beyond that, sludge is also deposited there (e.g., from waste treatment plants). The waste is dumped on landfills or into natural or excavated hollows in the land. Refuse contains soluble material (Table 32). The concentration of dissolved solids depends on the quality of the refuse, the method of dumping (above the water table or into the groundwater zone), the processes taking place within the waste, and the duration of the effect of the water dissolving the material (Bucksteeg 1969b, Klotter & Hantge 1969, Pierau 1967).

Table 33 gives data on the dissolved solids content in percolation water from different tips. The percolation water from site 1, the waste disposal site of a town of 110,000 inhabitants, seeps to the surface over impermeable layers. Site 2 receives the waste from a town of 180,000 inhabitants, and is sited over an impermeable base; percolate from it was probably diluted with certain amounts of laterally infiltrating groundwater.

In the percolation water from the third site, which is situated in a chalk pit, two sampling dates gave counts of 3500 and 2500 coliform bacteria per 100

Table 33 Composition of Percolation Water from Sanitary Landfills

		Site 1 (Klotter & Hantge 1969)		Site 2 (Klotter & Hantge 1969)		Domestic waste site (Davison 1969)
		range	average	range	average	
BOD_5	mg/l	105 – 236	151	70 – 88	80	90
Permanganate value	mg/l	586 –1310	850	382 – 484	470	135
pH		7.1 – 8.2		7.3 – 7.7		7.3
Spec. electrical conductance	µS/cm	7700 –9000	8350	3810 –4940	4050	–
Cl^-	mg/l	880 –1640	1340	500 – 600	550	1800
SO_4^{2-}	mg/l	369 – 558	428	1.65 – 30.0	13.7	–
NH_4^+	mg/l	270 – 310	293	138 – 190	169	1100
NO_2^-	mg/l	n.d.	n.d.	0 – 0.13	0.04	–
NO_3^-	mg/l	0 – 9	3	0 – 4.5	2.2	0
PO_4^{3-}	mg/l	1.90 – 2.56	2.30	0.48 – 1.53	0.81	–
Hydrol. phosphate	mg/l	0.32 – 1.60	1.15	0.51 – 1.58	0.91	–
Fe^{2+}	mg/l	0 – 6.5	2.2	8.0 – 88.8	41.7 $Fe^{2+(3+)}$	90
Mn^{2+}	mg/l	0 – 2.64	1.25	0 – 0.44	0.22	–
Ca^{2+}	mg/l	120 – 180	150	–	–	540
Mg^{2+}	mg/l	–	–	–	–	240
Na^+	mg/l	540 – 805	675	–	–	1540
K^+	mg/l	312 – 465	390	–	–	790
Phenolic compounds volatile with water vapor	mg/l	–	–	0.6	0.6	–
Area	hectares		9		12	6
Percolate	l/s	0.4 – 0.7	0.5 (1967–1968)	–	2	1.25
Disposal period		since 1956		since 1964		1963–1968
Amount of waste			900 000 m^3		600 000 m^3	4.10^8 kg

ml, respectively, and 160 and 25 *E. coli* per 100 ml, respectively (Davison 1969).

The results show that in the groundwater below the disposal sites considerable and long-lasting changes in quality have occurred. The topic has been investigated by Apgar & Langmuir (1971), Andersen & Dornbush (1967, 1968), Birk et al. (1973), Coe (1970), Denner (1951), Egger (1942), Emrich & Landon (1971), Exler (1972, 1979), Golwer et al. (1970, 1976), Hughes et al. (1969, 1971a, 1971b), Klotter & Hantge (1969), Klotter & Langer (1964), Kupke (1963), Lang & Bruns (1940), Langer (1963), Löhnert (1969), Matthess (1972), Mollweide (1971), Rössler (1951), Pettijohn (1972), Semmler (1958, 1960), Siebert & Werner (1969), and Zwittnig (1964). The changes appear as differences from the values for the groundwater quality upstream and downstream of the waste disposal site.

Table 34   Effect of Sanitary Landfills of Different Ages on Groundwater (After Knoll 1969)

			Effect of tip on groundwater downstream compared with upstream from tip			
Type of refuse	Age of tip (years)	Average depth of water table under tip (m)	Organic substances increase (%)	Alkaline earth[a] as Ca^{2+} (mg/l)	Sulfate SO_4^{2-} (mg/l)	Chloride Cl^- (mg/l)
Domestic and industrial refuse	15	5	+800	+614	+1500	+ 65
Domestic and industrial refuse	20	8	+350	+ 86	+ 79	+ 65
Domestic refuse	40	10	+560	+107	+ 300	+375
Domestic refuse	50	10	+500	+171	+ 310	+520
Milled and composted refuse	20	10	0	+ 21	+ 73	0

[a] Total alkaline earths calculated as Ca^{2+}.

Table 34, which gives the highest recorded test results in these series of observations, shows the chemical changes in the groundwater in areas of various old waste disposal sites in which the refuse was tipped 5–10 m above the water table. This table shows significant increases in alkaline earth and sulfate contents, but relatively low values associated with milled and composted wastes, where there was no increase in chloride and organic content. POssibly the low level of pollution is to be attributed to optimum filtration and breakdown effects in the compost fractions, in which the microbe count of 10^9–10^{10} microbes per gram exceeds that of the best culture medium (Knoll 1969).

Methods for sanitary landfills have been developed in recent years; these have proved to be quite a safe way to dispose of solid waste.

Whereas clear sludge or decayed sludge from municipal sewage treatment plants always must be considered to be infectious, on account of the presence of pathogenic bacteria and viruses as well as invading parasites, solid urban waste is to be regarded as potentially infectious only because it occasionally contains pathogenic agents (Knoll 1969).

Sulfate, the major pollutant from coal mine tips, is released by oxidation of pyrite in shaly coal and the various rocks associated with coal seams (Semmler 1958, Waterton 1969). Pulverized slag from a blast furnace has been reported as causing a rise in the calcium, iron, and manganese content in groundwater (Lang & Bruns 1940).

Blue Cross poison gas, which in 1918 had been buried or tipped into an old well in Germany, polluted the groundwater with arsenic. This was not discovered until 1958, when a woman died from arsenic poisoning after drinking water from a nearby private well. The polluted groundwater plume, which endangered a waterworks 650 m away, formed a tongue-shaped layer (60 m maximum width, 200 m long, 5 m deep) with a high arsenic content (1–10 mg As/l), which dipped downstream so that its point lay 5 m below the water table, a result of the more rapid groundwater movement in the lower and middle beds of coarse sands compared with finer grained sands in the top layer. The arsenic was released from the organic arsenic compounds in the Blue Cross gas by hydrolysis and was ultimately oxidized to As_2O_3, which slowly dissolved, forming arsenious acid (H_2AsO_3) and arsenic acid (H_3AsO_4) (Koppe & Giebler 1965).

About 30,000 m^3 of cyanide-containing waste from a chemical works, which from 1960 to 1963 was dumped into a gravel pit, polluted the groundwater in gravel deposits in the Cologne Bight. In a well 250 m away 0.6 mg/l cyanide was found in September 1962, and by November 1962 in the water of a waterworks 600 m away the figure was 0.2 mg/l. The highest value reached at the latter place was almost 0.5 mg/l. After removal of the cyanide-containing waste the cyanide concentration in the groundwater fell sharply. Detailed measurements showed that the pollution was caused by potassium hexacyanoferrate(III), from which no danger of acute toxicity originated (Effenberger 1964). The cyanide compounds, which are produced in metal plating works, steel works, blast furnaces, coke ovens, gasworks, carbide plants, and so on, have various degrees of toxicity (Knie 1966): the simple cyanides KCN and NaCN are highly toxic, but the complex salts $Na_2\{Zn(CN)_4\}$, $Na_2\{Cd(CN)_4\}$, and $Na_2\{Cu(CN)_6\}$ · $K_4\{Fe(CN)_6\}$, $K_3\{Fe(CN)_6\}$, $K\{Ag(CN)_6\}$, and $Na_2\{Hg(CN)_4\}$ are almost nontoxic, so that the actual state of the cyanide detected always must be checked.

A combination of man-made effects occurs in settlements and in large-scale animal husbandry. It has been known for a long time that well and spring waters within settlements with no sewerage systems exhibit an increase in dissolved solids, particularly chloride, sulfate, alkaline earth, permanganate value, iron, and manganese as indicators of reducing conditions, whereas geological causes can be excluded (Reichardt 1880, König, 1893, Luedecke 1899, 1901). It can be assumed that these substances originate from human and animal excreta, from the intensive use of fertilizers in domestic gardens, and from the disposal of a wide variety of waste materials.

In thin (3.5 m), shallow aquifer sediments beneath a village with no sewerage system, considerable changes in the water quality were observed (Table 35). The deterioration by the organic fraction of the pollution, shown by the raised permanganate value, resulted in a consumption of oxygen that was in-

Table 35   Changes in Water Quality Below a Village (Matthess 1958)

Pollutant	In groundwater beyond zone of pollution (mg(l)	In groundwater in village, maximum values (mg/l)
Cl^-	76	405.
SO_4^{2-}	143	650
Alkaline earths calculated as Ca^{2+}	165	523
HCO_3^-	388	837
Permanganate value	9.5	34.4

dicated particularly by the occurrence of ammonia and iron in the ground-water.

Effects Caused by Fertilizers.   The use of fertilizers raises the quantity of soluble salts in the ground; commercial fertilizers yield chloride, sulfate, ni-trate, phosphate, calcium, potassium, magnesium, ammonium, and sodium in various amounts. There are inorganic and organic compounds in stable and farmyard manure. All fertilizers may cause indirect effects by stimulating bacteria in the soil.

Vogel (1913) observed that after use of commercial fertilizers there was an increase in the sulfate and chloride content of groundwater. This was explained by the computations made by Nöring (1951b) on the basis of data obtained from the third decade of this century on the annual use of potash fertilizers in agriculture, paying exclusive attention to chloride-containing fertilizers. He found an annual addition of 1.3 g SO_4^{2-} and 4.1 g Cl^-/m^2 on ag-ricultural land of Germany and 0.8 g SO_4^{2-} and 2.6 g Cl^-/m^2 on the total sur-face of Germany. For a groundwater recharge of 10 mm this indicated for complete leaching of the sulfate and chloride from the fertilizer 130 mg SO_4^{2-}/l and 410 Cl^-/l, for 100 mm percolate 13 mg/l and 41 mg/l respec-tively. Less salt enters the groundwater because of loss by flushing away from the surface. In the cases of sulfate and nitrate an uptake by plants and microbial metabolism must be allowed for, particularly in the unsaturated soil zone (Corey & Kirkham 1965). The use of nitrogenous fertilizers, which has steadily increased during this century, has also made itself notice-able by raising the nitrate content.

In areas of intensive cultivation, such as vegetable growing and vineyards, nitrate contents of several 100–1000 mg/l have been measured in the groundwater (Harth 1969, Schneider 1964, Schwille 1969, Schmidt 1974). The amounts of fertilizer applied in wine-producing regions amounts to about 40–90 g N_2/m^2.

Raised nitrate content (up to 50 mg NO_3^-/l) was also observed by Harth (1969) for a single widespread application of balanced fertilizer in wood-lands.

Fertilizer application in the future may have a greater effect because liquid fertilizers (i.e., solutions of commercial fertilizer concentrates) and liquid manure from farmyards are being used in increasing amounts.

It should be noted that lime spreading causes a rise in the activity of soil bacteria, and a large amount of soluble sulfates can be found in the soil so treated. The reason is that the bacteria oxidize the sulfur, which previously occurred in an insoluble state (e.g., in organic compounds) to sulfate. After lime spreading Kurmies (1957) found a rapid rise in the amount of leached sulfate, and Pfaff (1937) even recorded about 100% increase in the sulfate leached from an acid soil.

The rise in alkaline-earth content in groundwater after addition of potash salts as a consequence of the liberation of Ca^{2+} ions by ion exchange has been described in Section 2.2.

Agricultural irrigation can lead to a rise in salt content and in nitrate, and to coloring of the groundwater (Eldridge 1963).

Effects Caused by Pesticides.   Pesticides are being brought into action in increasing amounts in the war against injurious organisms. These substances contain arsenic, cyanide, fluorine, silicon, and sulfuric acid compounds (copper sulfate), chiefly synthetic organic compounds of chlorine (organochlorines like DDT, HCH, dieldrin, aldrin, and heptachlor), and phosphate- or polyphosphate, ester-, or nitrogen-based, partly organometallic compounds (fungicides). The amounts of material brought into the environment in this way are considerable (Table 36). The world usage (excluding Eastern bloc countries and Korea) of mercury-based pesticides alone for agricultural purposes amounted to 52,946,000 kg in 1948–1967 (Man's Impact 1971).

Table 36   Pesticide Application (Man's Impact 1971)

	Amount (g/ha)
Japan	10,790
Europe	1,870
United States	1,490
Latin America	220
Oceania	198
India	149
Africa	127

In general these pesticides are thoroughly retained by the soil and eventually broken down (Lichtenstein et al. 1966, Schinzel 1968, Sieper 1971). Thus the mobility of insecticides in the soil is directly proportional to their solubility in water (aldrin < dieldrin < lindane) and inversely proportional to the sorptive power of the soil. The use of very large doses under unfavorable subsurface conditions may pollute the groundwater, for example lindane in a sandy soil with very high amounts of precipitation (Beran & Guth 1965).

Pesticides have been detected occasionally in groundwater (Quentin et al. 1973). For example in Basle hexachlorocyclohexane was connected with the fight against cockroaches (Stundl 1956), and the herbicide selinon was found in concentrations of 0.1–0.5 mg/l in 12 wells in Saxony as the result of use in the immediate vicinity of shallow wells (Lingelbach & Kühn 1965). In general any pollution of this kind apparently occurs only where the pesticides have moved directly into the groundwater through cracks or other passages, avoiding the protecting soil cover, or via percolating surface water and at wastewater disposal points. Thus Faust & Aly (1964) have reported pollution of groundwater by pesticides entering via waste-loaded river and lagoon waters. Bonde & Urone (1962) have noted pesticides entering via wastewater infiltration basins, and Ströhl (1966) via disposal points of sewage sludge and industrial solid waste. Pesticide pollution can also occur as the result of irrigation (Eldridge 1963).

Pesticides generally cause troublesome odor, taste, or color in the water; and in certain cases a pesticide content in potable water is toxic for humans (Schuphan 1971). Before new pesticides are introduced, therefore, their behavior in the soil and their leaching and mobility must be tested (Drescher 1971).

Pollution Caused by the Use of Salt on Roads.   The mineral content of groundwaters can also be raised by the use of salt on roads in winter (Agie 1974, Dowst 1967, Golwer 1973, Toler & Pollock 1974), and this effect is particularly noticeable in groundwater bodies with small catchment areas, which are crossed by roads heavily strewn with salt.

Migration of Pollution.   Man-made effects operate in a varied way in three dimensions. Polluting gaseous and aerosol-forming substances spread out over very large areas, which then become more or less severely affected. It appears that the main bulk of the emitted material is deposited in the immediate vicinity of pollution sources. Thus a comparison of the measured average sulfate supply with the atmospheric precipitation at the observation station at Hürth (about 100 g $\cdot$ m^{-2} $\cdot$ a^{-1}) with the estimated annual SO$_4^{2-}$ emission from the Goldenberg power station 1.5 km away (approximately 65.6×10^6 kg) shows that about 6.56 km^2 is sufficient to account for the total amount of SO$_4^{2-}$ emitted (Matthess 1961).

The area of influence of such linear sources of pollution as rivers and canals depends on the hydraulic gradient between the surface water and the

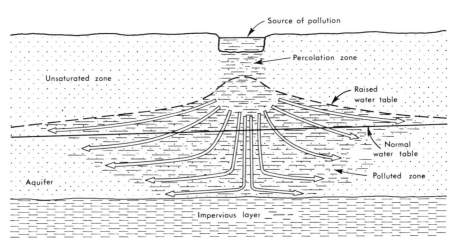

Fig. 38   Entry and disposal of polluted water in groundwater body (after LeGrand 1965 and Deutsch 1965).

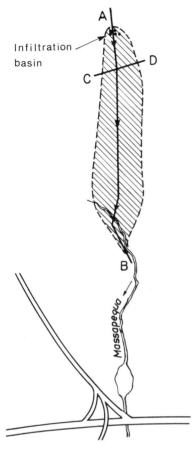

Infiltration
basin

Fig. 39   Lateral dispersal of chromium and calcium pollution in groundwater (after Perlmutter
et al. 1963).

0   500   1000 m

aquifer concerned. The affected zone is dependent on the hydraulic gradient, the hydraulic conductivity of the aquifer, the groundwater inflow from inland, and the period during which the hydraulic gradient is directed from the river toward the land (Nöring 1954). Special importance should be attached to the flooding of wide areas.

Road salting also constitutes a type of linear pollution. Most groundwater pollution arises from small source areas, which can be situated in the aquifer or above the water table (Fig. 38). The polluted water forms a more or less well-defined body, movement of which is determined by the hydrologic conditions of the aquifer (Deutsch 1965).

Figures 39–41 show the spatial dispersal in a polluted aquifer of waste containing cadmium and hexavalent chromium from a metal plating works at South Farmingdale, Nassau County, New York (Perlmutter et al. 1963).

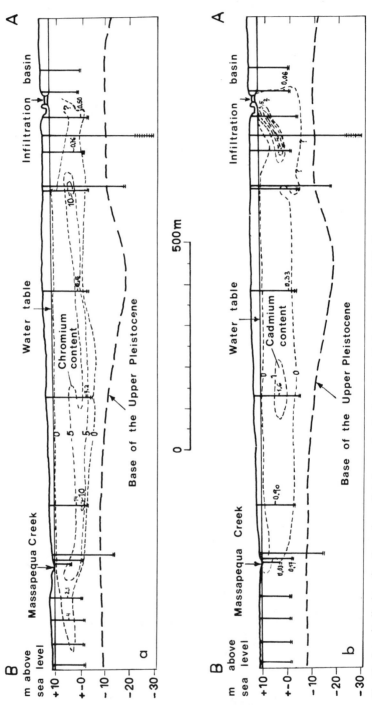

Fig. 40 Longitudinal section through a polluted groundwater body (after Perlmutter et al. 1963). Concentrations in mg/l.

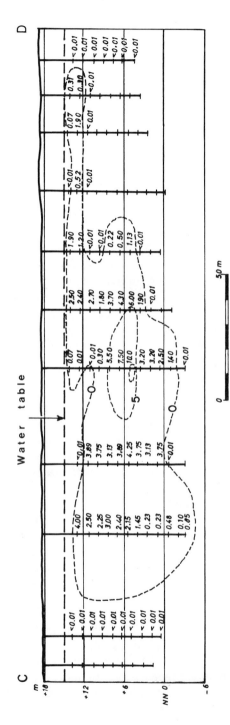

Fig. 41 Cross section across the polluted groundwater body (after Perlmutter et al. 1963). Chromium (VI) concentration in mg/l.

Since 1942 maximum values of 3.7 mg Cd/l and 14 mg Cr(VI)/l had been observed. The polluted groundwater moves through Upper Pleistocene glacial deposits 24–39 m thick, chiefly highly permeable fine-grained sands, gravels with some silt content, and clay and silt lenses. Because of their higher clay content and their poorly sorted condition, the underlying fluvial Upper Cretaceous deposits are less permeable than the Pleistocene deposits. A clayey dividing bed is absent from the profile; however the silt and clay lenses have an effect on local concentrations of cadmium and chromium ions.

Measurements on approximately 90 wells in 1962 detected a cigar-shaped pollution plume 1260 m long, with a maximum width of 300 m, between the infiltration basin and the creek. The lateral and vertical spread obviously results from dispersion and is modified by the groundwater movement (Pinder 1973) and retention of chromium and cadmium by the aquifer materials (Ku et al. 1978). The depth to water table in the area of the source of the pollution amounts to about 4.5 m in the north, the average groundwater gradient is 2.2 m/km. The polluted groundwater flows partly into Massapequa Creek and partly below its surface (Fig. 40).

Research on groundwater pollution by petroleum products (Schwille 1964, 1966, Schwille & Vorreyer 1969), and by solid wastes (Nöring et al. 1968, Farkasdi et al. 1969, Golwer et al. 1970, 1971, 1972, 1976, Golwer & Matthess 1972) shows that the materials introduced by man spread out in accordance with hydrologic principles (Deutsch 1963, 1965, LeGrand 1965) and are modified by the same natural processes to which the natural substances in the water are subject, namely, dilution, precipitation and coprecipitation, adsorption, ion exchange, gas exchange, microbial breakdown, and radioactive decay. The dilution by intermixing of pure seepage water and groundwater only has the effect of lowering the concentration below the maximum permissible concentrations for potable water, while the other processes remove pollution from the groundwater. Thus the extension of polluted groundwater zones results from temporarily changing dynamic equilibrium between polluting and self-purification processes in the ground. Their boundaries can, according to Golwer et al. (1971), be defined as the lines at which the concentrations of all pollutants have fallen below the maximum permissible concentrations for potable water, or where all water properties have taken on the normal values of the environment concerned. The dispersion of pollutants may be described by dispersion models (Fried 1975).

The extent of the pollution can be very variable, from a few tens and hundreds of metres to several kilometres; for example the picric acid pollution described by Lang & Bruns (1940) extended 4–5 km in an area with a groundwater flow velocity of about 2 m · d^{-1} in unconsolidated sediments.

If the fraction of the polluted seepage and groundwater is very small compared with the total amount of the groundwater, as in rapidly flowing karst groundwater occurrences, the change can be so small that pollution is almost undetectable. Conditions of that kind explain the insignificant changes and the absence of bacterial pollution in well waters near waste disposal points that are installed in old stone quarries in dolomite limestone. In one case in Sunderland, England refuse and mining waste from coal deposits were dumped from 1961 to 1963. Immediately at the edge of the quarry a well yields 20 l · s^{-1}; another well at a distance of 730 m yields 50 l · s^{-1}. In another case in the same locality disposal continued until the middle 1930s. A well 460 m away yields 30 l · s^{-1} (Waterton 1969).

Thews (1971) has pointed out that the balance between supply of pollutants and natural self-purification processes can be disturbed by additional pollution in such a way that undesirably·large areas are affected. This means that when reviewing any new factors that can lead to an occurrence of groundwater pollution, the state of equilibrium resulting from other existing pollution must be taken into account.

There are often risks that arise only from the combined effect of individual impacts. Therefore it is feasible to designate as potable water conservation zones all areas in which some harmful activity is taking place, so that further potential hazards can be prohibited.

Chapter 3 Groundwater

3.1 ORIGINS OF GROUNDWATER

Groundwater as freely mobile water fills pores and fractures in rock and sediments below the ground surface. It discharges at the surface after variable residence times underground, forming springs or swamps or feeding directly into surface water (rivers, seas, lakes). It is recharged by percolation of precipitation into the ground or — with favorable hydraulic gradient — by stream flow infiltrating water remains for some time in the unsaturated zone above the water table. The water in the unsaturated zone exhibits some characteristic properties.

3.1.1 Atmospheric Precipitation

Precipitation contains small amounts of dissolved and suspended substances, of which the amount and composition and the pH value ($3.0 \leq pH \leq 9.8$) vary with time and place (Carroll 1962).

Concerning the deposition of gases and the most minute suspended particles from the atmosphere it is necessary to distinguish between removal by solution in raindrops and washing out by entrainment. Solution includes the processes by which trace substances in a cloud are incorporated in its water droplets. Washing out is adsorption and collection of aerosols and gases by falling raindrops below the cloudbase (Georgii 1965).

The gaseous constituents of the atmosphere can dissolve in the precipitation. Corresponding to the various solubilities related to the prevailing air temperature there are deviations from the normal composition of air: oxygen and carbon dioxide are increased relative to nitrogen (Table 37).

The CO_2 dissolved in rainwater forms carbonic acid, which like dissolved oxygen acts as a weathering agent on the rocks at the Earth's surface.

Richards (1917) has noted that the oxygen content of atmospheric water diverges from the saturation value, and indeed more so in summer (deficit average 15%, maximum up to 25%) than in winter (average 7%).

Dissolved and Suspended Solids. The most important components in precipitation are the anions Cl^-, SO_4^{2-}, SO_3^{2-}, NO_3^-, NO_2^-, and HCO_3^-, and the

Table 37   Dissolved Gases in Rainwater (from Schoeller 1962 and Vogel 1956)

	Temperature (°C)	Content (cm³/l at 1013.24 mbar)			Fraction (vol %)		
		$N_2 + Ar$	O_2	CO_2	$N_2 + Ar$	O_2	CO_2
Rainwater	0	18.99	10.19	0.52	63.94	34.31	1.75
	10	14.97	7.87	0.36	64.53	33.92	1.55
	20	12.32	6.36	0.26	65.05	33.58	1.37
Air					79.03	20.93	0.03

cations Na^+, K^+, Ca^{2+}, Mg^{2+}, and NH_4^+; most of these are listed in Table 38. The agreement in many cases between calculated electrical conductance values and measurements shows that the known values account for the major proportion of the inorganic substances (Junge 1963).

Aerosols can also return to the ground dissolved or suspended in the precipitation but can fall out in dry form, as well. The leaching and entrainment effects depend on the type of precipitation (hoarfrost, dew, rain, heavy rain, snow) and the amount of precipitation (Table 39).

Sulfate concentration in snow is 2.9 times higher than in rain; for sodium, chloride, and ammonia the concentration is twice as high. The increased sulfate content probably can be attributed to sorption of SO_2 with consequent catalytic oxidation in addition to the uptake of aerosols (Georgii 1965).

With low rainfall the range of the deviation in the concentration of trace substances in rainfall is quite considerable. The spread decreases with increasing rainfall. High concentrations occur only with rainfall amounts

Table 38   Average Value of Dissolved Constituents in Precipitation

	Rainwater[a] (mg/l)	Precipitation,[b] Central Europe (mg/l)	Precipitation[c]	
			USSR (mg/l)	Northwestern part of USSR (mg/l)
Na^+	1.1	0.2	5.1	2.3
K^+	0.26	0.2	n.d.	n.d.
NH_4^+	n.d.	0.2	0.2	n.d.
Ca^{2+}	0.97	1.0	4.8	3.2
Mg^{2+}	0.36	0.2	1.7	1.3
Cl^-	1.1	0.5	5.5	3.6
SO_4^{2-}	4.2	3.0	9.2	4.7
NO_3^-	n.d.	0.3	1.7	n.d.
HCO_3^- (+ H_2CO_3)	1.2	n.d.	n.d.	n.d.

[a] After Wedepohl 1967.
[b] After Riehm 1961.
[c] After Chilingar 1956b.

Table 39 Mean Values of Trace Concentrations (mg/l) for Various Amounts of Precipitation (Georgii 1965)[a]

	Precipitation (mm)					
	≤ 1.0	1.1–3.0	3.1–7.0	7.1–11.0	11.1–20.0	>20.0
NH_4^+	2.9	1.9	1.6	2.1	1.3	0.7
NO_3^-	3.5	2.6	1.9	2.8	1.4	0.7
SO_4^{2-}	9.9	7.4	4.0	5.4	4.4	2.8
Cl^-	2.5	2.1	1.4	2.0	1.6	1.1
Number of cases	4	12	17	13	8	8

[a]Data from Taunus Observatory, West Germany.

below 2 mm. It is possible that the large deviations for low rainfall depend mainly on washing out and evaporation effects, whereas with increasing amount of rain the removal by solution in raindrops becomes more and more significant (Georgii 1965). Aerosols include soil particles that are blown into the atmosphere by wind, particularly in arid regions. They consist of tiny particles (< 1 μm radius) of various origins, rock dust from the land surface, salt particles from dried sea spray from the surface of the ocean, particles of very varied sizes from volcanic eruptions, or solid matter of man-made origin (smoke, soot from combustion). In addition, particles form by reaction between gases in the atmosphere: there are many possibilities involving reactions between SO_2, H_2S, NH_3, or hydrocarbons with oxygen or ozone.

In arid climates HCl can form when SO_2 in the atmosphere is oxidized to SO_3, which displaces the chloride in the NaCl particles in aerosols while forming sulfates (Eriksson 1958).

The fractional composition of aerosols varies widely from one region to another, and therefore the content in rainfall is similarly variable.

Terrigenous dust material is widespread above large landmasses and over desert regions, where sandstorms carry small particles into the atmosphere. Thus it appears that in the USSR 75–80% of all salts in the precipitation are of continental origin and that salts of marine origin make up less than 25% (Chilingar 1956a). This statement is supported by the observation that the ratio of the average SO_4^{2-} content to total salt content in atmospheric precipitation compared with the same ratio in the salt composition of the ocean water is twice as high as the respective ratios of chloride contents (Chilingar 1956b). Salt particles remain behind in the atmosphere after the evaporation of sea spray. Because these substances participate in the hydrologic cycle and eventually return to the sea, they are known as cyclic salts. Their composition over the sea or near coasts diverges very little, but it differs significantly from that of the salt content of seawater; the ratios Cl^-/Na^+, K^+/Na^+, Mg^{2+}/Na^+, and Ca^{2+}/Na^+ are higher. Wilson (1960) and Junge (1963) suggest that this is caused by an ion separation process occurring in films of organic matter on the surface of the sea; furthermore Junge (1963) noted the effect of sulfuric acid on salt particles when HCl escapes into the atmosphere. In addition, NaCl is removed from the atmosphere faster than KCl because with the higher sodium concentration in the spume blown off the sea, NaCl crystals form earlier and are larger than KCl crystals (Larson & Hettick 1956, Emanuelson et al. 1954).

With increasing distance from the sea there is a greater proportion of terrigenous material (K^+, Ca^{2+}, Mg^{2+}). The dissolved solids content of the precipitation varies rapidly from the coast landward and is dependent on wind direction and speed. This is apparent from studies made by Bätjer & Kuntze (1963) at the observation stations in East Friesia and Oldenburg. The highest sodium value occurred with onshore northwest winds, the lowest with offshore southerly winds. There was also a significant increase in the sodium content with increasing wind velocity (Tables 40 and 41). The magnesium content obviously originated from the sea, while calcium and potassium, and to a greater extent sulfate, and ammonia allow no such undoubted derivation of the salt content from the sea. Sulfate and ammonia are of predominantly terrestrial origin.

Table 40 Cation and Anion Concentration in Precipitation Water (mg/l) (after Bätjer & Kuntze 1963): Average values, November 1960–November 1962

Distance from coast (km):	0.1	9	18	28	75
Ion	Norderney	Grossheide	Abelitzmoor	Friedeburg	Friesoythe
Na^+	18.4	5.5	4.3	4.4	3.9
K^+	2.2	1.8	1.7	1.6	n.d.
NH_4^+	2.3	n.d.	n.d.	2.2	n.d.
Ca^{2+}	2.2	1.6	1.2	1.1	1.2
Mg^{2+}	2.2	0.4	0.4	0.4	0.2
Cl^-	41.4	8.2	5.6	5.4	n.d.
NO_3^-	2.0	n.d.	n.d.	1.6	n.d.
S (calculated as SO_4^{2-})	20.4	n.d.	n.d.	10.8	n.d.

Table 41 Dependence of Sodium Content (mg Na/l) on Northwesterly Wind Speeds (after Bätjer & Kuntze 1963)

	Wind speed (m · s⁻¹)		
	0–2.4	2.5–7.7	> 7.7
Norderney	53.2	70.9	102.3
Grossheide	6.1	12.5	22.6
Friedeburg	4.9	6.3	10.2
Friesoythe	5.2	7.0	9.6

Table 42   Dissolved Solids Concentrations During Storm Surge in Comparison with Average Values for Northwesterly Winds (After Bätjer & Kuntze 1963)

		Concentration during storm surge (mg/l)	Average concentration for northwesterly winds (mg/l)	Ratio of two concentrations
Na$^+$	Norderney	> 1480	74.9	> 20
	Grossheide	296.0	18.1	16
	Friedeburg	170.7	8.8	19
	Friesoythe	79.6	8.8	9
Mg^{2+}	Norderney	> 377	5.2	> 72
	Grossheide	27.4	0.7	39
	Friedeburg	11.8	0.5	24
	Friesoythe	6.8	0.4	17
Ca^{2+}	Norderney	135.8	9.1	15
	Grossheide	17.2	3.6	5
	Friedeburg	15.0	2.1	7
	Friesoythe	7.1	1.9	4

Distinct increases in the dissolved solids content are also related to storm centers, as shown by the values consequent on the storm surge of February 16 and 17, 1962, on the East Friesian coast (Table 42). The Na$^+$ and Mg^{2+} contents show mostly an increase of about 15 times and more, in contrast to the normal values for northwest winds, while the rise in Ca^{2+} values is lower except for Norderney, possibly in consequence of differential behavior of the salt particles in the atmosphere.

The most significant factor is the decrease of Na$^+$ and Cl$^-$ with increasing distance from the sea, which is the source of these typical cyclic salts (Table 43). This can also be seen in the distribution of the average Cl$^-$ content in the precipitation (Fig. 42).

Table 43   Na$^+$ and Cl$^-$ Contents of Rainwater as Function of Distance from Sea (After Riehm 1961)

	Distance (km)	Na$^+$ (mg/l)	Cl$^-$ (mg/l)
Westerland	0.2	22	49
Schleswig	50	2.5	4.8
Braunschweig	450	0.8	2.2
Augustenberg	800	0.4	0.9
Hohenpeissenberg	950	0.2	0.6
Retz	1250	0.2	0.5

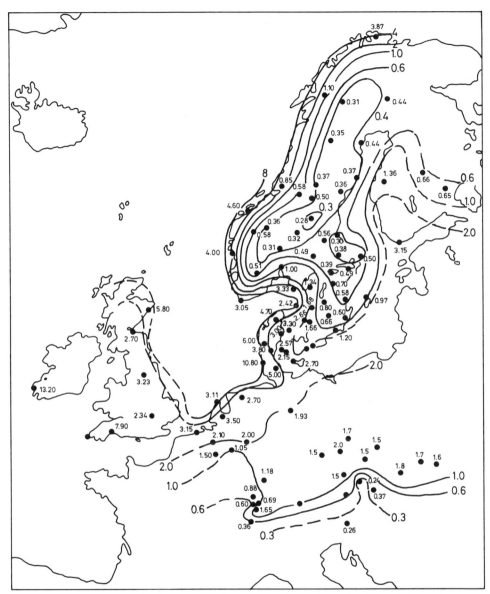

Fig. 42 Average Cl⁻ concentration (mg/l) in rainwater in Europe for the period 1957–1958 (after Junge 1963).

The salts derived from sea spray, which at the coast on stormy days can cause Cl⁻ contents of several thousand milligrams per litre, are more and more washed out of the atmosphere with increasing distance from the coast. Added to this there is a mixing of the more or less elutriated air masses com-

Table 44   Relationship Between Distance from Sea and Plateau Value of CL⁻ Content in Precipitation (After Junge 1963)

	Distance from sea (km)	Plateau value (mg/l)
England	30	7
Holland	50	3.5
Australia	150	1
Sweden	300	0.4
United States	600	0.15

ing from the sea with continental air. The decrease in Cl⁻ concentrations with increasing distance from the coast finally ends at a limiting value. The plateau values are different for various regions, with morphological features and the distribution of land and sea as controlling factors (Table 44).

Aerosols from Gas Reactions.   The oxidation of SO_2 from volcanic eruptions and from man-made combustion gases (Section 2.6) takes place only slowly by reaction with atmospheric oxygen and ozone according to Coste & Wright (1935). Sunlight, especially its ultraviolet content, mist, and catalyzers such as heavy metal oxides, platinum, selenium, and glass so greatly accelerate this reaction (Gmelin, 1953, Liesegang 1933) that in the precipitate sulfuric acid is detectable, or, in the presence of bases, sulfate and less often sulfurous acid (Liesegang 1927). In particular also the hydrogen sulfide entering the atmosphere from tidal zones is extensively oxidized there and is seldom detectable in the precipitation.

The distribution of average sulfate concentration over Europe and Asia (Fig. 43) reveals the connection with the sea (sulfate as a cyclic salt), overlapped by sulfate from other sources (man-made SO_2 output, sulfate from aerosol and sulfate from oxidiation of H_2S from coastal zones).

Ammonia as a volcanic gas, as a cyclic salt, as a terrestrial aerosol component, and as a product of combustion and of gaseous emanations (degradation product of biological substances or in increasing amounts from liquid ammonia fertilizers) is oxidized to nitrate photochemically. This oxidation process, which occurs in the presence of water droplets, provides the bulk of the nitrate found in precipitation (Junge 1958). The oxides of nitrogen occurring as intermediate products (N_2O and NO are detected, but NO_2 is commonest) can also come from nitrogen in the soil (Junge 1958) or from industrial flue gases, or it can be formed from nitrogen by electrical discharges. The importance of lightning has been described by Reiter (1964); its effect produces, according to the isotopic composition of the nitrate, only a small part (Hoering 1957, Gambell & Fisher 1964) of the measured concentration in rainwater—at the most 10–20% according to Junge (1958).

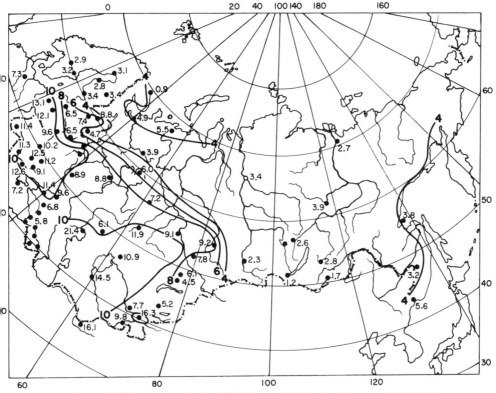

Fig. 43 Average SO_4^{2-} concentration (mg/l) in the precipitation in Europe and Asia (Eriksson 1970, after Selesneva 1966).

Aerosols also contain organic protein ammonia (albuminous ammonia) in its organic fraction, which can come from the soil, from seawater, or from human activity. On average there is 0.4–2.6 mg/l (Junge 1963), in an individual case (Mount Vernon) up to 28 mg/l (Rankama & Sahama 1960), and it can be liberated by oxidation.

Nitrogen dioxide in rain can be transformed into nitric acid and nitrous acid:

$$2NO_2 + H_2O \rightarrow HNO_3 + HNO_2 \tag{96}$$

The fact that compared with NO_3^-, NO_2^- is found only in minor amounts in the aerosols and in atmospheric waters shows that the oxidation is quickly completed in clouds and rainwater. Another form of aerosol can originate by the formation of $(NH_4)_2SO_4$ particles from the reaction of gaseous ammonia with photochemically produced sulfuric acid in the presence of water droplets; in any case, however, NH_4^+ and SO_4^{2-} conforming to stoichiometric proportions are found in rainwater monitoring stations on mountain tops (e.g., on the Zugspitze, West Germany) almost unaffected by man-made influences (Georgii 1963, 1965).

Some other substances besides man-made pollutants as described in Section 2.2, under "Dissolution and Precipitation During Mixing Processes," can be found in rainwater (Table 45).

Table 45   Trace Substances in Rainwater

Substance	Concentration	Source[a]
P_2O_5	0.1–1.0 mg/l	(1)
I^-	0.001–0.01 mg/l (average 0.0035 mg/l; Holland)	(1)
	0.0014–0.0146 mg/l (New Zealand)	(4)
H_2O_2	0.08–0.86 mg/l (Japan)	(1)
Formaldehyde	0.1–1.0 mg/l (average 0.5 mg/l)	(1)
	0.15–1.2 mg/l	(3)
Organically bound carbon	1.7–3.4 mg/l (average 2.5 mg/l; rain, Sweden)	(1)
	0.8–1.9 mg/l (average 1.3 mg/l; snow, Sweden)	(1)
^{226}Ra	<0.37–55.5 mBq/l	(2)

[a] After Junge 1963 (1), Banjeri & Chatterjee 1966 (2), Dhar & Ram 1933 (3), and Dean 1963 (4).

The substances carried down with the precipitation play an important part in plant nutrition, for example nitrogen, sulfur, and phosphorus (Buchner 1958, Riehm 1961, Ottermann & Krzysch 1965). Now and then the dissolved solids content of surface waters are without doubt derived from the materials entrained in the precipitation (Gorham 1961); for example the average chloride content in hoarfrost (3.6 mg/l), rain (3.4 mg/l), and snow (2.1 mg/l) at the Haldde Observatory in northern Norway sufficed to explain the Cl^- content of the lake and rainwater there (1.1-2.7 mg/l) (Köhler 1937). The bulk of the Cl^- content of the surface drainage in Israel appears to be derived from the chloride content of the precipitation (Loewengart 1958, 1961).

3.1.2   Percolation Water

Percolation water is the gravitational water that as a rule has reached the Earth's surface as precipitation; after a longer or shorter stay in the unsaturated subsurface zone in the form of drops or films of water it percolates to the water table and becomes groundwater. The unsaturated zone includes the soil and, if occasion arises, the surface layers lying under it down to the water table. In this zone solid material coexists with water in the form of adhesion, capillary, and percolating water, and ground air. The precipitation water has its chemical properties changed if it comes into contact with the soil surface and the vegetation and picks up soluble and suspended substances, which it carries below the surface during percolation. Soluble salts are particularly prominent at the surface in arid and semiarid regions. The processes taking place in the unsaturated zone, described in Chapter 2, critically determine the chemical properties of the percolation water. The duration of stay of the percolating water in the unsaturated zone, the composition of the ground air, and the physical and chemical properties of the soil and the near-surface strata down to the water table all play important roles.

Residence Time in the Unsaturated Zone. In fine-grained soils and other subsurface materials such as fine sand, silty sediments, and the products of rock weathering (e.g., loess and loam), the large surface area of contact and the long contact period may make possible the establishment of chemical equilibrium between water, ground air, and solid constituents. With hard-jointed and karst rocks the contact area and time are significantly lower, likewise for rapid water movement through animal burrows and cracks in the soil.

The percolate behaves as a solvent and carries dissolved or suspended solids into the groundwater. The movement of water in the unsaturated zone above the water table is controlled by gravity and capillary forces. The viscosity of water and the friction at the water-solid phase boundary counteract these forces. In the unsaturated zone percolation occurs within the intergranular spaces only when the soil moisture exceeds a certain limiting value, the field capacity. The field capacity in coarse underground materials is significantly smaller than in finer materials. The percolation velocity in rocks with low field capacity is therefore high (order of magnitude up to several metres or tens of metres per day), in skeletal and sandy soils for example, as well as in clayey formations that contain deep shrinkage cracks because of desiccation. Percolation velocities of 2.7 and 2.9 m $\cdot$ d^{-1} were for example observed in sandy weathered zones in the Bunter Sandstone of the Saarland (Heitele 1968).

In subsurface materials with a high field capacity the vadose water can move downward only very slowly, as observations with tritium-labeled percolation water have shown (Zimmermann et al. 1966, 1967a, 1967b, Matthess et al. 1979). With a velocity of 1 m/year, which according to the measurements just noted can be taken as a guide for sandy-loamy materials in humid climates, it is years before the water table is reached, and this is dependent on the thickness of the unsaturated zone. The percolate develops age stratification. Recently percolated precipitation water pushes the lower, older, water layers downward, so that the percolate that enters the water table at about 2–3 m depth a few hours or days after a rain event, is "old" water from the water layer directly above the water table. Besides the dominant intergranular percolation water movement, certain amounts of local water can move rapidly through cleavage and fracture systems, as can be demonstrated in a water profile in the English Chalk (Smith et al. 1970). This observation is important for low permeability clay profiles in which shrinkage cracks occur as a result of summer droughts.

Dissolution of the Ground Air. The ground air has a different composition from the atmosphere. The important difference compared with the atmosphere is the reduction in oxygen content and an increase in carbon dioxide. The nitrogen content as the third constituent is only slightly different from the amount in the atmosphere.

The average CO_2 content in the free atmosphere is 0.03 vol %. It varies over rural environments between 0.021 and 0.044 vol % and can be higher locally in the vicinity of industry, towns, hot springs, and volcanic emanations (Rankama & Sahama 1960). Because of the effect of acids such as nitric or sulfuric on carbonates, as well as biogenic CO_2 in soils as described in Section 2.5, the actual CO_2 content in the ground air is generally 10–100 times higher. Values between 0.2 and 5 vol % are common (von Pettenkofer 1871, 1873, Schloesing 1873, 1889, Smolensky 1877, Von Fodor 1875, Audoynaud & Chauzit 1880, Boussingault & Lewy 1853, Boynton & Reuther 1939, Fleck 1876, Russel & Appleyard 1915, Appleman 1927, Lundegårdh 1924). Occasionally even higher concentrations have been observed: up to 11.4 vol % (Schloesing 1889), 13 vol % (Boynton & Reuther 1939), about 20 vol % directly above the water table (Kristensen & Enoch 1964). Higher values occur in the vicinity of natural escapes of carbon dioxide, such as volcanic emanations, carbonation gases, and in areas of severe man-made subsurface pollution, for example at refuse tips, where values of up to 26 vol % have been recorded (Golwer et al. 1970, Bishop et al. 1966).

The average oxygen content of the free atmosphere is 20.95 vol %. The oxygen content in the ground air is generally below this amount, often by only an insignificant amount, and in general the values are between 8.9 and 21% (Schloesing 1873, 1889, Audoynaud & Chauzit 1880, Boussingault & Lewy 1853, Fleck 1876, Russel & Appleyard 1915, Boynton & Reuther 1939); in zones of very high CO_2 content it can however decrease to nil (Kristensen & Enoch 1964, Golwer et al. 1970). The five soil air analyses recorded by Russel & Appleyard (1915) showed high oxygen contents (21.01–21.71 vol %) and are explained by them as caused by the addition of oxygen dissolved in rainwater. Because oxygen is consumed during biological processes the oxygen content generally behaves inversely toward carbon dioxide and shows annual and daily variations tending in contrary directions (Boynton & Reuther 1939, Russel & Appleyard 1915), and a decrease with depth, that is, with increasing distance from the large reservoir of oxygen in the atmosphere. Oxygen is also used in the formation of sulfate and nitrate in the soil as well as in CO_2 production.

Besides nitrogen, oxygen, and carbon dioxide, other gases occur in the ground air. Von Fodor (1875) has detected gaseous ammonia at 2 m depth in the range 0.048–0.082 mg in 100 l of soil air (1013.24 mbar, 0°C), and this small amount is attributed to sorption on soil particles. N_2O is also found in the ground air with a concentration dependent on depth. It appears to have no seasonal changes in quantity. In pararendzina soil at 45 cm depth 1.5–2.0 ppm N_2O was found as the average for the year, and at 75 and 90 cm depths 2.0–2.2 ppm. For the three soils investigated (two pararendzinas, one deeply cultivated soil in a vineyard), there was an estimated average N_2O yield of 10^{-8} g · m^{-2} · s^{-1}, which agreed with the value of global N_2O production (Albrecht et al. 1970). Hydrogen sulfide is usually absent and at most occurs in quite small amounts as a product of putrefaction. Methane is likewise found as a putrefaction product (Von Fodor 1875). The ground air in deeper layers below the surface is saturated with water vapor at all times, and as a rule tends to be so in the upper rock layers and in the soil (Giesecke 1930).

Solution of the ground air components in the percolate is governed by Henry's law (cf. Section 1.2). With the significantly higher CO_2 partial pressure compared with the atmosphere, the percolate can have a higher CO_2 content: the most frequent CO_2 values between 0.2 and 5 vol % correspond to 10°C equilibrium concentrations of 4.6–115.9 mg/l, the CO_2 value of 25 vol % to 579 mg of CO_2 per litre. The CO_2 can dissolve carbonates in accordance with the theoretical principles described in Section 1.2.3.1.

The solution of oxygen determines the redox conditions (Section 1.2.3.4). According to Russel & Appleyard (1915) the gases bound in the soil moisture contain less oxygen (0–15.1%) and more CO_2 (6.3–99.3%) and nitrogen (0.7–84.8%) than the ground air.

The composition of the ground air is determined by the biological oxygen consumption, by production of CO_2, by solution and sorption processes in the subsurface media, and by gas exchange with the free atmosphere. Ground air and atmosphere compositions differ from each other more, the more difficult is the exchange of gases between the two.

Dissolution of Soil Material. Soils develop as a result of physical, chemical, and biological transformation of rock constituents. In addition to ground air and water soils contain living organisms and undecomposed organic substances.

Other things being equal, the type of soil that evolves depends chiefly on temperature and the amount of precipitation from the atmosphere. The most noticeable feature is the evolution of soil horizons that when fully developed represent a state of equilibrium between soil-forming forces and erosion at the locality concerned. The topmost soil levels are designated as topsoil or A-horizons as found in humid regions where there is a regular downward movement of percolate; the layers below this are called B-horizons and are often characterized by enrichment in various substances. The horizons below these comprise the weathered rock zone (C_v-horizon), and the fresh unweathered parent rock (C_n-horizon). With predominantly upward movement of water (as occurs in arid regions because of evapotranspiration) the horizon of mineral enrichment can lie on the surface.

The climatic zones can lead to the evolution of certain soil types (e.g., tundra soils, podzols and pseudogley, brown forest soils, red Mediterranean soils, caliche, chernozems, castanozems, and desert soils). Typical podzol and brown forest soil localities show a well-defined downward leaching and precipitation process related to the temperature and the precipitation, whereas with the other soils named above there is minimal and often seasonal upward movement of material depending on the precipitation conditions.

Other soil types, which depend on local peculiarities, are known; saline soils (e.g., in regions of saline shallow groundwater), gleys, and peat soils in regions of fresh groundwater near the surface, and rendzina soils (limerich) over limestones. Finally there are poorly developed soils, for example sierozems on dunes and immature pasture soils.

Soil contains appreciable amounts of living matter, such as plant roots, earth-dwelling animals (nematode, myriapods, insects, worms, and small mammals) and plants (ultimately the microflora) (edaphon). The higher organisms disturb the soil, which encourages gas exchange between the soil and the atmosphere, and favor infiltration of precipitation. The importance of microorganisms in the transformation of organic substances and in sulfurication and nitrification has been described in Section 2.5. Microorganisms metabolize the material of dead organisms into humus. The same happens with microbial slimes, polysaccharides, polyuronides, and other organic substances. Carbon dioxide, nitrous acid, nitric acid, and organic acids are further metabolic products, which in their turn attack inorganic materials below the surface.

Humus is a mixture of fulvic acids, humic acids, and humins (Scheffer & Schachtschabel 1976). Fulvic acids comprise numerous heterogeneous compounds that differ from humic acids by having a lower molecular weight and a higher content of carboxyl groups. They are, like their salts, soluble in water in the nonadsorbed form; they possess partly reducing properties and form complexes with metals (e.g., iron). The fulvic acids in the soil are mostly adsorbed onto iron and aluminum oxides, clay minerals, and organic compounds with higher molecular weight.

Humic acids are mostly highly polymerized, three-dimensional netted sphaerocolloids. Their acid nature and their potential for cation exchange depend chiefly on their COOH and phenolic OH groups. Their compounds with metals are often described as humates. Alkali and ammonium humates are easily soluble in water and then form colloidally dispersed solutions; the humates of higher valence cations (Ca, Mg, Fe, Al, etc.) and humic acids themselves are soluble in water with difficulty.

Humins are chemically heterogeneous, partly relatively little changed organic substances, partly humidified substances, which are bound to clay minerals or are included in other minerals.

The quantity of organic material in the soil depends on the ratio between its decomposition and the addition of new substances. High temperatures and humidity, among other factors, favor decomposition. The cultivation of soil is a further influence on its composition.

The soil moisture, containing the dissolved constituents, either transports these to the groundwater (leaching) or with upward water movement, as a result of prolonged evaporation from the surface, deposits them close to the surface. In humid climates and during wet seasons in other climates the downward water movement and leaching is predominant, but in arid regions there is a general upward movement and enrichment of salts by precipitation in the soil. There is a generally nonuniform movement trend of individual materials. Elements that are more easily released from plants than from water, for example silicon, can be enriched on the surface; other elements that are more easily removed from water than from plants, such as iron and aluminum, may precipitate in lower horizons. These processes are of importance in the evolution of the different soil types.

The extent of leaching of a substance depends on the climate and soil-dependent amounts of the percolate, on the quantity of the substance in the soil, on the solubility of its salts, and the bonding strength to the exchanger, and for organic substances on the resistance to microbial breakdown. Among anions, chloride, nitrate, and sulfate are easily leached; molybdate and phosphate ions less easily. Cations, which occur as water-soluble salts, are easily leached. With exchangeable cations the leaching depends chiefly on their quantity and on the bonding strength. However, because Ca^{2+} ions generally make up the largest part of the exchangeable cations in the soil, the Ca^{2+} fraction in the percolate is the highest in spite of higher bonding strength (cf. Section 2.2). The high sodium leaching in spite of the absence of sodium saturation is the result of the low bonding strength of the Na^+ ions. Potassium and ammonia are protected from leaching through bonding to clay minerals, likewise other substances that are incorporated into living or dead materials. This holds for nitrogen especially (Scheffer & Schachtschabel 1976), and to a lesser extent for sulfur as well.

By far the greatest part of the sulfur in soil is bound organically in amino acids (cysteine, methionine, cystine) that are derived from plant and microbe proteins, besides occurring in sulfuric esters, thiourea, glycoside, and alkaloids of biological origin. In the inorganic fraction, which makes up only 5–20%, typically less than 10%, the chief constituents are sulfates, sulfides, elemental sulfur, thiosulfate, and tetrathionate (Alexander 1961, Brümmer et al. 1971). The soil microbe activity—besides bacteria there are also mycorhiza fungi living in symbiosis with forest trees (Stremme 1950)—influences the leachable sulfur in such a way that the microorganisms use the organically bound sulfur partly for cell building and transform the remainder, eventually converting sulfides under aerobic conditions to sulfate ions so that in certain circumstances free sulfuric acid occurs (Heimath 1933, Kappen & Zapfe 1917) and the pH falls below 1 (Gmelin 1953, Porter 1950, Rankama & Sahama 1960). Under anaerobic conditions the microbes transform the sulfur compounds into H_2S, sulfides, and mercaptans.

Leaching of iron, manganese, copper, zinc, and cobalt is low in alkaline and weakly acid horizons because under these conditions they form only slightly soluble oxides, hydroxides, and phosphates. With decreasing pH the solubility rises, and with it the degree of leaching. Iron and manganese (Table 46) are preferentially removed under reducing conditions as divalent ions (Scheffer & Schachtschabel 1976).

The quality of the percolate can be described by analysis of soil solutions, lysimeter tests, and analysis of water of land drainage pipes.

Soil Solutions. The composition of soil solutions (Table 47) is clearly different from that of the precipitation. More or less high NO_3^- content is characteristic of soil solutions as a result of intensive biochemical nitrification

Table 46 Annual Leaching of Trace Substances from 61 German Soils (After Scheffer & Schachtschabel 1976)[a]

	Manganese	Copper	Zinc	Boron
Leachate (g/ha)	10–800	10–94	10–360	20–1040
Leachate (g/ha) (average)	250	30	100	250
Percolate (mg/l)	0.005–0.40	0.005–0.047	0.005–0.18	0.01–0.52

[a] On a basis of 200 litres of percolate per square metre.

Table 47 Compositions of Soil Solutions (mg/l) (After Schloesing 1870, and Schoeller 1962)

Ion	Samples[a]									
	1.1	1.2	2	3	4	5	6.1	6.2	7.1	7.2
Na^+	5.8	10.6	20.0	19.6	17.9	22.1	16.2	19.5	15.9	26.7
K^+	5.7	130	2.3	9.9	–	4.0	4.2	2.2	4.2	2.7
NH_4^+	–	–	–	–	0.32	0.0	0.0	0.85	0.49	–
Ca^{2+}	188.5	162	220	155	146	215	172	253	94.2	222
Mg^{2+}	8.1	12.1	12.5	11.0	9.0	11.4	10.7	13.1	9.9	20.1
Cl^-	7.4	6.7	5.6	5.1	12.2	35.2	12.6	26.0	17.3	31.7
NO_3^-	295	321	55	14.5	130.6	351	96.6	448	80.4	537
SO_4^{2-}	56.7	73	48.9	13.6	11.3	38.2	27.2	47.7	21.7	38.9
PO_4^{3-}	0.54	1.9	–	–	–	0.34	1.0	0.47	0.54	0.13
SiO_2	29.1	32.0	26.0	52.9	48.4	22.6	31.0	29.3	31.6	23.3

[a] 1.1, Tobacco field unfertilized for 10 years, Boulogne (Seine); 1.2, same tobacco field fertilized with KNO_3, ashes, and compost; 2, cereal field at Issy (Seine), 1869 harvest; 3, field at Neauphle-le-Château; 4, cereal field at Neauphle-le-Château (1869 harvest); 5, oat field at Neauphle-le-Château (1869 harvest); 6.1, 6.2, cereal field at Neauphle-le-Château (1869 harvest); 7.1, 7.2, field at Neauphle-le-Château.

and fertilizer application. Together with the carbon dioxide from the ground air (CO_2 content measured was 0.49–2.55 vol %) the nitric acid attacks the parent rock-forming minerals and Ca^{2+}, Mg^{2+}, K^+, and Na^+ are liberated in various ratios depending on the mineral chemistry. K^+ is chiefly bound to soil colloids and clay minerals and is therefore often present in small quantities in the soil solution analysis (Matthess et al. 1977). The sulfate content is additionally enriched by biochemical processes in the soil.

Low concentrations of NH_4^+ that originate in the soil by biological processes in the nitrogen cycle, as well as low concentrations of phosphate ion, were observed, the latter as a result of the low solubility of phosphate and its sorption on clay minerals or hydrous metal oxides in the soil. Finally the SiO_2 content is noteworthy; is produced by the weathering of silicates (Pekdeger 1977, Matthess & Pekdeger 1980).

The soil solutions within individual horizons possess different compositions, which M. Schoeller (in Schoeller 1962) has proved in aqueous extracts from various podzols (at Haillan, Martinac, and Caupian), and from brown forest soils (Baron and Pompignac) in the Bordeaux region as shown in Tables 48–50.

Table 48   Soil Solutions (mg/l) in Podzols (After Schoeller 1962)

Horizons:	Caupian, December 1955				Haillan, February 1954					
Ion	A_1	A_2	B	C	A_0	A_1	B_1	B_2	C_vC_n	Groundwater
Na^+	143	489	407	269	6.9	11.3	7.1	9.2	8.3	29
K^+	238	0	0	0	4.7	8.2	1.9	42	3.5	5.9
Ca^{2+}	130	24	0	0	0	0	24.4	20.6	9.8	41.2
Mg^{2+}	48	47	49	59	0	0	6.4	0	8.8	5.3
Cl^-	218	344	360	239	27	20	18	19	25	45
HCO_3^-	146	146	152	207	0	0	0	21	21.5	195
NO_3^-	0	53	56	0	–	–	–	–	–	–
SO_4^{2-}	0	0	0	0	30	111	108	100	44	34

Table 49   Equivalent Ratios and Index of Base Exchange (I_{BA}) in Soil Solutions in Podzols (After Schoeller 1962)

Horizon:	Caupian, December 1955				Haillan, February 1954					
	A_1	A_2	B	C	A_0	A_1	B_1	B_2	C_vC_n	Groundwater
SO_4^{2-}/Cl^-	0	0	0	0	0.832	4.053	4.328	4.053	1.300	1.180
Mg^{2+}/Ca^{2+}	0.622	325	0	0	0	0	0.435	0	1.48	0.174
K^+/Na^+	0.984	0	0	0	0.288	0.323	0.161	0.268	0.261	0.119
I_{BA}[a]	−2.73	−4.05	−0.89	−3.45	+0.315	−0.035	+0.306	−0.392	+0.344	−0.209

[a] See Section 2.2

In podzols the silica-rich parent material, which contains only a few soluble substances, and the strong leaching effect of percolating water, cause only small differences in the chemical composition of soil solutions in the various soil horizons. This applies particularly to the Cl^- content. The low calcium and magnesium content increase with depth. During the evapotranspiration periods the concentration gradient can reverse from time to time, so that short-term enrichment of material comes about. The ratio of SO_4^{2-} to Cl^- is also subject to horizontally controlled variations during the same periods. The ratio of K^+ to Na^+ is affected by base exchange processes (Matthess et al. 1977). The concentration of hydrogen carbonate generally increases with depth, the more so immediately after a period of leaching. After long rainy periods however the hydrogen carbonate concentrations in the A-horizons are raised because of increased biogenic CO_2 (Schoeller 1963).

Table 50   Soil Solutions (mg/l) in Brown Forest Soils,[a] Baron, May 1956 (Schoeller 1962)

	Horizons								
Ion	A_1	B_1	B_2	C	A	B_1	B_2	C_v	Groundwater
Na^+	21.6	31.8	22.3	21.9	15	35	30	6	7
K^+	14.5	8.5	4.9	8.1	10	9	7	2	1
Ca^{2+}	167.8	124.6	144.8	156	119	137	196.8	40.6	60
Mg^{2+}	35.9	15.2	15.29	14	25	17	21	64	5
Cl^-	64.5	70.5	71.2	78	45	77	97	20	23
HCO_3^-	1000	610	564	156	710	669	767	40	380
NO_3^-	7	47.5	96	188	51	522	131	50	0.9
SO_4^{2-}	76.3	56.3	104	−	54	61	142	−	11
SiO_2	109.2	31.5	31.8	57.9	77.4	34.8	43.2	15	9.6

[a] "Brown forest soils" defined in a broad sense.

The chemical quality of the soil solutions (Tables 50 and 51) for the brown forest soils investigated showed more distinct differences in the various horizons than the podzols, undoubtedly because of the calcareous parent rock and the clay content of the brown forest soils, which bring about solution of mineral substances and reactions with the subsurface materials generally. (Throughout this discussion the term "brown forest soils" is used in a broad sense.) The slow percolation in the clayey brown forest soils makes for a prolonged contact time between soil water and the subsurface materials, so that more mineral matter can be dissolved. Thus equilibrium between water and the subsurface can be established. This is why the different chemical compositions of the soil solutions in the various horizons for a soil and in different soils are much more distinct in brown forest soils than in podzols (Schoeller 1963).

Table 51 Equivalent Ratios and Index of Base Exchange (I_{BA}) in Soil Solutions in Brown Forest Soils,[a] Baron, May 1956 (Schoeller 1962)

	Horizons				
	A_0	B_1	B_2	C_v	Groundwater
SO_4^{2-}/Cl^-	0.875	0.586	1.084	–	0.367
Mg^{2+}/Ca^{2+}	0.356	0.205	0.166	0.133	0.136
K^+/Na^+	0.392	0.161	0.123	0.215	0.092
I_{BA}	0.273	0.192	0.453	0.475	0.049

[a] "Brown forest soils" defined in a broad sense.

The calcium and magnesium contents in the upper and lower horizons of brown forest soils are generally higher than in the middle layer. The concentration variations of sodium in the individual horizons show no uniform trends, only the ratio of K^+ to Na^+ exhibits a connection with the base exchange in the clayey fractions of the soils. The chlorides apparently pass through brown forest soils in the same way as through podzols without any noticeable effect. Vertical movements of salts through biological and meteorological processes lead to a certain homogeneity of the chemical composition of the soil solutions with increasing depth in spite of certain horizon-specific solid phase accumulations, and this is most clearly seen in podzols because of their high permeability. The comparison between podzols and brown forest soils in the Bordeaux region shows the dependence of the chemical composition of the soil solutions on soil type. The total concentrations and the concentrations of most elements are less in the soil water of podzols than in brown forest soils. The Ca^{2+} concentrations of less than 1.7 meq in podzols is in contrast with the 2–18 meq content in the brown forest soils. Magnesium concentration shows the same tendency. The hydrogen carbonate concentration is always definitely higher in the carbonate-containing brown forest soils than in the carbonate-free podzols. In podzols the concentrations of Ca^{2+} and Mg^{2+} deviate little from those of Na^+ and Cl^-, while the solution of carbonates in the brown forest soils leads to a relative decrease of Cl^- and Na^+. Hence the equivalent ratios of Ca^{2+}/Na^+, Ca^{2+}/Cl^-, CO_3^{2-}/Na^+, and HCO_3^-/Cl^- of the soil solutions in the brown forest soils investigated are always considerably higher than unity in contrast to near unity or less in podzols (Schoeller 1963).

Kretzschmar (1964) reported the following ion contents in soil solutions in loess loam soils above loess deposits, and in dune sand soils above the floodplain loam in the area of the lower river terrace at Cologne, West Germany: NO_3^-, 1–200 mg/l (average 22 mg/l); Cl^-, 7–7380 mg/l (average 388 mg/l); and SO_4^{2-}, 4–1160 mg/l (average 106 mg/l). The highest values were found in areas under cultivation and are attributable to materials added with fertilizer applications.

Table 52 Aqueous Solutions from Saline Soils in Guadalquivir Valley, Spain[a] (After Garcia et al. 1956)

	Sapillo West soil horizon					
	0–20 cm		20–40 cm		40–70 cm	
Ion	mg/l	meq/l	mg/l	meq/l	mg/l	meq/l
Na^+	204	8.89	455	19.80	724	31.49
K^+	59.7	1.53	35	0.87	35	0.87
Ca^{2+}	67.0	3.35	214.2	10.71	364.2	18.21
Mg^{2+}	16.6	1.38	115	9.57	138	10.51
Cl^-	324	9.14	1285	36.21	1228	34.60
HCO_3^-	48.8	0.80	26.8	0.44	29.3	0.48
NO_3^-	126	2.03	93	1.50	39.8	0.64
SO_4^{2-}	174	3.63	229	4.77	1272	26.52
pH	8.1	–	7.9	–	7.9	–

	Sapillo Northeast soil horizon							
	0–10 cm		10–20 cm		20–40 cm		40–60 cm	
Ion	mg/l	meq/l	mg/l	meq/l	mg/l	meq/l	mg/l	meq/l
Na^+	120	5.20	260	11.30	419	18.20	141	6.12
K^+	17.5	0.45	183	4.70	82	2.10	720	18.20
Ca^{2+}	46.0	2.30	90.0	4.50	55.6	2.78	46.0	2.30
Mg^{2+}	69.6	5.80	56.4	4.70	69.6	5.80	61.2	5.10
Cl^-	256	7.20	426	12.00	674	19.00	280	7.90
HCO_3^-	32.3	0.53	40.9	0.67	4.33	0.71	31.7	0.52
NO_3^-	99	1.60	149	2.41	56	0.90	90	1.45
SO_4^{2-}	206	4.30	341	7.10	293	6.10	221	4.60
pH	7.6	–	7.6	–	7.8	–	7.7	–

	Cologne soil horizon					
Ion	0–10 cm, meq/l	10–20 cm, meq/l	20–30 cm, meq/l	40–60 cm, meq/l	60–80 cm, meq/l	80–100 cm, meq/l
Na^+	0.85	1.05	2.05	3.10	4.24	5.20
K^+	0.12	0.12	0.12	0.12	0.17	0.17
Ca^{2+}	0.96	1.05	0.78	0.63	0.60	0.59
Mg^{2+}	0.32	0.28	0.19	0.20	0.18	0.16
Cl^-	0.37	0.53	0.63	0.87	1.55	2.20
HCO_3^-	0.80	0.70	0.90	1.18	1.08	1.12
SO_4^{2-}	0.35	0.46	0.51	1.84	1.64	1.70
pH	7.8	7.7	7.8	8.3	8.1	8.1

[a] Ratio of water to soil, 5:1.

Table 52 shows the distribution of soluble salts in saline soils. Saline soils are developed in the floodplain of the Guadalquivir, south of Seville. The heavy clay soils Sapillo, west and northeast, are strongly saline and slightly alkaline. The zone at Cologne is formed from a stiff clay, the salt content of which is considerably lower and only increases at some depth.

Lysimeter Measurements. Lysimeter measurements indicate the quality of the percolate. Table 53 gives some examples of compositions found in this way. The ammonia, nitrite, and nitrate contents that in heathland soils reach only very small values result from microbial metabolism in the soil.

Table 53 Analysis Obtained by Lysimeters (mg/l) for North German Soils (After Höll 1970b)

Ion	Soil types[a]							
	1		2		3		4	
	Mar 1, 1958	Oct 31, 1962	Feb 1, 1959	Dec 15, 1961	Mar 1, 1958	Sept 30, 1958	Sept 6, 1960	Feb 14, 1961
Na^+	—	10.0	—	—	—	—	—	8.2
K^+	—	13.5	—	—	—	—	14.0	10.2
NH_4^+	0.15	3.8	0.4	1.35	0.04	0.01	0.03	0.05
Ca^{2+}	21.4	22.8	98.1[b]	216.0[b]	6.9	5.1	34.3	35.7
Mg^{2+}	11.3	67.4	—	—	7.7	9.0	10.8	4.8
Fe_{total}	3.05	0.65	0.01	2.25	1.5	0.03	0.35	n.d.
Mn^{2+}	0.75	1.0	3.8	4.5	0.55	0.75	0.3	0.15
Cl^-	7.0	139.0	86.5	234.0	6.0	12.5	15.2	6.7
HCO_3^-	18.3	12.2	24.4	8.7	12.2	30.5	54.5	81.9
NO_2^-	0.88	0.9	2.0	1.1	0	0	1.6	2.0
NO_3^-	60.0	245.0	148.0	285.0	0.3	0.3	46.0	105.0
SO_4^{2-}	48.0	24.5	38.0	121.0	28.0	27.0	78.8	41.2
PO_4^{3-}	0.03	0	0.02	0.001	0	0.01	0	0.015
Free CO_2	20.0	66.0	30.0	66.0	16.6	18.8	61.5	—
pH	5.8	6.9	6.9	5.4	6.51	—	6.55	4.74

[a] 1, Light arable soil, birch forest region; 2, heavy arable soil; 3, light heathland soil, Calluna heath; 4, light sandy soil, coniferous forest region.
[b] Alkaline earths calculated as Ca^{2+}.

In arable soil regions (1 and 2 in Table 53) nitrogenous fertilizers had been added, in the coniferous forest example (region 4) only natural nitrification processes had been working. The importance of the nitrification processes is confirmed by the observations of Kretzschmar (1964) that the highest nitrate concentrations do not occur until the end of rainy periods.

Periodic Changes in the Composition of Soil Solutions. The chemical composition of soil solutions varies seasonally. In the podzolized brown forest soils of Baron the concentrations of all elements except for chloride ion in November at the end of the evapotranspiration period is higher than in May at the end of the period of percolation. The decrease of chloride ion in autumn is attributed to late summer rains. Because of this the equivalent ratio of SO_4^{2-} to Cl^- increases in November. The equivalent ratio $(Na^+ + K^+)/(Ca^{2+} + Mg^{2+})$ decreases with depth in November, but on the other hand increases with depth in May. As a result of exchange processes the alkalis can migrate upward more easily during the evapotranspiration period than the alkaline earths (Schoeller 1963).

Seepage from podzol lysimeters at Talence near Bordeaux immediately after the dry period gave much higher concentrations than in the rainy and recharge periods. Here the concentrations of all elements increased with the exception of HCO_3^-, as shown clearly in Table 54.

Table 54 Lysimeter Measurements of Ratios of Concentration During Dry Period to Concentration During Wet (Leaching) Period (After Schoeller 1963)

	Lysimeters[a]		
Ion	D_1, 0–0.20 m	D_2, 0–0.38 m	D_3, 0–10.5 m
Cl^-	23.8	44.9	48.5
SO_4^{2-}	18.7	14.7	39.4
Ca^{2+}	7.8	4.6	2.2
Na^+	4.1	2.8	1.2
HCO_3^-	0.56	0.64	0.71

[a] D_1, A-horizon; D_2, A- and partly B-horizon; D_3, A + B-horizon.

Concentrations of the Cl^- and SO_4^{2-} ions increased more rapidly than those of cations, Cl^- more than SO_4^{2-} and Ca^{2+} more than Mg^{2+}. During evaporation periods the index of base exchange, which was negative during the period of leaching, became positive because of the comparatively faster rise of Ca^{2+} than the Na^+ concentrations. These seasonal changes in the chemical composition of the soil solution also appear in the percolate, which after a long evapotranspiration period shows very high concentrations of mineral matter. For example the Cl^- concentration increases quickly and after a few hours reaches a maximum of 100–800 mg/l and possibly even more. After 15–30 days the values stabilize and do not rise again until the start of the dry season (Schoeller 1963).

Comparisons Between Percolate and Groundwater Quality.   The composition of the percolate, as can be deduced from the soil solutions and lysimeter measurements, is very similar to that of the groundwater (Matthess & Pekdeger, 1980). The equivalent ratios that can be read from Fig. 44 — Mg^{2+}/Ca^{2+}, SO_4^{2-}/Cl^-, $Na^+/(Ca^{2+} + Mg^{2+})$, Na^+/Mg^{2+}, Cl^-/Na^+ — are not all exactly equal because in these low concentration waters small changes in the absolute values can lead to marked changes in the relative values expressed as ratios. The generally much higher hydrogen carbonate concentrations in soil moisture are to be regarded as important differences between water in the vadose zone and the groundwater below, as mentioned below.

The moisture in brown forest soils generally contains 9–20 meq HCO_3^-/l, the groundwater below only 6 meq HCO_3^-/l. Soil moisture in podzols also

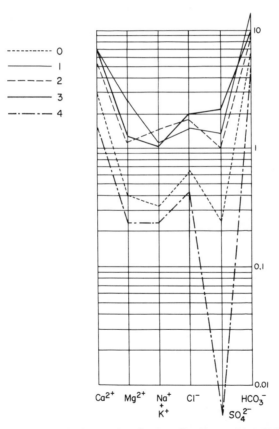

Fig. 44 Chemical properties of pore solutions in the soil at Baron, May 1956); (meg/l); solutions represented by curves 1–3 are aqueous extractions made up to standard volume, corresponding to pore volume). 0, groundwater below; 1, pore solution from A_1-horizon; 2, A_2-horizon; 3, B_1-horizon; 4, C_v-horizon (after M. Schoeller 1963).

shows higher values (0.1–1.9 meq HCO_3^-/l) than the associated groundwater (normally near 0.1 meq/l). In the seepage from lysimeters the values generally recorded were 2.5–4 meq HCO_3^-/l, and 1.4–2.5 meq HCO_3^-/l in the associated groundwater (Schoeller 1963). This observation was also confirmed by lysimeter measurements at Versailles, where values of 4.8–12 meq HCO_3^-/l (Bastisse 1951–1953) contrast with only 4 meq HCO_3^-/l (average) for the groundwater.

Furthermore, the NO_3^- content of the soil moisture is always much higher than in the groundwater lying below. In the soil solutions in brown forest soils concentrations occur between 0 and 1.5 meq/l; in the associated groundwater the range is 0.01–0.26 meq NO_3^-/l, and in the soil solutions of podzols 0.1–0.18 meq NO_3^-/l, compared with zero in the associated groundwater (Schoeller 1963). In the soil solutions described by Schloesing (1870) the NO_3^- content varied between 0.23 and 8.7 meq/l. The average given is 3.7 meq/l. Compared with this 95% of all groundwaters contain less than 0.84 meq NO_3^-/l. In the lysimeter measurements 32 meq NO_3^-/l was reached, whereas in the adjacent groundwater only 1 meq NO_3^-/l was observed, and in a polluted well 1.5–5 meq/l (Schoeller 1963). The fact that in most cases the NO_3^- content measured by lysimeters with bare soil surface is many times greater than the NO_3^- content recorded by plant-covered lysimeters shows that the nitrate is extensively taken up by vegetation. Schulz (1970) has found that denitrification results from reaction of nitrogen compounds with dissolved organic substances in the unsaturated zone, so that the concentrations of nitrate and organic materials decrease while that of CO_2 increases.

The equivalent ratios K^+/Na^+ for the soil solutions are always higher than in the groundwater below (Table 55).

The soil solutions and the seepage of lysimeters also show an increase in the equivalent ratio K^+/Na^+ in other soil types and in other regions. The soil solutions described by Schloesing (1870) show K^+/Na^+ ratios of 0.06–7.2 with an average of 0.97 (nine analyses). In seepage from lysimeters the ratios were found to be between 0.05 and 16.28. In general these K^+/Na^+ ratios are distinctly higher than the ratios reported for the groundwater, which according to Schoeller (1963) mostly lie below 0.18. The cause of this is the uptake of potassium by plant roots and sorption by soil colloids. Evidence for this is the smaller K^+/Na^+ ratios as recorded by seepage from lysimeters under plant cover in contrast with those measured away from vegetation. Only a small part of the potassium enters the groundwater (Matthess et al. 1977).

The connection between the composition of the soil solutions and that of associated groundwater is seen most clearly in saline soils. As Schoeller (1963) has shown the soil moisture in solonetz soils is similar to that in the associated groundwater below, both having high equivalent ratios SO_4^{2-}/Cl^-

between 4 and 10, Mg^{2+}/Ca^{2+} about unity, and a significantly smaller Cl^- content than Na^+. The water in solonetz-castanozem soils and the associated groundwater on the other hand have an equivalent ratio SO_4^{2-}/Cl^- below unity; the ratio Mg^{2+}/Ca^{2+} is very low, and the Cl^- content is greater than the Na^+ content.

The discrepancies between the quality of the water in the unsaturated zone and the groundwater lying below are explained by several factors: the retention of nutrients like nitrate, potassium, and sulfate by plants; redox reactions between the nitrogen compounds and the dissolved organic substances, particularly humates, which result in the production of nitrogen and carbon dioxide (Schulz 1970); and the effect of carbon dioxide in the strata between the surface and the water table, as well as hydrolysis of silicate minerals, sorption of potassium, and the dilution of soil solutions by the addition of fresh percolate, which leads to groundwater recharge. The percolate that reaches the water table is, as clearly shown by the concentration measurements by lysimeter tests, especially from those not covered by vegetation (Bastisse 1951, Schoeller 1963), most highly concentrated after dry periods. With a shallow water table and highly permeable surface strata, the addition of highly concentrated percolate can be observed in the quality of the groundwater (Buchan 1958).

The frequently found, but relatively low deviations of the concentrations of the unsaturated zone water from those of the groundwater below suggest that in the active soil zone and in the rock strata lying below, the percolate

Table 55 Ratio of Potassium to Sodium in Soil Solutions and Groundwater (After Schoeller 1963)

	Soil solutions	Groundwater
Brown forest soils, podzolized		
Baron		
November 1954	0–0.263	0.16
May 1956	0.161–0.392	0.092
Pompignac	0.115–0.158	0.038
Podzols		
Haillan		
February 1954	0.157–2.68	0.119
April 1955	1.64–3.9	0.43
Martignas	0.36–0.84	0
Caupian	0–0.89	0
Lysimeters (see Table 54)		
D_1	0.15–0.55	0.11–0.16 (wells)
D_2	0.1–0.35	
D_3	0.1–0.32	

takes on the characteristic properties of the groundwater, which are only slightly changed in its subsequent underground flow path.

3.1.3   Inland Surface Waters

Surface waters can influence groundwater quality when water level gradients cause the infiltration of river water in karst regions, or percolation of direct surface runoff in areas of flooding or by bank filtration to water supply wells. The quality of the infiltrating water is changed by geochemical processes (precipitation, dissolution, oxidation-reduction, ion exchange, and biological processes).

The surface runoff water can be considered to be a mixture of direct surface runoff and reappearing groundwater. The true surface runoff that results from storms or from glaciers and snowmelt waters over frozen ground has only a short period of contact with soil or vegetation, but has a generally significantly higher dissolved solids content than did the original precipitation. However, the base flow – that part of the runoff which is exclusively fed by groundwater – matches the dissolved solids in the latter, which is in general higher than direct runoff in its natural, unpolluted state. Hence it follows that the dissolved solids content of the surface water decreases with increasing total runoff. In addition, the water quality is influenced by reactions between surface water and rock and suspended material, water loss by evaporation, transpiration, and biological effects caused by water plants. Finally the surface runoff is especially subject to man-made factors because it serves as an important transporter of dissolved and suspended waste material.

The average dissolved solids content of surface water can locally and periodically show wide deviations, as shown in Table 56. Table 57 gives ex-

Table 56   Average Composition of River Water

Ion	After Clarke 1924 (mg/kg)	After Livingstone 1963 (mg/kg)
Na^+	5.8	6.3
K^+	2.1	2.3
Ca^{2+}	20	15.0
Mg^{2+}	3.4	4.1
Cl^-	5.7	7.8
SO_4^{2-}	12	11.2
$HCO_3^- + H_2CO_3$	35	–
HCO_3^-	–	58.4
SiO_2	17.3	13.1

amples of the composition of river waters (no average values). In addition, Table 58 gives collected data on the minor elements contained in North American rivers based on work by Durum & Haffty (1961) and Durum et al. (1960). These values reflect the different geological, pedological, and climatic conditions in the catchment areas.

Relationships between surface water quality and that underground have been investigated by Schmidt (1956) in the Fulda-Eder area. He found conductances significantly below 100 μS at heights above 400 m in the Rhön (basalt), the Vogelsberg (basalt), and the Rothaargebirge (Lower Carboniferous and Devonian slates).

At lower levels, in rocks of the Muschelkalk and Bunter Sandstone formations, the conductances range from 200 to 300 μS; in the area influenced by the Upper Bunter Sandstone (Röt) conductances are between 300 and 500 μS, and in the Zechstein even greater than 1000 μS. The pH values generally lie between 7 and 8, only extremely saline-poor waters show values between 6 and 7. Hydrogen carbonate and alkaline earth contents follow conductance values closely. Considerable fractions of alkaline earths, which can be correlated with sulfate and chloride ions, occur in the region of the gypsiferous and salt-bearing Zechstein. Very low Cl$^-$ contents ($<$ 10 mg/l) are, like the corresponding Na$^+$ contents, confined to high altitudes. The regional potassium distribution is similar to that of sodium, but the potassium content is always significantly lower than the sodium or chloride content. The natural distribution of substances in the surface waters of the region is locally distorted by municipal or industrial waste.

In a similar way Davis (1961) has correlated the properties of the river waters on the eastern slope of the southern Coast Range, California, with the rock chemistry in the catchment area. Explanations have been found for some peculiarities, such as the emergence of mineralized water. In the water at Little Panoche Creek and at Salt Creek (Merced County) the sodium and chloride contents are the major constituents, and unusually high boron values occur (1.4–19 mg/l in 3820 mg/kg total dissolved solids in the Little Panoche Creek, and 25 mg/l in 5600 mg/kg total dissolved solids at Salt Creek), which are attributed to the emergence of highly mineralized oil field brines. High magnesium content was traced back to ultrabasic intrusive rocks; for example more than 40% of the cation equivalent found in the waters of De Puerto, Orestimba, Cantua, and Los Gatos creeks.

Surface waters coming from a catchment area that is exclusively composed of one hard rock type exhibit a quite uniform chemical composition independent of the size of the catchment. Research into the average concentrations of waters in the quartzite, granite, and sandstone areas in the Sangre de Christo Range, New Mexico, showed a ratio of 1:2.5:10 in which the relatively high dissolved solids content of the runoff from the sandstone areas is attributed to the carbonate cement in the rock, and thin intercalated beds of limestone (Miller 1961).

Table 57 Chemical Composition of some of the Earth's Rivers (After Alekin 1962, Bösenberg & Lüttig 1959, and the International Commission for Prevention of Rhine Pollution 1967).

River and location	Test-period	Concentration							Total ions	Anions	
		HCO_3^-	SO_4^{2-}	Cl^-	NO_3^-	Ca^{2+}	Mg^{2+}	$(Na^+ + K^+)$		HCO_3^-	SO_4^{2-}
Amazonas (Obidos)	-	18.1	0.8	2.6	-	5.4	0.5	3.3	30.3	55.8	30.8
Parana (mouth)	-	33.9	9.8	15.6	-	7.0	2.7	18.6	90.6	46.4	17.0
Rio Negro (Mercedes)	-	105	16.2	5.9	-	23.5	2.6	13.5	166.7	77.8	15.0
Colorado(Austin)	-	108.4	199.0	159.5	-	105.8	9.5	102.7	684.9	17.0	39.8
Mississippi (New Orleans)	-	118.0	25.6	10.3	-	34.1	8.8	13.8	210.6	70.0	19.3
Missouri (mouth)	-	180.3	117.2	13.5	-	52.6	18.2	38.0	419.8	51.2	42.2
Northern St. Lawrence (Montreal)	-	133.7	16.6	3.6	-	30.6	9.5	7.2	200.2	83.0	13.2
Rio Grande (Lagero)	-	185.5	38.0	171.3	-	108.6	24.0	123.5	650.9	34.9	9.0
Colorado	-	196.6	40.1	56.3	-	49.6	16.5	42.0	391.1	55.8	15.2
Columbia River (Western Canada)	-	67.9	12.5	2.6	-	16.5	4.1	9.3	112.9	77.2	18.0
Yukon	-	91.9	10.5	0.4	-	21.8	4.6	6.0	135.2	87.2	12.2
Rhein	Average	150.2	79.3	149.1	10.5	81.2	10.2	85.9	566.4	29.0	19.5

(Emmerich/Lobith) 1963—1965
(24 readings)

Mosel (Koblenz) 1963—1965 (24 readings)	Average	147.7	94.2	252.4	11.1	128.0	15.4	85.1	733.9	20.6	16.7
Weser (Bremen)	Average	162	136	312	–	86	23	184	903	18.6	19.8
Klar—Älv (Sweden)		20.0	1.9	0.6	–	0.3	0.1	3.1	28.7	85.2	10.4
Neva (Ivanskoye)	9.7.1946	27.5	4.5	3.8	–	8.0	1.2	3.8	48.8	69.2	13.8
Moskva (Tatarowo)	Average (1914—1926)	250.7	5.6	2.3	–	61.5	14.2	23	358.5	94.8	2.6
Ob (Nowosibirsk)	21.8.1940	85.6	13.0	–	–	24.3	5.4	0.4	129.0	84.0	16.0
Lena (Kyuskyur)	8.9.1940	66.4	21.2	15.2	–	18.0	3.8	18.8	143.0	55.6	22.4
Jenissey (Krasnoyosk)	20.9.1936	73.2	4.0	2.6	–	19.3	4.0	1.5	104.6	88.8	6.0
Don (Aksaiskaya)	4.7.1939	260.0	112.	44.0	–	82.0	18.0	52.2	568	54.4	29.6
Wolga (Wolsk)	21.8.1940	210.4	112.3	19.9	–	80.4	22.3	12.5	458	54.4	36.8
Machanudi (S.E.India)		47.3	0.9	1.8	–	13.6	4.0	5.7	73.3	92.4	2.0
Seraja (Java)		63.5	18.0	7.0	–	14.3	4.2	15.3	122.3	64.6	23.0
Nile (Cairo)		84.6	46.7	3.4	–	15.8	8.8	11.8	119.1	56.5	39.6

Table 58 Minor Element Content in North American Rivers (After Durum & Haffty 1961 and Durum et al 1960)

Location	Date of Collection	B /ug/kg	Ba /ug/kg	Cr /ug/kg	Cu /ug/kg	Li /ug/kg	Mo+ /ug/kg	Ni+ /ug/kg	Pb /ug/kg	Sr /ug/kg
Colorado at Juma	16.9.1958	52.	152	24	8.8	35	6.9	30	8.0	802
Ariz. USA	20.1.1960	34	128	10	8.5	37	6.5	0	16	715
Columbia at	11.6.1958	11	33	9.2	3.0	-	0.x[a]	x.0	1.2	30
The Dalles	1.12.1958	3.9	48	18	3.8	3.9	2.1	10	5.0	112
Oreg., USA	8.9.1959	5.0	44	20	27	0.59	1.0	36	3.4	60
Hudson at	29.10.1958	9.0	28	30	8.6	2.2	0	12	2.9	106
Ford Motor Co.	22.1.1959	15	46	15	15	1.1	0.65	8.6	11	84
Powerstation at	17.8.1959	20	60	40	44	0.28	1.7	71	9.4	72
Green Island N.Y.,USA	19.10.1959	10	49	19	4.5	0.89	0.55	5.5	6.6	106
Mississippi	10.5.1958	29	78	2.6	10	-	0	0.x	4.0	34
at Baton Rouge,	14.10.1958	22	127	8.5	6.9	4.9	1.1	19	7.8	105
La., USA	13.3.1959	15	72	4.6	9.0	1.8	0	13	4.0	61
	18.5.1959	6.1	84	84	74	2.4	2.6	33	9.4	97

Sacramento at Sacramento Calif., USA	1. 5.1958	22	9	0.72	1.4	–	0	0.x	0	6.3
	25.11.1958	10	31	4.4	2.9	2.1	0.43	7.1	4.5	46
	1. 5.1959	9.8	25	2.4	7.0	0.77	0	6.6	4.2	15
	16. 9.1959	25	56	7.0	14	0.66	0.47	20	4.4	61
Susquehanna at Conowingo Md., USA	7. 5.1958	10	24	1.9	5.1	–	0	x.0	3.6	12
	10. 6.1958	12	38	4.5	4.0	–	0.x	0.x	1.1	40
	11. 9.1958	16	25	3.7	5.3	3.5	0	11	2.1	25
	5. 6.1959	4.3	37	1.3	105	3.4	0.54	3.9	7.2	74
Yukon at Mountain Village, Alaska, USA	28. 5.1958	11	26	2.3	6.3	–	0	x.0	1.5	15
	7. 1.1959	13	109	7.0	2.5	2.0	1.2	17	8.6	129
Fraser River at Mission City, British Columbia Canada	1. 7.1958	2.6	17	2.9	1.6	0.50	0	5.0	0.62	16
	1.10.1958	11	18	6.0	2.5	0.18	0	12	1.8	18
	12. 2.1959	2.6	20	23	3.0	0.40	0.79	13	3.9	40
	12. 6.1959	1.4	14	2.8	0.83	1.3	0.21	2.6	1.3	23

Table 58 (Continued)

Location	Date of Collection	B /ug/kg	Ba /ug/kg	Cr /ug/kg	Cu /ug/kg	Li /ug/kg	Mo+ /ug/kg	Ni+ /ug/kg	Pb /ug/kg	Sr /ug/kg
Mac Kenzie above delta, Northwest Territory Canada	24. 7.1958	13	65	12	11	1.1	0	36	2.9	96
	2.10.1958	8.6	70	5.7	3.9	1.0	0	8.2	6.5	228
	9. 6.1959	4.7	85	11	4.2	2.5	0.80	22	7.6	125
Nelson River at Amery, Manitoba Canada	5.10.1958	2.5	45	5.7	2.3	2.9	0	5.9	5.3	121
	29. 1.1959	7.6	50	2.5	1.0	4.8	0.87	4.1	0	73
	9. 3.1959	21	58	1.9	2.5	7.5	1.3	5.5	0	105
	9. 4.1959	3.6	56	4.7	4.2	8.1	0	7.8	22	86
	1. 5.1959	7.8	57	3.4	6.0	3.9	0	8.8	4.9	107
	3. 6.1959	58	53	21	11	1.7	0.53	55	5.0	55
St. Lawrence River at Levis, Quebec, Canada	25. 8.1958	13	30	12	4.3	0.41	1.7	13	3.7	66
	3. 9.1958	6.3	26	9.6	4.7	0.40	2.4	11	4.9	92
	24. 9.1958	11	46	23	59	0.60	2.6	22	55	78
	3.12.1958	11	36	5.8	4.7	1.1	1.1	13	3.2	86
	20. 4.1959	7.8	39	6.8	-	0.27	0.34	13	7.3	45
	23. 4.1959	5.8	40	9.7	-	0.36	0.37	12	3.6	58
	16. 6.1959	4.1	29	7.0	8.1	0.55	0.60	16	3.6	69

[a] X signifies semiquantitative result as the order of magnitude.

3.1.4 Seawater

Seawater can enter the underlying rocks of the mainland where permeable rocks occur adjacent to the sea. In these cases the groundwater is chemically related to the seawater, or deviates by mixing with fresh water or as a result of geochemical processes (precipitation, solution, oxidation, reduction, sorption, and ion exchange, gas exchange, and biological effects, cf. Sections 2.1–2.5) affecting seawater. The less dense fresh groundwater tends to form a layer above the denser seawater, and indeed so much so that a state of equilibrium is reached. The position of the boundary layer between the two is a dynamic function of the differences between densities and the velocity and the amount of the fresh groundwater flowing to the sea.

Along tidal coasts the equilibrium is furthermore dependent on tidal range. Because it is a dynamic equilibrium it is understandable that a brackish water zone, at most only a few metres thick, forms as a boundary layer. Any thorough mixing by diffusion is prevented by the steady movement of both water bodies in opposing directions, by which the mixed waters flow as well. Only under static conditions does appreciable diffusion occur (Jacobs & Schmorak 1960, Kohout 1960, Ødum & Christensen 1936, Todd 1960a, Richter & Flathe 1954). The hydrologic conditions described above are especially well developed on many small oceanic islands. In these cases the fresh groundwater from the precipitation accumulates as freshwater lenses above the saltwater layer deeper underground. During extraction of groundwater near the coast, entry of seawater may take place; with excessive demand a deterioration of the freshwater quality can result. Finally, seawater can also move inland beyond the coast by flooding, penetrating the subsurface of the interior and mixing with inland groundwater.

The salt content of seawater amounts to approximately 35 g/kg. Its average composition (Table 59) varies very little in the open sea; however in coastal waters significant variations in the percentage contents of individual ions can occur. The minor element content of seawater is shown in Table 64, below. The salt content increases when the rate of evaporation exceeds the water inflow. Water losses occur through evaporation and formation of ice, water inflow comes from the land, precipitation, and melting ice (Turekian 1969).

The pH of seawater is apparently determined by the system $CaCO_3$-CO_2-H_2O and lies close to 8. The free dissolved oxygen content in most oceans is close to saturation, based on the prevailing atmospheric conditions and the water temperature. The deviation ranges from 120% (oversaturation) in water close to the surface, as a result of heating and production of oxygen by phytoplankton, to approximately 70% and even lower at greater depths, because of biological oxygen consumption. For an oxygen partial pressure of 200 mbar and of pH of 8 the redox potential of oxygen-saturated seawater is calculated to be 0.85 V at 20°C. This value has not yet actually been

**Table 59   Concentration of
Major Dissolved Solids
Content in Seawater (Culkin
1965)**[a]

Ion	Concentration (g/kg)
Na^+	10.76
K^+	0.387
Ca^{2+}	0.413
Mg^{2+}	1.294
Sr^{2+}	0.008
Cl^-	19.353
SO_4^{2-}	2.712
HCO_3^-	0.142
Br^-	0.067
F^-	0.001
B	0.004

[a] Salt content 35‰.

measured, probably because other redox pairs likewise affect the system or because of difficulties in measurement techniques (Turekian 1969).

3.2   GROUNDWATER PROPERTIES AND CONSTITUENTS

Various estimates have been made of the total planetary groundwater volume. Such estimates are based on the assumed depth and porosity of the rock formation saturated with groundwater. Assuming an average porosity of 10 vol % down to average depths of 9654 m (beneath the continents) and 8045 m (below the oceans) Slichter (1902) estimated the total amount of the water as approximately 4.3×10^7 km³. This figure is too high because of the excessive estimate of porosity at 10%.

Van Hise (1904) assumed, starting from an effective porosity of about 1.4% in the strata near the surface and approaching 0% at about 10 km depth, that there is an average 0.69% voids in fissures and cracks below continents. According to Kalle (1943) the continents cover 148.94×10^6 km². Thus the total amount of groundwater can be calculated to be 1.026×10^7 km³. In 10 km this is equivalent to a water layer 69 m thick. Because in porous media the porosity varies between zero and more than 50%, Kalle surmised that in this connection the estimated value should be 2–3 times higher. Hence an equivalent total thickness of a layer of water 69 m thick would be too low.

Kalle (1943) has stated that there is in the Earth's crust, which as a geochemical concept defined by Clarke (1924) descends 16 km below and rises up to 10 km above the Earth's surface, 272.3 kg · cm^{-2} of movable water. Of that the groundwater fraction is 0.018% = 0.049 kg · cm^{-2}. The total amount for the whole global surface (510.1 × 10^6 km^2) works out to

$$\frac{0.049 \text{ kg} \times 10^{10}}{\text{km}^2} \times 510.1 \times 10^6 \text{ km}^2 = 25 \times 10^{17} \text{ kg}$$

This corresponds to about 2.5 × 10^5 km^3 for a density of 1.

Nace (1960) has calculated the total groundwater below continents to be 8.312 × 10^6 km^3, assuming an average effective porosity of 4% for depths down to 805 m and 0.7% over the zone 805–4023 m. This corresponds to a water column of 56 m. The average effective porosity for the whole column comes to 1.4%.

Nöring (1966) has taken a groundwater content of 1% as a basis for the first 800 m and 0.25% for 800–4000 m, which corresponds to an equivalent water column of 16 m. The total groundwater amount below the continents on this assumption amounts to 148.94 × 10^6 km^2 × 0.016 km = 2.38 × 10^6 km^3.

3.2.1 Groundwater Temperatures

Groundwater temperature at certain places and times is the result of various heating processes. It can be almost constant; however this is generally the exception. The temperature of groundwater is the result of the heat exchange on the Earth's surface under the control of incoming and outgoing radiation, heat conduction, convection, evaporation, and chemical and thermonuclear processes in the rocks.

The temperatures on the surface and in the uppermost rock strata of the crust are almost exclusively determined by energy exchange between sun, soil, and atmosphere. Energy is received as short-wave visible light radiation; heat moves up from the depths of the Earth's interior and is radiated out as long-wave heat radiation. The average temperature of the Earth's surface is determined by the total energy reaching it by these ways.

The energy reaching the Earth from the sun is on average 1.3 kW · m^{-2} (solar constant). This value is however reduced significantly by absorption and reflection to an average value of 0.75 kW · m^{-2}. Geographical latitude, altitude, exposure, type of surface, seasonal and daily variations lead to spatial and temporal deviations from this figure.

The heat flow q from the Earth's interior, which comes from natural radioactive decay and other deep-seated heat sources, shows regional variations (Table 60). The average heat flow from the Earth is given by Schoeller (1962) as (5.4 ± 0.4) × 10^{-2} W · m^{-2}. Hence for surfaces perpendicular to the flow direction and in unit time some 20,000–30,000 times as much energy is received from the sun as from the interior. The effect of terrestrial heat flow on the temperature at the Earth's surface is accordingly less than 0.1°C (Kappelmeyer 1968, Kappelmeyer & Haenel 1974).

Temperature varies with depth, and indeed the more so the greater the heat flow q and the smaller the thermal conductivity λ of the subsurface material.

The thermal conductivity λ is defined by equation 97

$$q = -\lambda \frac{d\delta}{dx} \tag{97}$$

in which $d\delta/dx$ is the temperature gradient in $K \cdot cm^{-1}$ and q is the heat flow $(W \cdot m^{-2})$.

Table 60 Natural Heat Flow q in Nonvolcanic Regions of the Earth (After Birch 1954)[a]

Region	q $(\times 10^{-3} W \cdot m^{-2})$
South Africa	46
Canada	40
Iran	36
England	56
United States	
California	54
Colorado	71
Michigan	39
Atlantic	41
Pacific	61

[a] Average from measurements at various stations.

Thermal conductivity depends on mineral composition, subsurface structure, porosity, and water content, together with the effects of various fissures and fault zones, bedding planes, jointing, and cleavage. Thus thermal conductivity is locally always under the simultaneous control of several factors. Thermal conductivities of the major rock types are given in Table 61 as material constants for steady thermal processes.

The thermal diffusivity a is related to the thermal conductivity λ by the expression

$$a = \frac{\lambda}{c \cdot \rho} \tag{98}$$

in which c is the specific heat and ρ the density of the material. Temperature changes on the Earth's surface decrease faster, the smaller the thermal dif-

Table 61 Thermal Conductivities of Some Rocks

Material	Thermal conductivity $(W \cdot m^{-1} \cdot K^{-1})$	Source[a]
Earth's crust, average	1.67	(2)
Soil, dry	0.14	(2)
Clay (dry)	0.84–1.26	(1)
Clay (wet)	1.26–1.67	(1)
Shale	1.67–3.34	(2)
Gneiss	2.09–2.51	(1)
Granite	1.67–3.34	(1)
Basalt	2.18	(2)
Chalk	0.84	(2)
Limestone	2.09–3.34	(1)
Marble	2.97	(2)
Marl	2.09–2.93	(1)
Rock salt	3.34–6.28	(1)
Gypsum	1.3	(2)
Sand (dry)	0.33–0.38	(1)
Sand (10% moisture content)	1.26–2.51	(1)
Sandstone, dry	0.84–1.26	(1)
Sandstone, wet	2.09–2.93	(1)
Coal	0.13–0.3	(1)
Water	0.59	(1)

[a] After Kappelmeyer 1961 (1) and Hodgman et al. 1958 (2).

Table 62 Thermal Diffusivity of Some Rocks and Soils (After Kappelmeyer 1961)

Material	Thermal diffusivity $(\times 10^{-6} \, m^2 \cdot s^{-1})$
Sand (dry)	1.1–1.3
Sand (wet)	0.9
Loamy sand	0.8
Sandstone	2.3
Granite	1.9–2.1
Peat	2 –1
Peat (dry)	1

fusivity a of the material. Indications of the order of magnitude of the thermal diffusivity are given in Table 62.

Inward and outward radiation is subject to daily and seasonal changes, sunspots and other fluctuation, and regional and local climatic peculiarities on the Earth's surface (e.g., polar and equatorial regions). Temperatures therefore vary at the Earth's surface, in the near-surface air layers, and in the zones immediately below the surface down to certain depth limits.

The limits of temperature range originate from the effect of incident and outward radiation at the actual surface of the Earth. These temperatures are further variable depending on the irregularities of the surface, such as exposure, slope direction and gradient, and absorption and reflection coefficients of the surface. The hottest climatic regions are found between latitudes 20° and 40° north and south (desert regions of the Sahara, the Arabian peninsula, western Pakistan, southeastern Iran, southwestern United States, and the semiarid and desert regions of northwestern Argentina, southwestern Africa, and central Australia). In these regions at least once yearly a temperature of 40°C is reached or even exceeded. Low-lying parts of the Lut Desert in Iran, the northern and western Sahara, and Death Valley in California show average annual maxima of more than 50°C. The lowest average temperatures are found in the north and south polar regions: in Antarctica below −70°C to nearly −90°C; in Alaska, northwestern Canada, and Siberia −50°C; and on the Greenland icecap −65°C (Hoffmann 1963).

The heat that is added to the soil by solar radiation, which varies strongly during the course of the day and the year, moves slowly into depth. The amplitudes of temperature variations decrease with depth and show a phase shift (Kappelmeyer & Haenel 1974) (Fig. 45). The deep zone, at which the annual variation of temperature is nil, is called the neutral zone [stratum of invariable temperature (Chang 1957)]. The disappearance of temperature changes at a certain depth depends critically on the thermal conductivity of the subsurface material. Correspondingly there occur zones with an annual temperature range below 0.01°C at depths between 15 and 39 m, for example in limestones between 24 and 27 m, and in granite between 34 and 39 m (Schoeller 1962). If 0.1°C, which corresponds to the accuracy of thermometers in general use, is chosen as the limit, the neutral zone lies substantially higher. With reference to this limiting value Chang (1957) has stated from interpretation of results from 780 recording stations around the world that the depth of the neutral zone in the tropics is at approximately 10 m, in temperate zones at 15 m, and in cold regions (Alaska and northern Canada) at 20 m.

Kappelmeyer (1968) has observed at 1 m depth below forests annual variations of 4.6°C, at 14 m depth 0.1°C, and below grassland 5.5°C at 1 m depth and 0.1°C at 12 m. A similar trend is shown by research into temperature variations at different depths (down to approximately 26 m) in a shallow water table at about 1.50 m depth), above a groundwater body approximately 27 m thick in Long Island, in which the temperature variations below unforested land were higher than below forest cover (Pluhowski & Kantrowitz 1963). The average annual temperature in the soil depends on latitude, exposure, moisture, and on the plant cover. At 1 m depth below woodland temperatures approximately 1–1.5°C lower than under grassland can be observed (Kappelmeyer 1961, 1968). Correspondingly the average

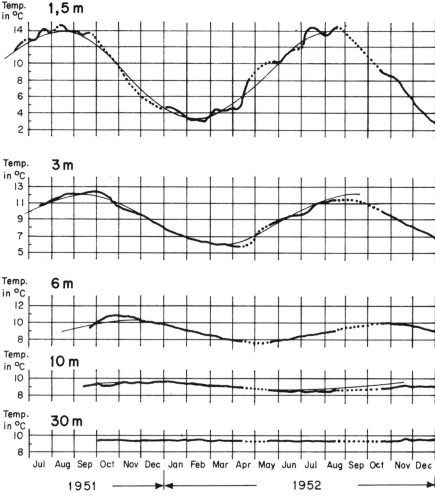

Fig. 45 Seasonal temperatures at various depths below surface at Hannover. The thin line shows ground temperature waves determined by use of harmonic analysis (after Kappelmeyer 1961).

annual temperature of the Long Island groundwater body mentioned, below nonforested land, was about 0.8°C higher than below forested land (Pluhowski & Kantrowitz 1963).

The changes of the temperature field of the subsurface as a consequence of the periodic temperature variations on the Earth's surface can be comprehensively described by considering the heat flow from the Earth's interior as given in equation 99 (Kappelmeyer 1968, Kappelmeyer & Haenel 1974):

$$\vartheta(z, t) = \bar{\vartheta} + \frac{q}{\alpha} \left(1 + \frac{\alpha \cdot z}{\lambda}\right) + \sum_{n=1}^{n=\infty}$$

$$\vartheta_n \cdot \exp\left(-\sqrt{\frac{n \cdot \omega}{2 \cdot k}} \cdot z\right) \sin\left(n\omega \cdot t - \sqrt{\frac{n \cdot \omega}{2 \cdot k}} \cdot z + \varphi_n\right) \qquad (99)$$

in which $z =$ depth (cm)

$t =$ time (s)

$\vartheta =$ average temperature on the Earth's surface, neglecting heat transfer from below (°C)

$q =$ terrestrial heat flow (= quantity of heat flow from the depths to the Earth's surface, $W \cdot m^{-2}$)l on continents $4.2 \times 10^{-3} - 8.4 \times 10^{-3} W \cdot m^{-2}$, average $6.3 \times 10^{-3} W \cdot m^{-2}$

$\alpha =$ coefficient of heat transfer at the Earth's surface ($W \cdot m^{-2} \cdot K^{-1}$); the value for α at the Earth's surface depends on moisture content, temperature, and wind velocity; as a rule values lie between 8 and 420 $W \cdot m^{-2} \cdot K^{-1}$. This has a marked effect on soil temperatures at 1 m depth and below only in regions with extremely high terrestrial heat flows.

$\lambda =$ thermal conductivity of the ground material ($W \cdot m^{-1} \cdot K^{-1}$) the thermal conductivity is low in very dry ground materials, it increases with its cohesiveness and moisture content. In general thermal conductivities of ground materials lie between 0.8–2.4 $W \cdot m^{-1} \cdot K^{-1}$ (cf. Table 60).

$\vartheta_n =$ amplitude of the nth harmonic temperature wave at the Earth's surface (°C)

$n =$ whole number

$\omega = 2\pi/T$ (s^{-1})

$T =$ wave period of the ground wave (usually 1 year) (s).

$k =$ thermal diffusivity of the ground material ($m^2 \cdot s^{-1}$); this value is also strongly dependent on water content and surface tension of water in the ground material; one can assume for different ground materials a range between 0.5×10^{-6} and $1.3 \times 10^{-6} m^2 \cdot s^{-1}$.

$\varphi =$ phase of the nth harmonic temperature wave in the ground material

The amplitudes of daily and annual temperature changes at the Earth's surface decrease exponentially and undergo a phase shift proportional to the depth, depending on the period of the oscillations and on the thermal diffusivity below the surface. For example values of thermal diffusivity the short-duration daily temperature variations decrease at half the surface amplitude as $1/\sqrt{365} \simeq 1/19$ of the depth; the long-wave annual temperature change is only half as great (Kappelmeyer 1961).

The diminution of the amplitude of temperature oscillations can be described approximately by equation 100 proposed by Geiger (1950):

$$\delta_2 = \delta_1 \cdot \exp (x_1 - x_2) \sqrt{\frac{\pi \cdot \rho \cdot c}{T \cdot \lambda}} \qquad (100)$$

in which δ_1, δ_2 = temperature amplitudes corresponding to depths x_1 and x_2, respectively

x_1, x_2 = depth (cm)

ρ = density of the soil ($Mg \cdot m^{-3}$)

c = specific heat of the soil ($J \cdot K^{-1} \cdot Mg^{-1}$)

λ = thermal conductivity of the soil ($W \cdot m^{-1} \cdot K^{-1}$)

T = period of oscillation (s)

The daily temperature variation in soil is significantly affected by the soil cover. The greatest amplitude (at 10 cm depth) and the greatest penetration depth were found in bare sandy soils; the lowest amplitude (at 10 cm depth) and a very low penetration depth were observed under woodlands, where the daily temperature variation detected at 50 cm depth was less than 0.1°C. The values under grassland lie between these limits (Kappelmeyer 1961). The same laws hold for the annual temperature ranges; however, the penetration depth is greater on account of the longer periods involved.

The effect of plant cover under central European conditions is allowed for in the empirical equations 101–103, which give the deviations from the annual average (Schubert 1930) for $T = 1$ year. These relationships were observed during long-term investigations at 16 field stations.

Open land: $\qquad \delta = 9.06 \sin \left(\frac{2\pi}{T}t + 251°\right) + 0.98 \sin \left(\frac{4\pi}{T} t + 80°\right)$ (101)

Pine forest: $\qquad \delta = 7{,}40 \sin \left(\frac{2\pi}{T}t + 244°\right) + 0.90 \sin \left(\frac{4\pi}{T}t + 54°\right)$ (102)

Beech forest: $\qquad \delta = 7.10 \sin \left(\frac{2\pi}{T}t + 244°\right) + 0.48 \sin \left(\frac{4\pi}{T}t + 64°\right)$ (103)

By use of equation 100 the depth of the neutral zone, in which the annual temperature variations are negligible, can be determined to a first approximation. A straight line can be drawn on semilogarithmic paper, using the observed values, which terminates on the abscissa at 0.01°C at the appropriate depth for the neutral zone (Fig. 46). Below the level known as the neutral zone the temperature increase depends on the local geological conditions and the heat flow. The unit of temperature increase with depth is known as the geothermal gradient. This is defined as the distance between two vertical points between which there is a temperature difference of 1°C. The geothermal gradient is smaller the greater the heat flow and the smaller the thermal conductivity. Generally the geothermal gradient is 33 m · °C^{-1}; in Germany values between 25 and 40 m · °C^{-1} have been found. Unusually low values

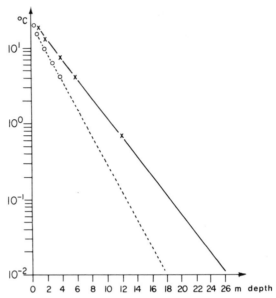

Fig. 46 Approximate determination of depth of the neutral zone from measurements of decreasing amplitude of temperature variations below ground level (after Wilhelm 1956).

occur in the Upper Rhine Valley (20 m · °C^{-1}) and in the Urach volcanic region (11 m · °C^{-1}). The greatest geothermal gradients are found in South Africa, where in kimberlite pipes gradients of 100 m · °C^{-1} occur (Kappelmeyer 1961).

Water has very variable temperatures when it enters the Earth's surface. The quicker the water table zone is reached and the more water infiltrates, the less will be the change in the temperature of the percolating water and the greater the heating or cooling effect of this change on the groundwater already there. This is particularly the case with very permeable strata and aquifers, especially with fissured rocks and karst, and with steady percolation of large amounts of bank-filtered river water and artificial groundwater recharge. The loss of heat from the groundwater into the solid rock of the aquifer remains low on account of the relatively low surface of contact. Other relationships occur when the residence time below the surface before reaching the groundwater body is long, especially when little water is percolating. In this case the temperature of the water can be close to equilibrium with that of the surrounding rock. In places where water is percolating through unconsolidated rocks, for example fine sand, the surface of contact is very high and the heat exchange is greatly accelerated. In heat exchange between water and rock, variations caused by differences in thermal conductivities can be expected (Table 60). Because of the low thermal conductivity of water and the very low rate of mixing at low flow velocities, temperature stratification occurs and different zones of temperature can exist between adjacent groundwater bodies laterally as well as vertically.

The evaluation of the changes in temperature and the periodic time lags of the occurrence of extreme values in groundwater can be useful for statements on flow velocity and water budget (Schneider 1962, Trueb 1962). Of importance here is the statement that flowing water carries heat with it and below the surface it generally effects a relatively rapid heat exchange, in places where otherwise only very long drawn out processes are working, dependent on conductivity alone. Furthermore, large amounts of heat can be moved through mass movement in the Earth's crust. The most striking example is that in volcanic regions, but besides that, water rising from the depths also brings up heat locally, and gases as well (Kappelmeyer 1961). Conversely, descending cooler water will cool the subsurface (Schneider 1964). Some heat exchange results from evaporation processes, particularly at the water table, during which heat is consumed, and with condensation, during which heat is liberated (cf. Section 1.1.3).

Below the neutral zone the water reaches different temperature zones depending on its path underground, some warmer, some cooler. Schoeller (1945, 1949, 1962) has investigated several models of water movement in connection with the shape of isotherms and has calculated the changes of groundwater temperature in a groundwater body that moves from shallow depths on flow channels into deeper strata, reascends afterward, and finally flows slowly under an assumed horizontal Earth's surface. At the beginning the shallow groundwater has the same temperature as the Earth's surface (neutral zone). In the descending groundwater stream there is a disequilibrium between rock and groundwater temperature because the heat flux is slower than groundwater flux. Therefore the rock temperature is rising faster than the groundwater temperature. When the groundwater has passed its deepest point and starts to reascend, it is still cooler than the surrounding rocks. Therefore the heating of water continues whereas the rock temperatures are diminishing with decreasing depth. At last a point is reached where rock and groundwater temperatures are equal. In the remaining part of the ascending branch the temperature diminishes quickly, but in the uppermost zone still remains above the soil temperature so that in this case the emerging water has a temperature corresponding to that of the neutral zone.

A hot spring needs the presence of very good, if possible vertical, flow channels, which allow the fast upward movement of water that has been heated at great depths, so that there is little heat loss, hence very small temperature change on the way to the surface. When calculating the depth of origin by using the geothermal gradient, a lower value is obtained always because heat loss on the way to the surface must always be considered.

Other temperature changes can be caused by further heating processes. In this category belong frictional heating during water flow through rocks (Schoeller 1962), heat of dilation of water, which—compressed under high pressure at depth—expands during ascent (Schoeller, 1962), and exothermic or endothermic chemical reactions such as oxidation or reduction, heat of solution, heat or sorption, and heat of dilution. Although generally of secondary importance, these phenomena can give rise to local thermal anomalies.

The oxidation of pyrite to iron sulfate and sulfuric acid according to equation 76 (Section 2.3) releases 1303 kJ/mole. This amount of heat is considerable; however, 2.3 kg pyrite per cubic metre per year must be transformed in this way to achieve a 1°C rise in temperature at 1.5 m depth. Notable temperature rises can be observed in the oxidation zone in the Bavarian pyrite deposits at Waldsassen (Paul 1939, Kappelmeyer 1961).

During the solution of salts heat energy is generally absorbed, and in the case of solution of rock salt this has an important effect; the solution of 353 g of salt in 1 litre of water reduces the temperature of the liquid about 3.16°C (Schoeller 1962).

Anomalies can also be of human origin: the temperature distribution in karst, very permeable Cenomanian-Turonian carbonate rocks in central Israel, which exhibit a very low vertical gradient as a result of homogenizing through vertical groundwater movements, is locally disturbed by intensive water extraction in areas where distinctly cooler water flows out of shallow water tables in the coastal plain (Schneider 1964).

Cooling effects are brought about by the ventilation air pumped through the galleries of mines and by drilling fluids in boreholes; refrigeration plants have a similar effect.

Intensive water circulation can keep groundwater temperatures unusually low. This effect is seen in the deep (ca. 450 m) freshwater wells in the city of Hamburg, where unexpectedly cool water at only 12°C is extracted.

In urban areas groundwater temperatures 2–3°C higher than that of the surroundings have been measured, a consequence of heating of the subsurface by the general accumulation of heat below the houses. The percolation of irrigation water warmed by the sun in hot climates (Eldridge 1963, Mink 1964), and the injection of heated industrial cooling water and wastewater into the subsurface, can heat the groundwater and the rocks.

The recycling of warmed cooling water after a period of cooling underground is one of the processes used in industry. This is based on heat exchange with the Earth's surface, involving loss of heat to rocks and mixing with cooler groundwater (Nöring 1954).

Spring Water Temperatures. The temperatures of perennial springs are often near the temperature of the neutral zone, which in its turn corresponds almost exactly to the average annual temperature of the locality. Some anomalies do occur, particularly at low yielding springs where water remains for quite long periods near the Earth's surface. The heat exchange processes are the same as those occurring in groundwater recharge. With slow shallow water movement the surface influences are quite marked, especially climate, altitude, attitude with respect to north, and soil cover (Fig. 47).

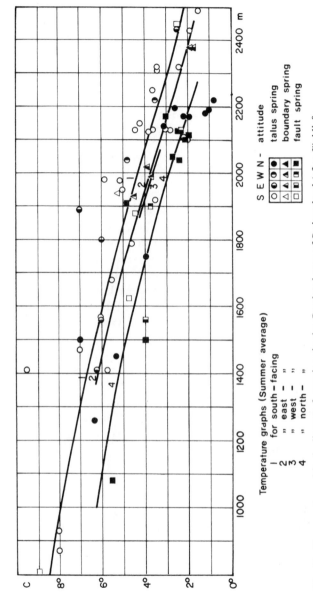

Fig. 47 Temperature-altitude diagram for springs in the Gotthard area of Switzerland (after Jäckli & Kleiber 1943).

Table 63   Variations in Spring Water Temperatures in the Jura
and in the Spring at Vaucluse (After Schoeller 1962)

	Range (°C)	Variation (°C)
Spring at Arcier	7.7–11.6	3.9
Spring at Billecut	10.0–12.9	2.9
Spring at Mouillère	10.5–15.0	4.5
Spring at Orbe	3.5–14.7	11.2
Spring at Vaucluse		
Average year	8.0–13.5	5.5
In 1903	8.0–15.0	7

In the northern hemisphere the water temperature should be significantly higher (order of magnitude about 2°C) on the south exposed slopes than on the north exposed slopes, and areas higher in open than in forest areas (order of magnitude about 1.5°C) (Schoeller 1962).

In fractured and karstic aquifers with high groundwater velocity the particular groundwater recharge periods can sometimes be discriminated with help of the temperature of the spring water. The subterranean flow distance shows up as a phase lag. A long flow period can for instance lead to the discharge of cooler water during the warmer seasons, because this water was recharged during the winter. Examples of temperature changes are given in Table 63.

An average spring water temperature lower than the average annual temperature is to be expected when all the groundwater recharge takes place during the cold period, or when the recharge area lies considerably higher than the spring, so that the lower average annual temperature there controls the groundwater temperature.

Schoeller (1962) cites as an example in this context a spring at the foot of the Djebel Ressas, Tunisia, in which the water has a temperature of 9°C instead of the expected 15.7–17°C, and a spring at Pont du Mort, in the Guil Valley in the Alps at Queryas, with a temperature of 6.8°C instead of 10°C.

Wilhelm (1956) in his research into springs in the Bavarian Alps and in the Alpine foreland was able to differentiate between two types of spring: a "radiation" type, the temperature behavior of which is essentially controlled by the heat radiation and absorption budget, and an "infiltration" type, in which the temperature is chiefly controlled by the infiltration. Figures 48 and 49 compare the temperature change with the dryness index $i = r/(T + 10)$ as a measure of the precipitation, in which i is the dryness index, r the amount of precipitation (mm), and T the air temperature. With "radiation-type" springs (Fig. 48) the snowmelt and the summer heating effect control the temperature of the spring water, whereas with the "infiltration-type" temperature minima, besides being associated with snowmelt, are also affected by summer precipitation (Fig. 49). The dominant peak value is attributed to the

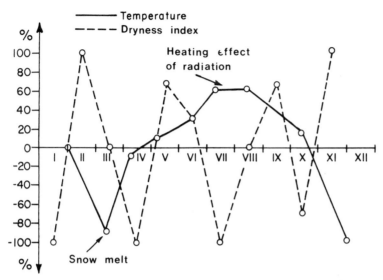

Fig. 48 Monthly distribution of the percentage frequency of maxima and minima for "radiation-type" spring (after Wilhelm 1956).

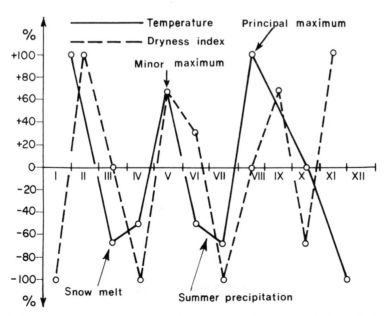

Fig. 49 Monthly distribution of the percentage frequency of maxima and minima for "infiltration-type" springs (after Wilhelm 1956).

summer insolation and lack of precipitation. Strong temperature changes in spring water can indicate a short residence time underground, hence can point to a chance of contamination in the groundwater body.

On a basis of the average annual air temperature or the soil temperature, Schoeller (1962) has classified as thermal (hyperthermal) spring waters with temperatures more than 4 and 2°C, respectively, above these parameters; as orthothermal those that lie up to 4 and 2°C, respectively, above; and as hypothermal those that have a temperature below the annual average air temperature, at least 2°C below the soil temperature. The majority of workers in this field define thermal water as groundwater with a temperature above 20°C. This definition is commonly used in Germany (Quentin 1969).

In spite of the complexity of the factors that control the temperatures of groundwater and springwater, observation of temperature may have a key function in the evaluation of a groundwater system.

Determination of Groundwater and Springwater Temperature. For temperature measurement, so far as is possible, a thermometer of accuracy to 0.1°C is needed. The thermometer can be incorporated into a sampling cup or bucket so that the mercury bulb is in a sufficiently large quantity of water to maintain the temperature until the thermometer can be read. Other types are straight stirring, maximum-minimum, and tilt thermometers. Thermometers based on electrical resistance or thermocouples are used for continuous recording of temperatures (Pfeiffer 1962, Kappelmeyer & Haenel 1974).

While measurement at springs generally presents no difficulties, during measurement in wells, particularly in those with large diameters, the temperature distribution may be disturbed by vertical flow movement of the water resulting from density differences caused by heating at depth or cooling at the surface, or from hydrodynamic potential distribution. General statements about the groundwater temperatures can sometimes be based on surface geothermal measurements (Kappelmeyer 1961).

3.2.2 Substances in Groundwater

The solids dissolved in groundwater form part of the geochemical cycle, which, starting from the stock of material in igneous rocks, proceeds via weathering, formation of sedimentary rocks, and evolution of terrestrial waters to eventual distribution in the sea. The important factor is the geochemical mobility of the elements in a given geochemical environment, in the rock, in the hydrosphere, and in the atmosphere. One important factor is their solubility, which is determined by the ionic potential, the ratio between ionic charge and ionic radius. The ionic potentials can be classified into three categories, according to Wickmann (1944):

1   0–3, cations that remain in true ionic solution up to high pH values, for

example Li^+, Na^+, K^+, Mg^{2+}, Fe^{2+}, Mn^{2+}, Ca^{2+}, Sr^{2+}, and Ba^{2+}; their hydroxides have ionic bonds and are therefore soluble.

2 3–12, elements that through their hydroxyl bonds tend to hydrolysis, for example Al^{3+}, Fe^{3+}, Si^{4+}, and Mn^{4+}.

3 Above 12, substances that form complex oxygen-bearing anions and as a rule form true ionic solutions, for example B^{3+}, C^{4+}, N^{5+}, P^{5+}, S^{6+}, and Mn^{7+}. These elements can form hydrogen bonds.

Besides solubility there is also a tendency for elements to be incorporated into new minerals, to be adsorbed on active surfaces, or to be coprecipitated; all these tendencies are important in the mobility of elements. As a measure of the mobility of an element X in water the mobility coefficient K_x proposed by Perel'man (1961) can be used. K_x is the ratio between the concentration of an element X in water and the concentration in the rock through which the solution is flowing. If the concentration of the element X in the water (M_x) is given in grams per litre and the content nx in the rock in wt %, then $K_x = (M_x \cdot 100)/a \cdot nx)$, in which a is the sum of the dissolved mineral substance in water in grams. Large values of K_x indicate intensive leaching from the rock and mobility in the solution. With the help of K_x substances can be grouped into different degrees of mobility:

	Mobility	Value of K_x	Examples
I	Very mobile substances	$n \times 10$ to $n \times 100$	Cl^-, SO_4^{2-}
II	Moderately mobile substances	n to $n \times 10$ ($n < 2$)	Ca^{2+}, Mg^{2+}, Na^+
III	Mobile substances	$0n$ to n ($n < 5$)	Si, P, K
IV	Slightly mobile and inert substances	$0.0n$ and below	Fe, Cr, Al, Pb

The differential geochemical behavior of the elements is reflected in the differences in the average composition that sedimentary rocks and seawater exhibit, compared with the initial material in igneous rocks (Table 64).

The commonest elements in igneous rocks, silicon, aluminum, and iron, have low mobility in the hydrosphere. Chlorine, which is relatively scarce in the Earth's crust, is very mobile and widespread in the hydrosphere. Calcium is mobile to some extent, and sodium is considerably more mobile than potassium, although both occur in almost equal amounts in the primary igneous material.

The mobility of an element in the hydrosphere is determined by the solubility of its various compounds, the tendency of its ions to deposit on rocks by sorption and ion exchange, and the degree to which it is bound in the biosphere.

The concentration variations of the important dissolved solids in groundwater are given in Fig. 50.

Table 64 Average Composition of Rocks (After Horn & Adams 1966 and Hem 1970) and of Seawater (after Turekian 1969) (mg/kg)

	Igneous rocks	Resistates (sandstones)	Hydrolysates (argillaceous rocks)	Precipitates (carbonate rocks)	Evaporites	Sea water
Si	285 000	359 000	260 000	33 900	386	2.9
Al	79 500	32 100	80 100	8 970	29	0.001
Fe	42 200	18 600	38 800	8 190	265	0.0034
Ca	36 200	22 400	22 500	272 000	11 100	411
Na	28 100	3 870	4 850	393	310 000	10 800
K	25 700	13 200	24 900	2 390	4 280	392
Mg	17 600	8 100	16 400	45 300	3 070	1 290
P	1 100	539	733	281		0.088
Mn	937	392	575	842	4.4	0.0004
F	715	220	560	112	24	1.3
Ba	595	193	250	30.1	173	0.021
S	410	945	1 850	4 550	26 800	904
Sr	368	28.2	290	617	234	8.1
C	320*	13 800*	15 300*	113 500*		28.5
Cl	305	15	170	305	525 000	19 400
Cr	198	120	423	7.08	10.6	0.0002
Rb	166	197	243	46.0		0.12
V	149	20.3	101	12.6	0.3	0.0019
Cu	97.4	15.4	44.7	4.44	2	0.0009
Ni	93.8	2.57	29.4	12.8	1.4	0.0066

Element						
Zn	80	16.3	130	15.6	0.6	0.005
N	46*	15	600*	5.16	30	16.17
Li	32.2	0.328	46.2	0.123	1.6	0.17
Co	23	5.87	8.06	2.69	0.9	0.00039
Ga	18.5	13.5	22.8	16.5	0.2	0.00003
Pb	15.6	3.94	80	2.01	1.2	0.00003
Th	11.4	90	13.1	16		0.0000004
B	7.5	2.15	194	0.771		4.45
Cs	4.3	0.258	6.2	0.175		0.0003
Be	3.65	1.01	2.13	2.2	0.2	0.0000006
U	2.75	0.115	4.49	0.166		0.0033
Sn	2.49	1.0	4.12	6.6		0.00081
Br	2.37	1.0	4.3	1.75	33	67.3
As	1.75	1.56	9.0	0.561		0.0026
W	1.42	0.881	1.92	0.363		0.000001
Ge	1.39	0.50	1.32	0.75	1.5	0.00006
Mo	1.25	3.75	4.25	1.59	1	0.01
I	0.45	0.0574	4.4	0.0456		0.064
Hg	0.328	0.0199	0.272	0.0476		0.00015
Cd	0.192	0.122	0.183	0.189		0.00011
Ag	0.151	0.525	0.271	0.315	0.2	0.00028
Se	0.050	0.00457	0.60	0.00179		0.00009
Au	0.00357		0.00345			0.000011

* after Hem (1970)

213

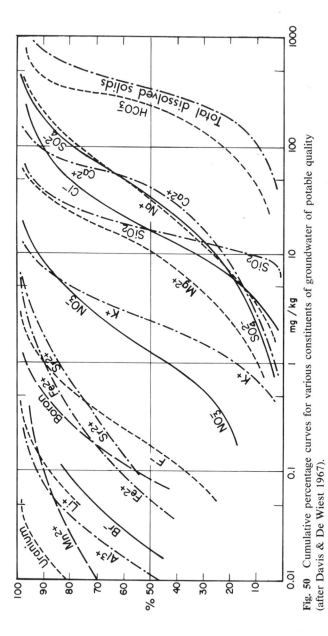

Fig. 50 Cumulative percentage curves for various constituents of groundwater of potable quality (after Davis & De Wiest 1967).

214

3.2.2.1 Dissolved Constituents: Total Dissolved Solids. The total dissolved solids comprise dissociated and undissociated substances, but not suspended material, colloids, or dissolved gases. This value is commonly determined by evaporation of a sample to dryness, and 1–2 hours drying at 180°C (formerly 110°C). This residue on evaporation is not quite the same as the solution content: the hydrogen carbonate ions are precipitated as carbonates and sulfate as gypsum; small amounts of magnesium, chloride, nitrate, and boron, as well as organic substances, can escape. The loss of CO_2 on the precipitation of carbonates is usually taken into account by adding half the hydrogen carbonate ion content to the residue on evaporation. In geochemical research this correction is frequently omitted. The result is in that case given as "measured value," "total," or "as reported."

The sum of the dissolved constituents, like the conductance, can be used for the control of chemical analyses of individual substances. In this way however only large experimental errors can be detected, because the calculated total from the individual determinations of the dissolved solid constituents can deviate from the directly determined values for contents between 100 and 500 mg/l $\pm$ 10–20 mg/l. Difficulties can arise in the comparison of results of different origin if the determinations have been carried out at different temperatures, or when it is not indicated whether they have been found by calculation or experimentally.

The total dissolved solids content can amount to less than 10 mg/l in rain and snow, less than 25 mg/l in water in humid regions with relatively insoluble rocks, and more than 300,000 mg/l in brines (Davis & De Wiest 1967, Hem 1970).

3.2.2.1.1 Major Constituents

Sodium. With a concentration of 28,100 mg/kg (Table 64), sodium is a major constituent of igneous rocks in which it chiefly occurs in plagioclase, the soda-lime feldspar with the sodium-rich end member albite, $NaAlSi_3O_8$. Of relatively less importance are sodium-containing silicates such as nepheline ($NaAlSiO_4$), sodalite ($Na_8Al_6Si_6O_{24}Cl_2$), jadeite ($NaAlSi_2O_6$), arfvedsonite [$Na_3Fe_4^{II}Al(OH)_2Si_8O_{22}$], glaucophane [$Na_2Mg_3Al_2(OH)_2Si_8O_{22}$], aegirine ($NaFeSi_2O_6$), and members of the zeolite group such as natrolite. Sodium is liberated during the weathering of these silicates. The distribution of this element and the rate of its liberation are reflected in the quality of spring water.

The high solubility of sodium salts and the limited degree of sorptive bonding of sodium onto clay minerals and other adsorbents lead to a considerably enrichment in the sea — there it is the most important cation at 10,800 mg/kg — and in evaporite deposits, in which it chiefly occurs as NaCl, rock salt.

In the resistates, particularly in sandstones, sodium has an average 3870 mg/kg concentration, either as a constituent of unweathered mineral grains, as a constituent of the cement, or as crystalline residue of pore solutions.

These contents can with prolonged or quantitatively lower groundwater movement be the cause of higher salt concentration, but are in general quickly leached out. The hydrolysates (argillites) are sodium rich at 4850 mg/kg, partly as a result of the lower permeability of this fine-grained material and partly because of the sorptive bonding of sodium to clay minerals. The salts can be liberated on the weathering of argillites; beyond that sodium can enter groundwater as a result of ion exchange with Ca^{2+}. This is indicated for example by the 857 mg/kg Na^+/l content together with very low calcium content (3 mg/l) and magnesium (7.4 mg/l) in well water from sandy and clayey beds in Richland County, Montana (Table 65, analysis 1). The numerous occurrences of $Na-HCO_3^-$ and $Na-SO_4^{2-}$ waters are likewise evidence of such ion exchange processes (cf. Section 2.2).

In precipitates, the carbonate rocks, sodium is only a very minor constituent at 393 mg/kg.

Sodium occurs as a cyclic salt or as part of the terrestrial dust in the precipitation in small amounts (of the order of 0.2 to a few mg/l), more in coastal areas (Section 3.1.1). Sodium is a bioelement and occurs in various chemicals used by man, so that it can enter into groundwater with man-made pollution (Section 2.6).

Sodium is generally present in fresh water as Na^+ ions; in concentrated solutions, however, complex ions and sodium ion pairs, such as $NaCO_3^-$, $NaHCO_3(aq)$, and $NaSO_4^-$ are present (Hem 1970). The solubility of sodium salts is high. A pure solution of sodium hydrogen carbonate ($NaHCO_3$) can contain 15,000 mg Na^+/l at room temperature. Occasionally the solubility limit of sodium salts is even exceeded, particularly in inland drainage basins, where it leads to precipitation of sodium salts, such as sodium carbonate, sodium sulfate, sodium nitrate (Langbein 1961, Toth 1966). The highest sodium concentrations occur in association with Cl^- ions. The saturation concentration in relation to NaCl occurs at approximately 150,000 mg Na^+/l; high concentrations of this order are barely reached in natural waters even in rock salt-bearing strata; the highly concentrated solutions originating on the contact surfaces are obviously removed continuously by the flow.

A highly sodium-rich groundwater, found in a borehole in rock salt-bearing strata (Salado Formation and Rustler Formation) in Eddy County, New Mexico, at 97 m depth contained 121,000 mg Na^+/l for a total dissolved solids content of 329,000 mg/l (Table 65, No. 2). Sodium is the dominant cation in most mineralized groundwaters with the exception of gypsiferous and many $Ca-HCO_3^-$ waters (Davis & De Wiest 1967). Excluding salt-bearing sediments and evaporites, the sodium content of groundwater in humid climates is mostly of the order of 1–20 mg/l.

Potassium. The average potassium content of igneous rocks is 25, 700 mg/kg (Table 64), only slightly less than that of sodium. In these rocks it occurs mostly in the potassium feldspars orthoclase and microcline ($KAlSi_3O_8$), the micas muscovite and biotite, and in other minerals such as leucite ($KAlSi_2O_6$), and – in smaller amounts – in nepheline [$Na_3K(Al_4Si_4O_{16})$]. In

spite of the greater resistance of the potassium feldspars and other potassium silicates to weathering, potassium ions are released by weathering.
However after a more or less prolonged migration they tend to become fixed
again, partly through sorption on clay minerals — for example incorporated
into the spaces between the lattice layers of illite, partly by taking part in the
formation of secondary minerals. This is reflected in its relative abundance
in clays (24,900 mg/kg) just as in its relative scarcity in seawater. The Na^+/K^+ ratio in igneous rocks is 1.09, in seawater on the other hand it is 27.84
(Rankama & Sahama 1960). The relatively high potassium content of 13,200
mg/kg in sandstones is caused by adsorbed potassium in the cementing material, and by unweathered potassium feldspar and micas (Hem 1970).

Evaporite deposits can bring about considerable potassium enrichment,
especially where there are occurrences of carnallite, $KCl \cdot MgCl_2 \cdot 6H_2O$, syl-
vite, KCl, kainite, $KMg(Cl)(SO_4) \cdot 3H_2O$, and other double and triple salts
with calcium, magnesium, and sodium. Deposits of potassium nitrate, saltpe-
ter, are of terrestrial origin.

Potassium occurs in rainwater in amounts from 0.1 to several mg/l (Section 3.1.1). This element is an important plant nutrient and thus takes part in
the biological cycle. Incorporation in vegetation draws potassium from the
hydrologic cycle; however the amounts involved can hardly be responsible
for the significantly lower potassium content in groundwater relative to sodium, because the materials incorporated into plants are once more returned
to the soil by plant decay and are there subjected to possible leaching.

Potassium is used as an agricultural fertilizer, and this can lead to significantly higher potassium concentrations in groundwater below cultivated
areas (Harth 1965). It also occurs as a biological element in oil field brines
(Krejci-Graf 1963b). High potassium values are also occasionally observed
as a result of human pollution (Section 2.6). The solubility of the potassium
salts is high (cf. Table 10). At 20°C water saturated with KCl can contain
approximately 133,000 mg K^+/l (Davis & De Wiest 1967).

Because of its lower geochemical mobility in fresh water, potassium is seldom found in greater or almost equal concentrations compared with sodium.
For one such exception, occurring in groundwater with low mineral content,
see analysis 23, Table 65. In general, however, in those occurrences of
groundwater with sodium content below 10 mg/l the potassium content only
amounts to 1–5 mg/l. Potassium ion concentrations of a few tens of milligrams per litre are unusual. Exceptions occur in basins with no outflow and
in mineral water occurrences. Knie & Gams (1960) found in the groundwaters of the Seewinkel area in Austria potassium values up to 305 mg/l,
even as much as 817 mg/l, and in the latter case there is certainly a suspicion
of pollution of the well, which is located in the village. In a mineral water in
Eddy County, New Mexico, Lang (1941) observed a potassium content of
16,600 mg/l. In general the K^+ concentration rises with increasing mineral
matter content more slowly, so that groundwater with 2000 mg/l total dissolved solids can for example contain less than 20 mg K^+/l (Davis & De
Wiest 1967).

Table 65 Chemical and Physical Data of Groundwater

Serial No.	1		2		3		4	
Sampling point	Richland County Montana, Borehole 167 m deep		Eddy County New Mexico, 97 m test borehole 3		Jumping Springs New Mexico		Midland, Michigan Borehole 1717 m deep	
Aquifer	Sandstone and shales (Fort Union Formation)		Salado and Rustler Formation		Gipsum (Castile Formation)		Sylvania Sandstone	
Date of sample	3 Oct. 1949		31 Jan. 1938		25 Nov. 1949		26 March 1952	
	mg/l	meq/l	mg/l	meq/l	mg/l	meq/l	mg/l	meq/l
Cations								
Na^+	857	37.28	121 000	5 263.50	17	0.75	28 100	1 222.35
K^+	2.4	0.06	3 700	94.61			11 700	299.17
Ca^{2+}	3.0	0.15	722	36.04	636	31.74	93 500	4 665.65
Mg^{2+}	7.4	0.61	2 490	204.83	43	3.54	12 100	995.35
Fe^{2+}	0.15	0.005					41	1.47
Mn^{2+}								
Others							Sr^{2+} 3 480	79.45
Total		38.11		5 599		35.3		7 263

Anions

	Torrey & Kohout 1956		Hem 1970		Hendrickson & Jones 1952		Scott & Barker 1962	
HCO_3^-	2 080	34.09	40	0.66	143	2.34	0.0	0.00
SO_4^{2-}	1.6	0.03	11 700	243.59	1 570	32.69	17	0.35
NO_3^-	0.2	0.00			18	0.29		
Cl^-	71	2.00	189 000	5 331.69	24	0.68	255 000	7 193.55
Others	CO_3^{2-} 57	1.90	CO_3^{2-} 63	2.10			Br^- 3 720	46.54
	F^- 2.0	0.11					I^- 48	0.38
Total		38.13		5 578		36.0		7 241
SiO_2	16.0		n.b.				130	
Others	B 0.4				29		Ra 720 pCi/l	
Residue on evaporation (calculated)	2 060		329 000		2 410		408 000	
Conductance μS at 25°C	2 960		225 000		2 510			
pH	8.3						5.29	
Temperature °C							46.1	
Literature	Torrey & Kohout 1956		Hem 1970		Hendrickson & Jones 1952		Scott & Barker 1962	

Table 65 (Continued)

Serial No.	5		6		7		8	
Sampling point	Siegler Hot Springs Lake County California		Eddy County Test borehole 1 47–65 m deep		Monticello Drew County Arkansas		Aqua de Ney, at Mt. Shasta California	
Aquifer	Serpentine and sedimentary rocks				Shales, sand and marls (Jackson Group)			
Date of sample	18 Oct. 1957		7 Febr. 1939		13 Dec. 1955		1958	
	mg/l	meq/l	mg/l	meq/l	mg/l	meq/l	mg/l	meq/l
Cations								
Na^+	184	8.0	59 000	2 566.60	416	18.10	10 900	474.10
K^+	18	0.46	2 810	71.85	11	0.28	135	3.45
Ca^{2+}	34	1.70	130	6.49	424	21.16	7.3	0.36
Mg^{2+}	242	19.90	51 500	4 236.39	194	15.96	1.6	0.21
Fe^{2+}					0.88	0.03	n.n.	0
Mn^{2+}					9.6	0.35	n.n.	0
Others			Al^{3+}		Al^{3+} 28	3.11	Al^{3+} 1.8	0.20
							Li^+ 3.2	0.46
							NH_4^+ 148	8.22
Total	30.1		6 881		58.99		487.00	

Anions

	(1)		(2)		(3)		(4)	
HCO_3^-	1 300	21.31	1 860 000	30.49	n.n.	0	n.n.	0
SO_4^{2-}	6.6	0.14	299 000	6 225.18	2 420	50.38	267	5.56
NO_3^-	0.2	0.00			3.1	0.05	n.n.	0
Cl^-	265	7.47	22 600	637.55	380	10.72	7 180	202.50
Others	F^- 1.0	0.05			CO_3^{2-} n.n.	0	CO_3^{2-} 5 560	185.30
					F^- 1.8		F^- 3.0	0.16
							Br^- 11.0	0.138
							I^- 5.7	0.045
							PO_4^{3-} 4.3	0.14
							OH^- 1 430	84.08
Total		29.0		6 893		61.24		477.92
SiO_2	175		n.b.		98	3 970	3 970	
Others					Ra 1.7 pCi/l		B 242	
					U 17 µg/l		H_2S 100	
Residue on evaporation (calculated)	1 580		436 000		3 990		31 200	
Conductance µS at 25°C	2 500				4 570		36 800	
pH	6.5				4.0		11.6	
Temperature °C	52.5						12.2	
Literature	Hem 1970		Hem 1970		Scott & Barker 1962		Feth et al. 1961	

221

Table 65 (Continued)

Serial No.	9		10		11		12	
Sampling point	Spring at the Southern foot of the active Yakeyama Volcano, Niigata Ken Honshu, Japan		Burra-Burra-Mine Ducktown Tennessee		Waukesha, Wisc. Well, 636 m deep		John de la Howe School, McCormick County, South Carolina 84 m deep Well	
Aquifer	Andesite/basalt		Copper deposit groundwater below mine		Sandstones Cambrian and Ordovician		Granite	
Date of sample	1 Febr. 1938		about 1911		2 May 1952		24 Nov. 1954	
	mg/l	meq/l	mg/l	meq/l	mg/l	meq/l	mg/l	meq/l
Cations								
Na^+	1 540	67.0	23	1.00	12	0.52	8.4	0.36
K^+	380	9.72	20	0.51	4.0	0.10	3.5	0.09
Ca^{2+}	2 010	100.3	68	3.39	60	2.99	13	0.65
Mg^{2+}	520	42.8	41	3.37	31	2.55	4.3	0.35
Fe^{2+}	970	52.1	2 178	77.99	0.37	0.01	0.18	0.006
Mn^{2+}			0.2	0.01	0.05	0.002	0.13	0.005
Others	Al^{3+} 340	37.8	Al^{3+} 433	48.65	Sr^{2+} 52	1.19	Al^{3+} 0.1	0.01
	H^+ 431	428	Cu^{2+} 312	9.82			Zn^{2+} 0.09	
			Zn^{2+} 200	6.12				
Total		738				7.36		1.47

Anions								
HCO_3^-	n.n.	0	n.n.	0	285	4.67	72	1.18
SO_4^{2-}	n.n.	0	6 600	138.65	111	2.31	6.9	0.14
NO_3^-	8 380	173.9			0.8	0.01	0.4	0.01
Cl^-	20 000	564			12	0.34	3.8	0.11
Others			F^- 0.1	0	F^- 0.5	0.03	F^- 0.2	0.01
							PO_4^{3-} 0.1	0.01
Total		738				7.36		1.45
SiO_2	640		56		8.7		35	
Others								
Residue on evaporation (calculated)	35 300		9 900		440		148	
Conductance μS at 25°C					658		150	
pH	0.4						7.0	
Temperature °C	88						18.1	
Literature	White et al. 1963		Emmons 1917		Lohr & Love 1954		White et al. 1963	

223

Table 65 (*Continued*)

	13		14		15		16	
Serial No.								
Sampling point	West Warwick, Kent County, Rhode Island, 37 m deep well							
Aquifer	Granite							
Date of sample	26 May 1955							
	mg/l	meq/l	mg/l	meq/l	mg/l	meq/l	mg/l	meq/l
Cations								
Na^+	5.9	0.26						
K^+	0.8	0.02						
Ca^{2+}	6.5	0.32						
Mg^{2+}	2.6	0.21						
Fe^{2+}	0.19	0.007						
Mn^{2+}	n.n.	0						
Others								
Total		0.82						

Anions			
HCO$_3^-$		38	0.62
SO$_4^{2-}$		0.9	0.02
NO$_3^-$		1.5	0.02
Cl$^-$		5	0.14
Others	F$^-$	0.5	0.03
Total			0.83
SiO$_2$		20	
Others			
Residue on evaporation (calculated)		82	
Conductance µS at 25°C		76	
pH		7.6	
Temperature °C			
Literature	White et al. 1963		

Table 65 (*Continued*)

	14 mg/l	14 meq/l	15 mg/l	15 meq/l	16 mg/l	16 meq/l	17 mg/l	17 meq/l
Serial No.	14		15		16		17	
Sampling point	Alsbach, Hesse Spring		Hochstädten, Hesse Spring		Jugenheim a.d.B., Hesse Spring		Waterloo, Howard County, Maryland 25 m deep well	
Aquifer	Granodiorite		Para-gneiss		Migmatite-gneiss		Gabbro	
Date of sample	1 April 1965		1 April 1965		1 April 1965		23 Dec. 1952	
Cations								
Na^+	10.9	0.47	11.6	0.50	9.8	0.43	6.2	0.27
K^+	0.8	0.02	1.6	0.04	1.3	0.03	3.2	0.08
Ca^{2+}	22.3	1.11	27.4	1.37	23.2	1.16	5.1	0.25
Mg^{2+}	4.7	0.39	5.9	0.49	5.4	0.44	2.3	0.19
Fe^{2+}							5.1	0.19
Mn^{2+}							0.19	0.007
Others								

	Diederich & Matthess 1972	Diederich & Matthess 1972	Diederich & Matthess 1972	White et al. 1963	
Total	1.99	2.40	2.06		0.99
Anions					
HCO_3^-	10.9	10.9	8.7	37	0.61
SO_4^{2-}	63.9	51.4	71.3	9.2	0.19
NO_3^-				0.3	0.01
Cl^-	12.3	32.7	12.0	1.0	0.03
Others					
Total	1.86	2.17	1.96		0.84
SiO_2				39	
Others					
Residue on evaporation (calculated)				112	
Conductance /uS at 25 °C				77	
pH				6.7	
Temperature °C					
Literature	Diederich & Matthess 1972	Diederich & Matthess 1972	Diederich & Matthess 1972	White et al. 1963	

Table 65 (*Continued*)

	18		19		20		21	
Serial No.								
Sampling point	Webster, Jackson County, North Carolina 83 m deep well		Oahu, Hawaii Well		Oberkalbach, Hesse Spring		Presberg, Hesse	
Aquifer	Peridotite		Basalt		Basalt		Quartzite	
Date of sample			6 March 1928		4 Febr. 1971		23 Sept. 1965	
	mg/l	meq/l	mg/l	meq/l	mg/l	meq/l	mg/l	meq/l
Cations								
Na^+	0.2	0.01	38	1.65	7.8	0.39	6	0.27
K^+	0	0	3.1	0.08	0.7	0.02	-	-
Ca^{2+}	2.5	0.12	17	0.85	} 30.7	} 1.53	5	0.25
Mg^{2+}	7.7	0.63	12	0.99	} 30.7	} 1.53	1	0.08
Fe^{2+}	0.08	0.003	0.08	0.003			n.d.	0
Mn^{2+}							n.d.	0

	White et al. 1963		White et al. 1963		Matthess 1971		Thews 1972	
Others								
Total		0.76		3.57		1.94		0.56
Anions								
HCO_3^-	44	0.72	84	1.38	54.5	0.89	24	0.39
SO_4^{2-}	0	0	15	0.31	29.2	0.61	2	0.04
NO_3^-			0.4	0.01			–	–
Cl^-	0.7	0.02	63	1.78	9.2	0.26	6	0.17
Others	F^- 0.3	0.02						
Total		0.76		3.48		1.76		0.60
SiO_2	16				18.3			
Others								
Residue on evaporation (calculated)			296					
Conductance /μS at 25 $^{\circ}C$								
pH	8.5							
Temperature $^{\circ}C$								
Literature	White et al. 1963		White et al. 1963		Matthess 1971		Thews 1972	

Table 65 (Continued)

Serial No.	22		23		24	
Sampling point	Forst Herzberg, Hesse S Breitenbach a.Herzberg Spring		Schlitz Forest W Rimbach, Hesse Spring		Krepbach spring Garmisch-Partenkirchen Bavaria	
Aquifer	Sandstone		Sandstone		Limestone	
Date of sample	11 Sept. 1958		11 Oct. 1958		3 Oct. 1946	
	mg/l	meq/l	mg/l	meq/l	mg/l	meq/l
Cations						
Na^+	28.0	1.22	2.4	0.10	Tr.	
K^+	6.8	0.17	8.4	0.22	Tr.	
Ca^{2+}	21.8	1.09	11.4	0.57	45	2.25
Mg^{2+}	4.5	0.44	2.0	0.16	7.8	0.64
Fe^{2+}					Tr.	
Mn^{2+}						

Constituent	Matthess & Thews 1963 (mg/l)	Matthess & Thews 1963 (meq/l)	Matthess & Thews 1963 (mg/l)	Matthess & Thews 1963 (meq/l)	Gerb 1953
Others					
Total		2.93		1.06	2.89
Anions					
HCO_3^-	6.5	0.11	6.5	0.11	171
SO_4^{2-}	120.0 [a]	2.50	28.8 [a]	0.60	Tr.
NO_3^-					Tr.
Cl^-	11.3	0.32	12.7	0.36	Tr.
Others					
Total		2.93		1.06	2.8
SiO_2					
Others					O_2 9.85
Residue on evaporation (calculated)					
Conductance /μS at 25 °C					
pH					
Temperature °C					
Literature	Matthess & Thews 1963		Matthess & Thews 1963		Gerb 1953

Table 65 (*Continued*)

	25		26		27	
Serial No.						
Sampling point	Murnau, Bavaria Well II		Amberg, Bavaria, Spring		Fürstenquelle, Eichstätt, Bavaria Well II	
Aquifer	Limestone (reduced water)		Jurassic Limestone		Jurassic, Limestone (reduced water)	
Date of sample	22 March 1949		15 Nov. 1946		11 Nov. 1950	
	mg/l	meq/l	mg/l	meq/l	mg/l	meq/l
Cations						
Na^+	Tr.		Tr.		Tr.	
K^+	Tr.		Tr.		Tr.	
Ca^{2+}	67.9	3.39	58.6	2.93	85	4.25
Mg^{2+}	19.6	1.61	27.4	2.25	17.4	1.43
Fe^{2+}	0.2	0.007	0.01	0.001	0.36	0.03
Mn^{2+}			n.d.	0	2.0	0.07

	Sample 1		Sample 2		Sample 3	
	mg/l	meq/l	mg/l	meq/l	mg/l	meq/l
Others						
Total		5.01		5.18		5.78
Anions						
HCO_3^-	238	3.89	303	4.97	298	4.89
SO_4^{2-}	48	1.0	Tr.	-	36	0.75
NO_3^-	-	-		0.178	n.d.	
Cl^-	Tr.		11		1.3	0.0357
Others						
Total		4.89		5.15		5.68
SiO_2						
Others O_2	0.13				0	
Residue on evapora-tion (calculated)						
Conductance /uS at 25 °C						
pH						
Temperature °C						
Literature	Gerb 1953		Gerb 1953		Gerb 1953	

Table 65 (*Continued*)

	Serial No. 28		Serial No. 29	
Serial No.	28		29	
Sampling point	Glen Springs at Gainesville, Alachua County, Fla.		Tuscumbia Colbert County, Alabama	
Aquifer	Miocene limestone		Dolomite	
Date of sample	16 Apr. 1946		1 Oct. 1952	
	mg/l	meq/l	mg/l	meq/l
Cations				
Na^+	3.2	0.14	2.0	0.09
K^+	0.6	0.02	0.6	0.02
Ca^{2+}	15	0.75	34	1.70
Mg^{2+}	6.7	0.55	14	1.15
Fe^{2+}	0.01	0.001	0.12	0.01
Mn^{2+}				
Others				
Total		1.46		2.97

Anions

HCO_3^-	74	1.21	160	2.62
SO_4^{2-}	2.6	0.05	3.7	0.08
NO_3^-	1.8	0.03	3.2	0.05
Cl^-	3.4	0.10	2.8	0.07
Others	F^- 0.4	0.02	F^- 0.1	0.01
Total		1.41		2.83
SiO_2	10		9.2	
Others				
Residue on evaporation (calculated)	118		230	
conductance µS at 25°C	143		259	
pH	7.0		7.5	
Temperature °C	22.2		16.7	
Literature	White et al. 1963	White et al. 1963	White et al. 1963	

[a] Calculated from total ions.

235

Potassium has a radioisotope, ^{40}K, which has an approximate distribution of 0.012% in natural potassium. Water with approximately 2.5 mg K^+/l therefore contains 3×10^{-4} mg $^{40}K/l$, corresponding to approximately 75 p Bq/l radioactivity. ^{40}K is the cause of 5–50% of the natural β-activity in water. It is taken up in the body and tends to be concentrated in the gonads. However, its low specific activity and its vanishingly small content in water greatly reduce any potential danger arising from its presence (Davis & De Wiest 1967).

Calcium. The average calcium content in igneous rocks is 36,200 mg/kg (Table 64); it is predominantly held in plagioclase feldspar, a solid solution series with end members anorthite ($CaAl_2Si_2O_8$), and albite ($NaAlSi_3O_8$); besides that there is an appreciable Ca content in the associated amphibole and pyroxene mineral groups, in the garnet group, in the epidote-zoisite series, and in wollastonite ($CaSiO_3$), apatite $[Ca_5(F,Cl)(PO_4)_3]$, and fluorite (CaF_2).

On weathering and deposition calcium is incorporated to a small extent only in resistates (22,400 mg/kg), or in hydrolysates, the clayey rocks (22,500 mg/kg). The chief concentration is found in precipitates, especially carbonate rocks (272,000 mg/kg). Calcium forms deposits of the carbonates, calcite and aragonite, $CaCO_3$, and dolomite, $CaCO_3 \cdot MgCO_3$, and the sulfates anhydrite, $CaSO_4$, and gypsum, $CaSO_4 \cdot 2H_2O$. Limestones contain besides calcite mixtures of magnesium and other impurities. Rocks with equivalent Ca/Mg ratios close to unity are classified as dolomites. Calcium carbonates and sulfates occur also as cementing material in psephitic and psammitic sediments.

Finally calcium is a constituent in some zeolites and montmorillonite and occurs as adsorbed ions on mineral surfaces in soils and rocks (Hem 1970).

Calcium ions are added to groundwater in small amounts of the order of a few milligrams per litre via atmospheric precipitation (Section 3.1.1). According to Schoeller (1962) the contents are mostly 0.8 – 10 mg/l, but locally there may be more: for example in Grignon 88.4 mg Ca^{2+}/l was observed. Calcium originates either as a cyclic salt from seawater or from the dust of calcareous rocks and industrial emissions. Calcium also plays a part in biology, incorporated in bones and scaly materials in animals. It plays an important part in the base budget soils and as a fertilizer and agent for improving the soil structure. Calcium can also appear as a component in man-made pollution (Section 2.6).

The calcium ion is relatively large (ionic radius 99 pm), and it can be hydrated. It can form complexes with some other inorganic ions. But they affect significantly the calcium concentration in natural waters only in specific cases. Thus the $CaHCO_3^+$ complex can make up approximately 10% of the dissolved calcium at HCO_3^- concentrations about 1000 mg/l, and in solutions with more than 1000 mg/l sulfate the ion pair $CaSO_4$ (aq), may comprise more than half the Ca^{2+} content. Strongly alkaline waters can contain hydroxide and carbonate ion pairs with calcium (Hem 1970).

Calcium is the most common cation in fresh water. Its content in most cases is determined by the equilibrium system $CaCO_3$-CO_2-H_2CO_3-HCO_3^--CO_3^{2-}, known as the calcium carbonate–carbon dioxide equilibrium, the principles of which were considered in Section 2.1. Pure water at 20°C can dissolve 14 mg $CaCO_3$ (calcite), approximately 5.6 mg Ca^{2+}/l, causing a pH of 9.9–10 in the solution (Garrels & Christ 1965, D'Ans-Lax 1967). In the presence of CO_2 it forms the carbonate, and the H^+ ions produced by the dissociation of carbonic acid accelerate the solution process (Fig. 13). It is of importance for the CO_2 content of groundwater that the CO_2 partial pressure in the ground air is consistently 10–100 times that of the atmosphere (Section 3.1.2).

Groundwater in silicate terrain often contains less than 100 mg Ca^{2+}/l. If calcium carbonate is present in the unsaturated zone and in the aquifer, groundwater may contain about 100 mg Ca^{2+}/l, which corresponds to the elevated CO_2 content in the ground air. In the presence of locally even higher CO_2 values, more $CaCO_3$ can be dissolved as in volcanic regions or in other endogenic CO_2 sources, where values of 200–300 mg Ca^{2+}/l are found dissolved in the presence of large amounts of hydrogen carbonate ion. With increasing pH carbonate ions increase in concentration, leading to the precipitation of $CaCO_3$ when saturation is attained. In groundwater it is to be expected that with sufficient residence time underground, the calcium carbonate–carbon dioxide equilibrium in carbonate-rich aquifers has become established (Weyl 1958); however, there are many anomalies to be found in this respect, not only oversaturation but also undersaturation (Adams & Swinnerton 1937). Thus Back (1963) has discovered in groundwater in Eocene limestones in central Florida that in spite of many years of contact with limestone, the water is not saturated with $CaCO_3$ and even at several hundred metres below the water table more limestone could have been dissolved than actually has been. (See also discussion of carbonate–carbon dioxide disequilibria, Section 2.1).

If the active H^+ ions in the $CaCO_3$ solution chiefly originate from dissociated sulfuric acid, or if the groundwater lies in contact with gypsum or anhydrite, then the Ca^{2+} content is controlled by the solubility equilibrium of the gypsum, which can be calculated by taking into account the solubility product $[Ca][SO_4]$ and the effects of the ion pair $CaSO_4$(aq) and the ionic strength. Excluding other dissolved solids the maximum concentration amounts to approximately 600 mg Ca^{2+}/l (Hem 1970). As an example of a groundwater of this type analysis 3 of Table 65 gives a water emerging from gypsiferous beds in the Castile Formation at Jumping Springs, Eddy County, New Mexico, showing 636 mg Ca^{2+}/l. In the presence of sodium chloride, which results in a higher ionic strength, the Ca^{2+} content can be a little higher.

Significantly higher Ca^{2+} contents can be observed in mineralized Ca-Cl waters. One example is the mineral water of the Wittekind spring at Bad Oeynhausen, which has the highest Ca-Cl content (76.4 eq %)in Germany, with a calcium content of 5993 mg Ca^{2+}/l (Harrassowitz 1935). The water from a 1717 m deep borehole at Midland, Michigan, showed a 93,500 mg Ca^{2+}/l content for a total dissolved solids value of 408,000 mg/l (analysis 4, Table 65). There is still some doubt about the origin of Ca-Cl water of this kind. The possible mechanisms involved include ion exchange (Valyashko & Vlasova 1965), particularly of Na^+ for Ca^{2+} ions, or a selective bonding of magnesium during reciprocal crystallization (e.g., solution as calcite and reprecipitation as dolomite), or during diagenetic processes at greater depths (at raised temperatures) in the presence of highly mineralized groundwater (> 290–350 g/l) (Pinneker 1968).

Ca^{2+} ions are removed from the groundwater by ion exchange for sodium and other ions. The reciprocal process, the displacement of Ca^{2+} by Na^+ ions, usually takes place, particularly in the boundary layer between Na-Cl water and fresh water (Section 2.2). The same process also comes about in soils irrigated by salt-bearing surface water (Hem 1970).

The reaction of soap with calcium, and also magnesium, iron, manganese, copper, barium, and zinc, has universally led to the concept of water hardness, which has however been defined in various ways. Generally this property is attributed only to calcium and magnesium dissolved in water, but it is usual to calculate it on the basis of calcium alone. Thus in three different systems 1° of hardness is defined as follows:

Germany: $1° = 10$ mg CaO/l = meq $(Ca^{2+} + Mg^{2+})$/l $\times$ 2.8
France: $1° = 10$ mg $CaCO_3$/l
Britain: $1° = 10$ mg $CaCO_3$/0.7 l $= 14.3$ mg $CaCO_3$/l

Instead of the common classification into degrees of hardness (°), Käss (1965) has proposed that the data be stated in milliequivalents per litre, and a new classification used, replacing that based on Klut-Olszewski's system (Table 66).

Table 66   Hardness Classifications Based on Klut-Olszewski and Käss

	Klut-Olszewski (1945)		Käss (1965)
	°German hardness	meq/l	meq/l
Very soft	0–4	0–1.43	0–1
Soft	4–8	1.43–2.86	1–3
Average hardness	8–12	2.86–4.28	3–5
Rather hard	12–18	4.28–6.42	
Hard	18–30	6.42–10.72	5–10
Very hard	>30	>10.72	>10

If only incomplete water analyses are available, the total hardness, which is often measured in waterworks practice, may serve as a measure of the total dissolved solids for most fresh water. As Thews (1972) has shown, the residue after evaporation increases linearly with increasing hardness (region below 26° German hardness) insofar as the predominant cations control the hardness. Highly alkaline groundwaters, for example Na-Cl and Na-SO$_4$ waters, attract attention through their Cl$^-$ and SO$_4^{2-}$ content and show by that fact that the connection between total hardness and residue after evaporation does not hold for them.

Magnesium. The average magnesium content of igneous rocks is 17,600 mg/kg (Table 64). The chief magnesium-containing minerals here are the olivine series (one of the end members is forsterite, Mg_2SiO_4), garnets, and cordierite, the pyroxene, amphibolite, and mica groups, chrysotile, sepiolite, talc, serpentine, the chlorite group and magnesium-bearing clay minerals. On weathering, magnesium is released. A part (8100 mg/kg) is retained in resistates, some (16,400 mg/kg) in hydrolysates. Important amounts of magnesium are contained in precipitates (45,300 mg/kg), the carbonates dolomite ($CaCO_3 \cdot MgCO_3$) and magnesite ($MgCO_3$), and mixtures of these in limestone. Finally magnesium occurs as a constituent of salt deposits of marine origin, bischofite ($MgCl_2 \cdot 6H_2O$), kieserite ($MgSO_4 \cdot H_2O$), hexahydrite ($MgSO_4 \cdot 6H_2O$), epsomite ($MgSO_4 \cdot 7H_2O$), and carnallite ($KMgCl_3 \cdot 6H_2O$). Magnesium salts are also present in terrestrial evaporites (Toth 1966). Magnesium phosphate is contained in guano (Rankama & Sahama 1960). The precipitates contain small amounts of the order of 0.1 to a few mg/l (Section 3.1.1). Schoeller (1962) quotes 0.4–1.6 mg/l as the commonest range of values; as a remarkably high value he mentions 4.3 mg/l from Grignon, France. Magnesium comes from the ocean (cyclic salt) or from the dust of magnesium-containing rocks and industrial emissions.

The magnesium ion has a smaller ionic radius and a greater charge density than calcium and sodium and therefore tends to form a sheath of six water molecules in octahedral coordination given by the formula $Mg(H_2O)_6^{2+}$. However, it is commonly cited without the water of hydration. The complex $MgOH^+$ is important at pH values below 10. The ion pair $MgSO_4$(aq) possesses approximately the same stability as $CaSO_4$ (aq), and the other magnesium complexes with carbonate or hydrogen carbonate show approximately the same stability as the corresponding calcium complexes. The sulfate and the hydrogen carbonate ion pairs are unimportant in solutions below 1000 mg/l sulfate or hydrogen carbonate ion (Hem 1970).

The solubility of magnesium carbonate is much harder to determine than that of calcium carbonate because of the presence of various carbonates such as magnesite ($MgCO_3$), nesquehonite ($MgCO_3 \cdot 3H_2O$), lansfordite ($MgCO_3 \cdot 5H_2O$), and basic carbonate hydromagnesite [$Mg_4(CO)_3$ $(OH)_2 \cdot 3H_2O$]. However the solubility of magnesium carbonate in pure water and in the presence of CO_2 is greater than that of calcium carbonate, so that under normal groundwater conditions $MgCO_3$ is not precipitated (Hem 1970). The dolomite and hydroxide in sediments are poorly soluble (Table 11). The solubility of magnesium chloride and sulfate is distinctly higher (Table 10).

In spite of the higher solubility of most of its compounds, the magnesium content in fresh water is generally below that of calcium, most probably because of the lower geochemical abundance of magnesium. Occasional exceptions occur in magnesium-rich aquifers, such as olivine-basalts, serpentines, and dolomite rocks; however the absolute Mg^{2+} contents in these cases are also low. Minerals with exchange capability adsorb magnesium only slightly more firmly than calcium so that low magnesium contents can occasionally be attributed to cation exchange (Davis & De Wiest 1967).

Fresh water generally shows values below 40 mg Mg^{2+}/l. Higher concentrations occur in magnesium-rich rocks, mostly in those with a generally higher salt content, for example 242 mg Mg^{2+}/l in a thermal water from serpentine, with 1580 mg/l total dissolved solids (Siegler Hot Springs, Lake County, California; Table 65, analysis 5), and 51,500 mg Mg^{2+}/l in a well in Eddy County, New Mexico, with 436,000 mg/l total dissolved solids (Table 65, analysis 6).

Iron. Iron is one of the most important elements in igneous rocks, with an average concentration 42,200 mg/kg (Table 64). It occurs especially in the dark-colored minerals, pyroxenes, amphiboles, biotite, magnetite (Fe_3O_4), and pyrite (FeS_2), in garnets (andradite, $Ca_3Fe_2Si_3O_{12}$ and almandite, $Fe_3Al_2Si_3O_{12}$), and olivine, a solid solution series of the end members fayalite (Fe_2SiO_4) and forsterite (Mg_2SiO_4).

The bulk of the iron freed by weathering is transformed into only slightly soluble stable Fe(III) oxide and oxyhydroxides, which are generally deposited again after erosion. The bulk of it (38,000 mg/kg) is found in hydrolysates, and (18,600 mg/kg) in resistates; the iron content in precipitates is lower (8190 mg/kg). The iron content in seawater (0.0034 mg/kg) reflects the element's relatively low mobility. In sediments it occurs as divalent iron in the polysulfide FeS_2 and in siderite ($FeCO_3$), in the trivalent state mixed with Fe(II) in magnetite and glauconite, and finally as trivalent iron in oxides and hydroxide, the most hydrated of which is ferric hydroxide [$Fe(OH)_3$], which forms as a gel and in the course of time changes into the oxihydroxide form (FeOOH), or the anhydrous oxide hematite (Fe_2O_3).

Iron plays an important biochemical role in the life cycles of plants and animals. Its occurrence in groundwater is influenced by microorganisms, which catalytically help either the oxidation to ferric iron (under aerobic conditions) or the reduction to divalent iron (under anaerobic conditions) (Section 2.5). By its indication of reducing conditions, an increase in the divalent iron content can be used as a criterion for the presence of pollution by organic substances (Section 2.6). The commonest type of iron dissolved in groundwater is divalent iron Fe^{2+}. The $FeOH^+$ complex can occur in very low CO_2 content waters; likewise an ion pair $FeSO_4(aq)$ can occur in waters with more than several hundred milligrams per litre sulfate content. Divalent iron builds numerous but not very well understood complexes with organic substances. Above the pH value of 11, seldom found in natural waters, the anion $HFeO_2^-$ can occur in considerable amounts (Hem 1970).

Trivalent iron can be dissolved in acid solutions as Fe^{3+}, $FeOH^{2+}$, and $Fe(OH)_2^+$, and in a polymerized state, the dominant species depending on pH. Above pH 4.8 the solubility of the trivalent iron compounds is less than 0.01 mg Fe/l. The polymerized forms occur chiefly in concentrated iron(III) solutions, for example $Fe(OH)_3(aq)$ has a solubility of less than 1 μg Fe/l (Hem 1970).

Trivalent iron can form complexes with chloride, fluoride, sulfate, and phosphate ions, and with organic substances; in groundwaters however, only the complexes with organic substances seem to be of practical importance. The organic complexes of trivalent iron are stronger than those of divalent iron. The exact complex-forming process is only partly known; besides a reduction of trivalent iron by organic agents there is a peptizing effect on ferric hydroxide suspensions attributed to it (Hem 1970).

The liberation of iron by weathering and its content in natural waters result from chemical equilibria, comprising oxidation and reduction, solution and precipitation of hydroxides, carbonates, and sulfides, formation of organic complexes, and biological metabolic processes (cf. Section 2.5). The combined effects of pH, *Eh,* and dissolved CO_2 and sulfur species can be best made clear by stability field diagrams (Hem & Cropper 1959, Hem 1960a, 1960b), the use of which is represented by an example for iron in Section 1.2 (Figs. 17–19).

The solubility diagram (Fig. 51) shows that small *Eh* or pH changes can have a powerful effect on the solubility. Within the usual pH field in groundwater (pH 5–9) relatively large amounts of divalent iron are soluble with a relatively low *Eh* (i.e., less than 0.20 mV and above −0.10 mV). Finally the iron solubility can be controlled in a reducing environment by the precipitation of iron carbonate, in which case the iron solubility depends on pH and the concentration of the dissolved CO_2 species. With addition of atmospheric oxygen however, the *Eh* rises and iron oxidizes and precipitates as ferric hydroxide (Hem 1970).

Percolation water usually brings with it oxygen from the atmosphere and ground air so that in addition to other oxidized substances iron sulfide is oxidized and Fe^{2+} and SO_4^{2-} ions go into solution. Iron sulfate water also occurs with leaching of alum shales and alum schists.

High iron content can thus be associated with the oxidation of reduced iron mineral such as iron sulfide in the boundary zone between reducing and oxidizing environments. The oxygen content of groundwater is completely used up sooner or later by oxygen-consuming reactions underground.

In the anaerobic environment the presence of organic matter can lead to reducing conditions, in which the oxidized iron compounds are reduced, so that the groundwater contains much divalent iron.

In most oxygenated groundwaters iron is not present or is detectable only in minor amounts. In "reduced" groundwaters dissolved iron often is measured between 1 and 10 mg Fe^{2+}/l. so that the content can be widely variable in place and time. Occasionally very high iron contents are observed: in oil field brines above 1000 mg/l (Hem 1970), in water heavily polluted by human activities up to 700 mg/l (Golwer et al. 1970). When taking samples it is necessary to remember that on contact with atmospheric oxygen $Fe(OH)_3$ is precipitated.

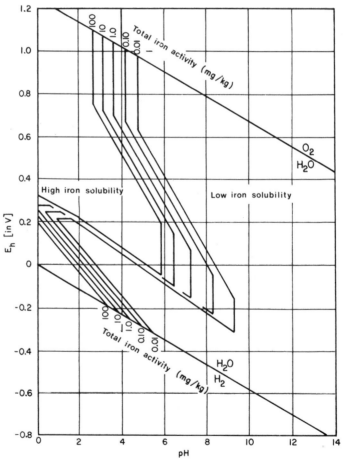

Fig. 51 Solubility of iron as a function of pH and Eh in an equilibrium system consisting of carbonate species with activities of 100 mg/kg (as hydrogen carbonate) and sulfur species with activities of 10 mg/l (as sulfate) (after Hem 1961a).

Manganese.   Manganese is a relatively minor constituent of igneous rocks (937 mg/kg – Table 64). Its relative rarity in seawater (0.0004 mg/kg) and its occurrence in resistates, hydrolysates, and precipitates reflect its low geochemical mobility. Nevertheless, manganese can cause considerable difficulties as a troublesome groundwater constituent.

In igneous rocks divalent manganese is chiefly incorporated in the mineral lattice, replacing other divalent ions of similar size, for example substitutions in ferromagnesian minerals such as biotite and hornblende. Manganese(II), (III), and -(IV) are widely distributed in sediments and soils as oxides and hydroxides and the carbonate rhodochrosite ($MnCO_3$).

Manganese participates in plant metabolism and can be considerably accumulated in plant material. After leaf fall or on the death of plants, manganese is liberated again. Microorganisms play a catalytic role in the oxidation and reduction of manganese (Section 2.5).

At about pH 7 Mn^{2+} is the dominant manganese ion. In waters containing HCO_3^- the manganese complex $MnHCO_3^+$ occurs, which can, at HCO_3^- concentrations about 1000 mg/l, account for half the manganese present. The ion pair $MnSO_4^0$ can be of importance in waters with more than 1000 mg SO_4^{2-}/l (Hem 1970).

Manganese equilibrium is not established quickly compared with iron. Nevertheless, the solubility of manganese can be shown on *Eh*-pH diagrams (Fig. 52). For the defined conditions, for pH values near 7 in oxygenated waters, a manganese content between 1 and 10 mg/l may be expected. At higher pH values the aqueous solution lies in the stability field of manganese oxides and manganese carbonate in which the pH and *Eh* controls are complex.

The presence of sulfur has not been considered because manganese sulfide minerals are scarce and have no large influence on the occurrence of manganese in natural systems. Figure 52 also shows that manganese at concentrations of the order of 1 mg/l will be precipitated only on the addition of oxygen at higher pH values. Oxidation and precipitation are accelerated with increased pH and by surface catalytic effects on manganese oxide and feldspar surfaces, and – as already mentioned in Section 2.5 – by manganese-absorbing bacteria (Hem 1970, Schweisfurth 1968).

Manganese is detectable in minor amounts in most groundwaters. In addition, under reducing conditions manganese concentrations above 1 mg/l are relatively rare, but values down to 0.05 mg/l will have an adverse affect on the potability of water. Higher concentrations can be observed in thermal waters, for example 42 mg/l at Arima, Hyogo District, Japan, and in highly mineralized waters, for example 30 mg/l in an oil field brine from a 3160 m deep borehole in West Bay, Plaquemines Parish, Louisiana, and 20 mg/l in the 1092 m deep borehole at Gevelsberg in the Ruhr area. Very high concentrations occur in acid mine waters, for example 841 mg/l in water in the center galleries of the Comstock Lode, Storey County, Nevada (White et al. 1963). Generally the manganese content of groundwater is lower than the

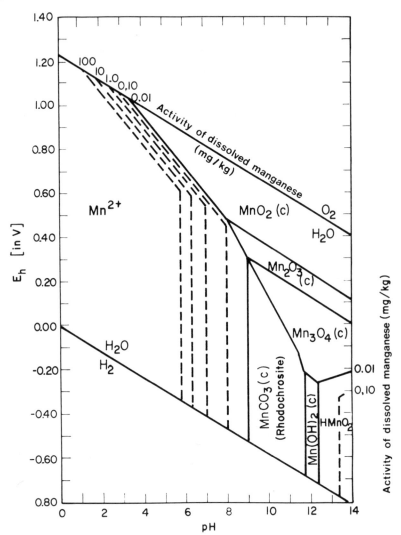

Fig. 52 Stability field diagram of manganese at standard temperature and pressure. The activity of the dissolved CO_2 species is 100 mg/1 (as hydrogen carbonate). Sulfur species absent; (c) = crystalline form (after Hem 1970).

iron content, possibly because of its more limited geochemical abundance (cf. Table 64). Occasionally manganese is the predominant ion, for example in bank-filtered waters on the Ohio River at Parkersburg, West Virginia, in which 1.3 mg Mn^{2+}/l has been recorded, compared with 0.04 mg Fe^{2+}/l (Scott & Barker 1962), and in a well at Monticello, Drew County, Arkansas, which gave 9.6 mg Mn^{2+}/l compared with 0.88 mg Fe^{2+}/l (Table 65, analysis 7). More often the common occurrence of troublesome iron and manganese concentrations is a phenomenon that is also observed in waters with low mineral content (e.g., the Breslau incident mentioned in Section 2.3).

Carbon Dioxide, Carbonate, Hydrogen Carbonate, Alkalinity, and Carbonate Hardness.   The geochemical importance of carbon, which with an average concentration of only 320 mg/kg is a relatively unimportant constituent of igneous rocks (Table 64), is due to its central role in all organic substances. In the lithosphere it is concentrated in caustobiolites and carbonate rocks (113,500 mg/kg), which have an origin in biological processes. Carbon can occur in considerable quantities in resistates (13,800 mg/kg) and hydrolysates (15,300 mg/kg).

The system $CO_2 + H_2O \rightleftharpoons H_2CO_3 \rightleftharpoons HCO_3^-.+ H^+ \rightleftharpoons CO_3^{2-} + H^+$ is very important in water, controlling the formation of various carbonate species that originate from carbonate in rocks and free carbon dioxide. The total CO_2 content in water is usually subdivided into free and bound CO_2. Free CO_2 comprises dissolved carbon dioxide and the undissociated H_2CO_3; bound CO_2 includes the hydrogen carbonate and carbonate ions.

Free carbon dioxide in groundwater has various origins. It may originate from atmospheric CO_2 that becomes dissolved in rainwater, or from ground air dissolving in percolating water, or directly from groundwater. Furthermore during transformation from peat via lignite, coal, to anthracite (carbonization process), large amounts of CO_2 are liberated (Krejci-Graf 1934). Free CO_2 is produced by anaerobic or aerobic oxidation of organic substances, fossilized organic deposits, by man-made organic pollution, and by the effect of organic acids (humic acids, naphthenic acids, Smith & Sutton Bowman 1930) or inorganic acids (sulfuric and nitric acids) on carbonates. Carbon dioxide is a component of endogenic emissions. Part of it is of primary magmatic origin, part of it is set free from carbonate rocks by thermal metamorphism in contact with magma or by hydrochloric acid in volcanic gases. According to Muffler & White (1968) carbon dioxide is released at depths below 300 m at temperatures between 150 and 200°C by conversion of dolomite and kaolinite into chlorite and calcite, and at depths below 900 m at temperatures between 300 and 320°C by the breakdown of calcite during the formation of epidote.

Finally CO_2 can also be liberated due to the mixing of different "hard" groundwaters (Fig. 14), and the subsequent precipitation of calcium carbonate during pH and pressure-temperature changes in the groundwater.

Schwille (1953a, 1955) has shown that with the incorporation of Ca^{2+} ions during exchange processes the resulting free CO_2 can act as aggressive CO_2 toward carbonates.

The definition of alkalinity (basicity) as a measure of the capability of a water to neutralize acids is based in most groundwaters on the quantity of carbonate and hydrogen carbonate ions. Alkalinity is defined in Germany by titration with hydrochloric acid to the methyl orange end point (pH 4.3) (*m*-value, total alkalinity), and to the phenolphthalein end point (pH 8.2) (*p*-

value, phenolphthalein alkalinity). In analytical practice alkalinity is usually determined by potentiometric titration. Any ion that reacts with a strong acid can contribute to alkalinity when the reaction takes place well above the titration point. Alkalinity mostly originates from anions or molecular types of weak, incompletely dissociated acids. The commonest weak acid is carbonic acid, which forms a buffer system with hydrogen carbonate ions (cf. Section 1.2.3.2.1). An approximation that is useful for most practical evaluations is given on Table 67.

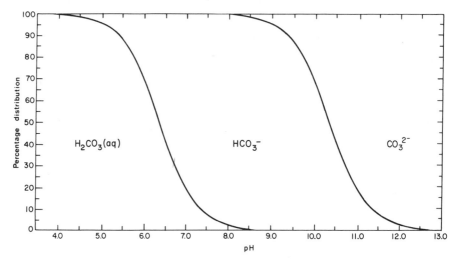

Fig. 53   Percentage of total dissolved CO_2 species in solution as a function of pH, at standard temperature and pressure (after Hem 1970).

Under equilibrium conditions the pH value of a water indicates the fractions of the different carbonate species. In Fig. 53 the percentages of the activities at STP of the undissociated carbonic acid and the hydrogen carbonate and carbonate species are plotted against pH. The diagram, which strictly applies to ideal solutions only, indicates that over the usual pH range of groundwater (6–8) CO_2 occurs chiefly as hydrogen carbonate ion and undissociated carbonic acid, while in the acid range (< pH 6) undissociated carbonate species and in the alkaline range (> pH 10) carbonate ions are dominant.

Sometimes it is wrong to attribute alkalinity to dissolved carbonate species exclusively; for example appreciable amounts of other weak acids such as silicic, phosphoric, and boric acids, and hydroxides [e.g., $Al(OH)_4^-$], organic acids, and other uncommon substances also occur dissolved or suspended in groundwater. An illustrative example is a spring in Custer County, Idaho, in which half the reported silica (75 mg SiO_2/l) at the prevailing pH of 9.4 is present as ionic silicate (Hem 1970). The effect of

Table 67   Approximate Hydroxide, Carbonate, and Hydrogen Carbonate Contents Calculated from p- and m-Values

	Sample contains		
Result of titration	Hydroxide (meq)	Carbonate (meq)	Hydrogen carbonate (meq)
$p = 0$	0	0	m
$p < \frac{1}{2} m$	0	$2p$	$m - 2p$
$p = \frac{1}{2} m$	0	$2p$	0
$p > \frac{1}{2} m$	$2p - m$	$2(m - p)$	0
$p = m$	m	0	0

1 meq/l corresponds to	17 mg OH^-/l
	30 mg CO_3^{2-}/l
	61 mg HCO_3^-/l

In the example calculation	p-value $= 0.4$; m-value $= 1.6$
Carbonate content	$2p = 0.80$
	$0.8 \times 30 = 24$ mg CO_3^{2-}/l
Hydrogen carbonate content	$(m - 2p)$
	$1.6 - 0.8 = 0.8$
	$0.8 \times 61 = 49$ mg HCO_3^-/l

other weak acids is however small in most groundwaters, so that generally the alkalinity can serve as a measure of the carbonate species present. This assumption does not hold for oilfield and landfill waters, in which organic acids may be very important in contributing to alkalinity (Golwer et al. 1976, Baedecker & Back 1980).

The usual range of free dissolved CO_2 in groundwaters is 10–20 mg/l, but more CO_2 can occur, and there is a transitional range that includes the high concentrations found in groundwaters used for medicinal purposes. The concentration of hydrogen carbonate ion in rainwater is below 10 mg/l, often below 1 mg/l. In groundwater concentrations between 50 and 400 mg HCO_3^-/l often occur. Concentrations above 1000 mg/l can occur in water of low alkaline earth content, especially at high CO_2 concentrations through endogenetic or diagenetic processes (cf. Foster 1950). When there is oversaturation with respect to the partial pressure of atmospheric CO_2, precipitation of limestone will occur at the contact with the atmosphere. pH values below 4.5 (at which the majority of hydrogen carbonate has been transformed into undissociated carbonic acid), and above 8.2 (at which carbonate ions are present) seldom occur in groundwater.

Natural water contains the stable isotopes ^{12}C and ^{13}C, also the radionuclide ^{14}C, which because of its half-life of 5730 years can under favorable conditions be used to determine groundwater ages up to 50,000 years (Münnich 1968). Radiocarbon is a β^--emitter. ^{14}C is produced in the atmosphere by cosmic radiation at a rate of about 2 atoms $\cdot$ cm^{-2} $\cdot$ s^{-1}. A global equilibrium quantity of about 4×10^6 mole ^{14}C is established by production and radioactive decay, of which the major proportion ($>$ 95%) is located in seawater (Haxel & Schumann 1962, Münnich 1963). Atomic weapon tests since 1950 have increased the total amount by about 4% (Roether 1970). The ^{14}C content of the troposphere can be assumed to have been constant in time and space up to 1950.

Groundwater dating is based on the theoretical decrease in radioisotope concentration from an initial value C_0 to a concentration C at the time of observation (p. 10). As a rule the ^{14}C content of recently recharged groundwater can be taken to be between 50 and 100% of that in the atmosphere (Münnich 1968). The CO_2 content of the ground air, which results from root respiration, microbial breakdown of organic materials, and gas exchange involved with the atmosphere, shows the same ^{14}C content as the atmosphere. By the interaction of this CO_2 with the generally ^{14}C-free carbonate below the ground, hydrogen carbonate ions originate in the water with a ^{14}C content of 50% of the atmospheric content. The generally higher ^{14}C content actually found is explained by the associated free carbon dioxide, an isotope exchange with surplus carbon dioxide (Münnich 1968), also possibly by a ^{14}C content worth mentioning in the dissolved limestone (Geyh 1970). Commonly the ^{14}C content of recently recharged groundwater is taken to be 85% of the atmospheric concentration, which in certain cases however introduces a possible age error of $\pm$1500 years (maximum) (Roether 1970). No new ^{14}C is added in the groundwater itself. The use of this method is made difficult by the geochemical state: that is, the carbonate-dissolving CO_2 can originate from the diagenetic and endogenetic processes described above, and in these cases it generally contains no ^{14}C. When carbonates are dissolved by CO_2 of that kind the ^{14}C content of the groundwater is therefore decreased, as noted at the beginning of Section 2.6. In simple cases an examination and consideration of those processes by determination of the total dissolved carbon inventory and of the ^{13}C content is possible (Pearson & Henshaw 1970, Reardon & Fritz 1978). Another cause of error can arise if isotope exchange with carbonates occurs in the aquifer (Thilo & Münnich 1970).

In favorable circumstances, particularly in carbonate-free aquifers and in the absence of other sources of CO_2, groundwater flow velocities, recharge rates, and mixing between multiple groundwater bodies can be determined (Münnich 1968). For the determination of flow rate — in the example in Fig. 54, 0.66 m/a — only the age differences are involved, not actual ages. In many ^{14}C determinations of groundwater a uniform flow rate in the aquifer is assumed. In very many cases however, a nonhomogeneous sample is measured; that is, the sample is derived from groundwaters of different ages, and the ^{14}C age is younger than that corresponding to the true average age of the groundwater. When for example half the water is approximately free of ^{14}C content, therefore at least 50,000 years old, and half of it is recent water, the true average age is in excess of 25,000 years. The ^{14}C content of the

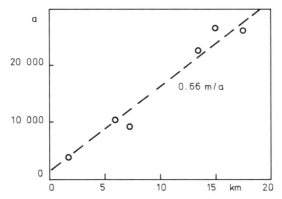

Fig. 54 ^{14}C ages in artesian groundwater adjacent to the South African coast as a function of distance from recharge area (after Vogel 1970).

sample tested however amounts to 50% of the normal initial ^{14}C content, which corresponds to an age of about 6000 years. Dispersion effects within the groundwater body must therefore be considered in each case (Münnich & Roether 1963, Vogel 1970, Lal et al. 1970).

Nitrate, Nitrite, Ammonia, and Other Nitrogenous Compounds. Most nitrogen occurs as a gas in the atmosphere, but there is an important fraction in the soil, and bound in organic substances. In the lithosphere nitrogen is most often found in hydrolysates, at 600 mg/kg (Table 64).

Nitrogen compounds are added with rainwater (Section 3.1.1). The ammonia content in precipitation is generally between 0.01 and 1 mg NH_4^+/l, most often in the range 0.1–0.2 mg NH_4^+/l (Junge 1963). Occasionally values up to 28 mg/l are observed (Schoeller 1962). The nitrate content of precipitation varies mostly between 0.3 and 2.5 mg/l (Junge 1963); however values up to 12 mg/l do occur (Schoeller 1962). Over the ocean and in coastal areas typical values are between 0.15 and 0.5 mg/l (Junge 1963). Nitrite is mostly present as traces in the precipitation. Schoeller (1962) gives the average as 2.5×10^{-4} mg/l; Rankama & Sahama (1960) quote values up to 1.35 mg/l. The NO_2^-/NO_3^- ratio is usually 0.01–0.1 (Junge 1963).

Nitrogen compounds are transformed by biological processes. In the presence of oxygen, organically bound nitrogen is oxidized via amino acids and ammonia to nitrite and finally to nitrate. Many organisms reduce nitrate to elemental nitrogen or ammonia (Section 2.5).

Plants extract nitrogen as the nitrate from the soil. Root nodule bacteria (Stainer et al. 1976) and other bacteria as well as algae can raise the nitrate supply by oxidation of atmospheric nitrogen. Any excess is available for leaching, as is the nitrate released by leaf fall or death of the plant.

Soluble nitrogen compounds can be concentrated in closed basins in semi-arid regions. The nitrate deposits in northern Chile were formed in this manner. Fertilizer application, animal husbandry, and liquid and solid wastes can locally increase the supply of soluble nitrogen compounds (Section 2.6). Various oxidation states of nitrogen occur: N^{3-} in ammonia, aqueous forms of which NH_4^+ and $NH_4OIH(aq)$; organically bound nitrogen in proteins and amino acids and urea (NH_2CONH_2) containing the amino group NH_2; N^- in hydroxylamine (NH_2OH). On oxidation, elemental nitrogen gas (N_2) forms, also trivalent nitrogen in nitrite NO_2^- and pentavalent nitrogen in nitrate NO_3^-.

Besides these species complex inorganic ions form during man-made pollution, such as cyanide ion CN^- (Section 2.6), which forms strong complexes with many metal cations (Hem 1970).

The concentrations of free nitrogen correspond approximately to the equilibrium concentrations of nitrogen content of the atmosphere and the ground air. High concentrations of nitrogen gas (up to 98 vol % N_2) are often found in highly mineralized, alkaline, silica-rich thermal waters in regions of recent tectonic and seismic activity (Barabanov & Disler 1968).

Ammonia is in general observed in groundwater only under reducing conditions, and mostly in small amounts. Its concentration is chiefly controlled under anaerobic conditions by sorption and under aerobic conditions by oxidation (Preul & Schroepfer 1968). In oil field brines ammonia concentrations above 100 mg/l are often observed, for example in Illinois up to 309 mg/l (White 1957b). The average concentrations are 40 mg/l in Illinois and 90 mg/l in California according to White, and according to Schoeller (1956) 10.6 mg/l in Hungary, 122 mg/l in Polasna-Krasnokamsk, USSR, and 155 mg/l in Romania. In springs affected by volcanic activity considerable concentrations of ammonia occur, which White (1957b) attributes to sedimentary organic origin. Hem (1959) mentions as an example Devil's Inkpot, Yellowstone National Park, at 769 mg NH_4^+/l. With man-made pollution very high values (> 1000 mg/l) have been observed (Golwer et al. 1970).

Nitrate concentrations mostly lie below 20 mg NO_3^-/l in 80% of cases in the groundwaters used in West Germany. Higher concentrations are found in areas under intensive cultivation, and values up to 700 mg/l have been reported (Schwille 1953b, 1962, 1969, Schneider 1964, Sturm & Bibo 1965, Ulbrich 1957, Young 1980). In dry regions high nitrate concentrations are quite common, for example values up to 1000 mg/l in the Seewinkel between Neusiedlersee and the Hungarian border (Schwille 1962). George & Hastings (1951) found in numerous cases in Texas groundwaters with abnormally high nitrate content (three water samples contained 1440, 1610, and 1950 mg/l), which only in some cases could have been caused by fertilizers or man-made pollution. Rather, the causes are aerobic microbial nitrate formation and—as Kreitler & Jones (1975) have proved by considering the $\delta^{15}N$ range of natural soil nitrate—the oxidation of organic nitrogen. High nitrate concentrations can however also arise by leaching of nitrate-bearing deposits in caves, caliche, and playas; but for their primary formation and preservation, some protection from leaching through the position in the cave or by a predominantly dry climate is necessary. After loss of this protection (say through falling in of the roof of the cavern or through flooding by groundwater), the nitrate deposits can be dissolved (Feth 1966).

Nitrite is found only seldom, and then in small amounts, in groundwater. In 1450 water analyses from waterworks in West Germany, of which 81% were supplied from groundwater including bank-filtered river water and artificial recharge, nitrate was reported in only seven cases, and five of these came between trace amounts and 1.5 mg/l (Giebler 1960). Nitrite is occasionally observed in trace amounts (up to 0.25 mg/l) in oil field brines (Krejci-Graf 1963a).

Sulfate. Sulfur, with an average concentration of 410 mg/kg (Table 64), is only a relatively minor constituent of igneous rocks. The presence of sulfur in resistates with an average of 945 mg/kg, in hydrolysates at 1850 mg/kg, in precipitates at 4550 mg/kg, and in seawater at 904 mg/kg reflects its geochemical mobility. Geochemical considerations lead to the conclusion that sulfur, which at present is distributed fairly uniformly in the sea, in evaporites, and in sedimentary rocks, originally came mainly from magmatic gases (Ricke 1961). Heavy metal sulfides and polysulfides occur especially in igneous rocks, locally concentrated in occurrences of sulfides, and in sedimentary rocks formed under reducing conditions. The oxidized form of sulfur is abundant as sulfate. Sulfate is contained in some feldspathoid minerals [nosean, $Na_8Al_6Si_6O_{24} \cdot SO_4$; hauyne, $(Na,Ca)_{4-8}Al_6Si_6O_{24} \cdot (SO_4,S)_{1-2}$]. For the most part however it occurs in gypsum ($CaSO_4 \cdot 2H_2O$) and anhydrite ($CaSO_4$) in evaporites, particularly in potash salt deposits, or as a cementing material, or enclosed within layers or cracks in sedimentary rocks. Calcium sulfate, and the sulfates of sodium, potassium, and magnesium, at times as double salts, occur in marine and terrestrial evaporite deposits.

Hydrogen sulfide, which originates from magmatic gases, or is formed by biochemical reduction of sulfates, occurs as a gas or in dissolved form as HS^- or S^{2-} ions or as undissociated H_2S. Under reducing conditions considerable quantities can occur. For example in highly mineralized oil field brines in the Permian Basin in western Texas the H_2S content is of the order of 500 mg/kg and in the Permian Kalinova Series (USSR) values of the order of 1000 mg/kg are known (Davis 1967). Finally, elemental sulfur can occur in nature; it originates by oxidation of sulfides and hydrogen sulfide and by reduction of sulfates.

Sulfur participates in biochemical reactions and is contained in organic compounds in plants and animals. Oxidation and reduction are frequently bound up with biological processes. Sulfate reduction in particular involves the assistance of microorganisms (Section 2.5).

Sulfur compounds are introduced into the sulfur cycle by man's activities, particularly SO_2 in flue gases and sulfate in fertilizers (Section 2.6). Atmospheric precipitation contains sulfate ions as cyclic salts, as terrestrial dust, and as man-made pollution; in rural areas less than 15 mg SO_4^{2-}/l, in nonindustrial urban areas between 15 and 30 mg/l, and in industrial areas 30–450 mg/l and above (Riffenburg 1925, Collins & Williams 1933, Drischel 1940, Junge 1954, Matthess 1961) (Section 3.1.1.).

Considerable amounts of sulfate are added to the hydrologic cycle with the precipitation from the atmosphere. This content comes from dried sea spray as a cyclic salt, from the dust of continents, and from the oxidation of H_2S that enters the atmosphere from coastal marshes, from volcanic emanations, and from man-made air pollution.

Further additions of sulfate to groundwater arise from the breakdown of organic substances in the soil (Section 3.1.2), from the addition of leachable sulfates in fertilizers, and from other human influences (Section 2.6).

Figure 55 shows the distribution in water of thermodynamically stable sulfur species SO_4^{2-}, HSO_4^-, elemental sulfur, HS^-, $H_2S(aq)$, and S^{2-}. The diagram is based on an assumed sulfur activity of 10^{-3} M, corresponding to 96 mg SO_4^{2-}/l. With increasing sulfur activity in the solution the only change would be an increase in the stability field for elemental sulfur, which would be correspondingly smaller at lower activities (Hem 1970).

Figure 55 gives only the general boundary conditions for the transformations of sulfur species. Because the individual reactions run to completion slowly, species that do not conform to the equilibrium can occur in groundwater over prolonged periods. According to Gundlach (1965) this can also explain the occurrences of short-lived anions such as thiosulfate (approximately 1–3 mg/l), pentathionate (1–4 mg/l), and tetrathionate (approximately 1–3 mg/l) in Greek sulfur springs, because the measured values exceed the equilibrium concentrations for the species that are stable on the Earth's surface, HS^-, S^{2-}, and SO_4^{2-} – in the case of $S_2O_3^{2-}$ by at least a factor of 10, and for $S_4O_6^{2-}$ by several orders of magnitude. The observed concentrations that are elsewhere exceeded in wells in volcanic regions in New Zealand (with concentrations > 50 mg/l) and in the United States (concentrations up to 10 mg/l) obviously indicate the equilibrium conditions of the temperature and pressure constraints in the environment of formation of these compounds. In the case described their origin is attributed to the hydrolysis of elemental sulfur.

The presence of the metastable sulfur species (polysulfide ions, colloidal sulfur, and thiosulfate) in a sulfide-rich water from Enghien-les-Bains, France, is reported by Boulegue (1977). Electrochemical and analytical data show that the water is in equilibrium with amorphous FeS formation. The relative concentration of the sulfur species is mediated by bacterial processes (*Desulfovibrio* and Thiobacteriaceae).

Sulfuric acid is not completely dissociated at low pH values; as a result HSO_4^- ions are predominant at pH below 1.9 (Fig. 55). The sulfate ion is a complex with a strong tendency to form further complexes. The most important complexes in natural waters are the ion pairs $NaSO_4^-$ and $CaSO_4^0$, which increase with increasing sulfate concentration. The strongest sulfate ion pairs form with di- and trivalent cations. At 100–1000 mg SO_4^{2-}/l there is an appreciable amount of $CaSO_4^0$ present, which affects the solubility of gypsum (Hem 1970). The solubility products of $SrSO_4$ and $BaSO_4$ (Table 11) suggest that in water with approximately 10 mg Sr^{2+}/l there should be at the most a few hundred milligrams of SO_4^{2-}/l, and with 1 mg of Ba^{2+}/l, only a few milligrams of SO_4^{2-}/l. Lower sulfate concentrations however only seldom result from the presence of strontium or barium; in most cases the cause is microbiological reduction of sulfate (cf. Section 2.5).

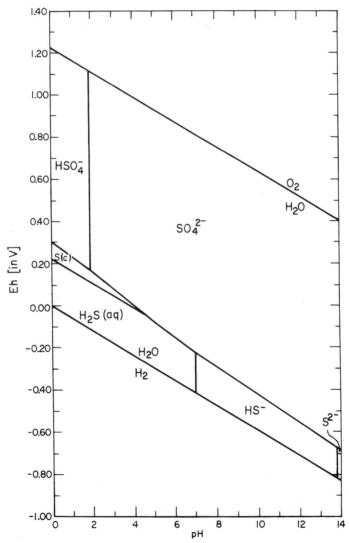

Fig. 55 Stability field diagram of sulfur species at standard temperature and·pressure. Total dissolved sulfur activity 96 mg/l SO_4^{2-} (after Hem 1970).

In reducing waters with pH < 7, undissociated H_2S may be expected, whereas in alkaline conditions HS^- is predominant. The S^{2-} ion requires such a high pH that it seldom occurs in nature (Fresenius & Quentin 1969).

H_2S and HS^- form insoluble heavy metal sulfides that are precipitated. Thus because of sulfate reduction waters with low sulfate contents usually contain very little dissolved sulfide or none at all (Hem 1970).

The sulfate content of groundwater in igneous rocks and psephitic and psammitic unconsolidated and solid rocks generally shows very low values (< 30 mg/l). Local sulfate and sulfide concentrations lead to higher SO_4^{2-} contents in groundwater. In gypsiferous rocks SO_4^{2-} concentrations ultimately reach the saturation point of gypsum (at 10°C, 1925 mg $CaSO_4$/l, corresponding to about 1360 mg SO_4^{2-}/l).

Even higher concentrations are observed in sodium and magnesium sulfate waters, for example an extreme value of 302,500 mg/l in a magnesium-rich brine in Eddy County, New Mexico (Lang 1941), and in the zone of oxidation of sulfide deposits, where thousands and tens of thousands of milligrams of SO_4^{2-}/l occur in mine waters. One extreme value of 209,100 mg SO_4^{2-} was reported in the water of the center galleries of the Comstock Lode (Lindgren 1933, Matthess 1961). Very low or even zero sulfate concentrations are typical for groundwaters in which microbiological reduction has been taking place.

The isotopic compositions of sulfur in sulfate in geologically different ancient evaporites vary very considerably from one to another. Thus the $\delta^{34}S$ values in evaporites of the same age are generally uniform across the world. It is possible to classify groundwaters in which the sulfate content stems from evaporites on the basis of the $\delta^{34}S$ value into distinct regions of origin (Nielsen & Rambow 1969). In this kind of study errors can arise from any process that can lead to isotope separation (solution, precipitation, bacterial sulfate reduction), as well as from mixing with sulfates from other sources that are dissolved in admixing groundwater or originate from oxidation of sulfides or from occurrences of secondary sulfate (e.g., in the cementing material in sandstones).

Chloride.   More than three-quarters of the known chlorine content in the topmost 16 km of the Earth's crust and in the hydrosphere is dissolved in the ocean as chloride, which amounts on average to 19,400 mg/kg. A small part of this amount owes its origin to weathering, because igneous rocks contain only 305 mg/kg, resistates even less (15 mg/kg), hydrolysates 170 mg/kg, and precipitates 305 mg/kg (Tab 64). This statement even holds when according to Billings & Williams (1967) the 170 mg/kg content in hydrolysates is too low because in the calculation the leached near-surface argillaceous rocks were overrepresented; the authors cited an average value of 1466 mg/kg for deeper argillaceous rocks. The bulk of the chloride in the ocean should however originate from degassing of the Earth's crust by volcanic emanations, which early in Earth's history gave rise to the chlorine in the primeval atmosphere and oceans (Goldschmidt 1937, Behne 1953, Correns 1956). In igneous rocks chlorine is found in scapolites, a solid solution series with end members marialite $[Na_4(Al_3Si_9O_{24})Cl]$ and meionite $[Ca_4(Al_6Si_6O_{24})CO_3]$, in sodalite $[Na_8(Al_6Si_6O_{24})Cl_2]$ and apatite $[Ca_5(F,Cl)(PO_4)_3]$, replacing (OH) in mica and hornblende (200–600 mg/ kg), and as liquid inclusions in minerals (Correns 1957, Goguel 1965). The chlorine content in feldspars is of the order of 50 mg/kg (Behne 1953).

Chloride ions are not retained in permeable rocks. In more or less impermeable argillites NaCl crystals or Na-Cl solutions are contained in pores (Hem 1970). Chloride ions are possibly held selectively because of their relatively large size and by osmosis in pore spaces in argillites, whereas other ions can escape, so that ultimately high concentrations can arise (Hem 1970). Possibly Cl^- ions can also be fixed in magnesium hydroxychlorides (Feitknecht & Held 1944). Chlorides are eventually concentrated in marine and terrestrial evaporite deposits. The Cl^- ion is cyclic; furthermore, it occurs in terrestrial dust, in volcanic emanations, and as a man-made airborne pollutant in the atmosphere. The chloride content of rainwater is about 1 mg/ l, rising to several tens of milligrams per litre near the sea; mostly however, it decreases rapidly inland. Local anomalies in this picture are the result of terrestrial and human causes (Section 3.1).

Chloride also enters into Cl^--containing liquid and solid waste material, and in Cl^--containing fertilizers, and highway salt, and thus joins the hydrologic cycle (Section 2.6).

Groundwater in chloride-poor igneous and sedimentary rocks mostly shows low Cl^- values (< 30 mg/l); higher values indicate the admixture of mineralized waters or man-made pollution (Section 2.6). In Cl^--bearing rocks, particularly in the vicinity of salt deposits, in the region of saline impregnation, contents range from several hundreds to thousands milligrams of Cl^- per litre. Hem (1970) quotes extreme cases with 189,000 mg Cl^-/l in a well in Eddy County, New Mexico (Table 65, analysis 2), and 255,000 mg/l in a well at Midland, Michigan (Table 65, analysis 4).

In acid groundwaters associated with active or recent volcanism Cl^- is occasionally the dominant anion (White et al. 1963).

The Cl^- content of groundwater and precipitation can be used in Cl^--free aquifers and near-surface strata, and in the absence of surface runoff, to determine groundwater recharge rates. By analysis of several years of observations Schoeller (1963) gave

$$A_u = \frac{Cl_p}{Cl_n} \qquad (104)$$

in which A_u = groundwater recharge
Cl_p = average Cl^- content of rainwater in the observation period
Cl_n = average Cl^- content of the groundwater and of percolate from lysimeters

Application of this relationship is not possible in regions of intensive agriculture with heavy fertilizer application. Of less importance is an occasional bonding of Cl^- ions in vegetation, because they are liberated again after death of the plant, and loss may occur when plant material is harvested. Observations should be made over several years because of the possibility of temporary retention of chloride ions in the soil during periods of high evaporation.

The radionuclide ^{36}Cl, which is produced in small amounts by cosmic radiation acting on ^{40}A, could be used for dating groundwater because of the geochemical mobility of the chloride ion and the long half-life of 4×10^5 years. However the natural concentrations need special identification and detection techniques. The recently developed ultrasensitive mass spectrometry using a tandem accelerator can use less than 18 mg of chlorine, thus requiring only small water samples (Elmore et al. 1979).

Silica.   Silicon, with an average distribution of 285,000 mg/kg, is the second most abundant element in igneous rocks after oxygen. Its relatively low geochemical mobility is shown by its abundance in weathering residues, resistates (359,000 mg/kg), and hydrolysates (260,000 mg/kg) (Table 64). The bulk of the silica occurring in groundwater comes from the weathering of silicate minerals; amorphous silica also contributes to total dissolved silica, while crystalline silica, particularly quartz, is almost insoluble (6 mg/l SiO_2 at 25°C).

The solubility of amorphous silica is up to 60–80 mg/l at 0°C, 100–140 mg/l at 25°C, and 300–380 mg/l at 90°C. It is in fact a true solution (molecular dispersion), in which the dissolved silicon is probably present for the most part, or even totally, as hydrated monomolecular silicic acid (H_4SiO_4). In acid, fluoride-containing waters with pH values above approximately 4 the occurrence of the fluor silicate complex SiF_6^{2-} has to be considered. The solubility of amorphous silica is hardly affected by pH changes from 0 to 9, but increases quickly above pH 9, in which region silicic acid is dissociated in appreciable amounts following the equation $H_4SiO_4(aq) = H^+ + H_3SiO_4^-$ (Krauskopf 1956a, Hem 1970).

With time, oversaturated silica solutions transform into colloids that flocculate in weakly basic solutions, or form gels in weakly acid solutions (Krauskopf 1956a).

The polymerization of true silica solutions to colloidal silica is affected by pH, temperature, degree of supersaturation, and presence of already formed colloidal and gelatinous silica (White et al. 1956).

The processes of solution and polymerization of silica in dilute solutions are very slow. The establishment of equilibrium between solid and dissolved silica, and between colloidal and truly dissolved silica can last for days or even years. This explains why not only oversaturated solutions but also colloidal silica below the solubility equilibrium can occur in natural waters (Krauskopf 1956a).

Colloidal silica can be precipitated by evaporation, by coprecipitation with other colloids, and by fairly concentrated electrolytic solutions, particularly by aluminum. The precipitation is optimal at pH $\sim$ 4.5, in contrast to silica that is molecularly dispersed at pH 8. True dissolved silica is not influenced by other electrolytes or colloids except at oversaturation or when precipitated as metal silicates (Krauskopf 1956a, Okamoto et al. 1957).

The bulk of the silica in natural waters should be present as a true solution (i.e., H_4SiO_4). The transition into the colloidal state can occur during cooling of high concentrations in volcanic and thermal waters or from increased concentrations by evaporation. Most waters are undersaturated with respect to silica because of the long period needed for solution of silica and silicates, especially in their crystalline form (Roy 1945, Krauskopf 1956a).

Experimental data given by Krauskopf (1956a) were recorded in thermal waters in which equilibrium existed between dissolved and amorphous silica, and solubility was measured at approximately 315 mg/l at 90°C and 110 mg/l at 25°C (White et al. 1956).

Many groundwaters on the other hand are undersaturated with respect to amorphous silica, probably because of the precipitation of almost insoluble quartz and chalcedony (White et al. 1956).

The SiO_2 concentrations of natural waters lie mostly between 1 and 30 mg/l; Davis (1964) has given 17 mg/l as an average value for groundwater.

Oil field brines mostly show 20–60 mg/l total silica, hence far below saturation with respect to amorphous silica, perhaps because of a tendency toward equilibrium with quartz and chalcedony. Occasionally however equilibrium concentrations can come about. For example at Wilbur Springs, Colusa County, California, a value of 190 mg/l total silica at 57°C has been attributed to weathering processes deep below the surface, particularly in serpentine; and pore water in the Santa Cruz Basin contained 68 mg/l at 4°C (White et al. 1956). The higher values are associated with rocks of particular types (volcanic rocks; and sedimentary rocks consisting of volcanic and granitic material), or with higher temperatures.

In thermal waters in active volcanic regions Bogomolov et al. (1966) reported values of 328–350 mg SiO_2/l in an acid well on the island of Paramurhir, Kuriles, and up to 300 mg/l in water from geysers in the Kamchatka peninsula. In addition considerable silica concentrations occur in thermal waters along deeply rooted active tectonic belts. Along these "thermal lines" there are up to 150–230 mg SiO_2/l in the Caucasus, 122–145 mg SiO_2/l in Chukotka, Daurskaya, and in the Pamirs (Bogomolov et al. 1966).

Very high silica concentrations are found in waters with pH values above 9, which very seldom occur in natural water. The water of a mineral spring at Mount Shasta, with the extraordinarily high pH of 11.6, contains 3970 mg SiO_2/kg (Feth et al. 1961) (Table 65, analysis 8).

3.2.2.1.2 *Minor Constituents and Trace Elements.*

Elements that because of their low level of distribution in the Earth's crust, or their low geochemical mobility, occur only in very low concentrations, are known as minor elements and trace elements. Modern analytical techniques (concentrating processes, atomic absorption, activation analysis, etc.) can be used to detect these elements and have proved their importance in allowing the estimation of the origin of the waters concerned, the location of mineral deposits (Hawkes 1957, Leutwein 1960, Ginsburg 1963), and the assessment of potable and irrigation water. However, for many minor and trace elements few or even no concentration values have been reported.

In this connection natural and artificial radionuclides that occur as trace elements in water also call for consideration. Among the natural radionuclides other than those already mentioned, (^{3}H, ^{14}C, and ^{40}K), there are the isotopes ^{87}Rb, ^{232}Th, ^{235}U, and ^{238}U and their daughter products ^{222}Rn and ^{226}Ra. The ^{238}U family and ^{40}K should be considered the most important radioactive elements in groundwater. The related thorium group has only local importance.

Radionuclides created in thermonuclear fission and fusion processes, which can enter groundwater flow systems, include ^{89}Sr, ^{90}Sr, ^{131}I, ^{3}H, and ^{103}Ru. With regard to the problem of use and removal of radionuclides, reference should be made to the literature (Sections 2.2 and 2.6).

Lithium. Lithium is a relatively uncommon element (Table 64) that can be taken up in igneous rock silicate lattices to replace magnesium and occurs in relatively high concentration in pegmatites and evaporites. It forms rather weak bonds by ion exchange; after release by weathering it stays principally in solution. The concentrations in groundwater range mostly from 0.001–0.5 mg/l, but values can exceed 5 mg/l in thermal waters and brines, for example 5.2 mg/l in a thermal spring (94°C) in Upper Geyser Basin, Yellowstone National Park, with 1310 mg/l total dissolved solids, and 8 mg/l in a well 248 m deep at Steamboat Springs, Washoe County, Nevada, having a bottom temperature 186°C, dissolved solids 2360 mg/l (Hem 1970).

Rubidium. The chemical behavior of rubidium resembles that of potassium. Correspondingly it is concentrated in resistates (197 mg/kg) and in hydrolysates (243 mg/kg) (Table 64). The groundwater content ought in general to be of a distinctly lower order of magnitude than that of lithium; however, measured values are scarcer than for lithium. Pentcheva (1967a) gives rubidium contents for nitrogen-containing thermal water of the order of 1–100 μg/l with a frequency maximum occurring a little above 10 μg/l; in Na-Cl waters she gives values between 0.1 and 1000 μg/l with a frequency maximum a little above 1 μg/l, and in alkaline-earth–HCO$_3$ waters values below 5 μg/l, chiefly under 1 μg/l. Krejci-Graf (1963a) reported oil field brines from the Zechstein Hauptdolomit at Kirchheiligen with 3 mg Rb$^+$/l and at Langensalza with 21 mg Rb$^+$/l.

Beryllium. Beryllium is a trace element in igneous rocks, with an average concentration of 3.65 mg/kg (Table 64); beryllium ions can replace silicon in silicate lattices, but the element also occurs in the mineral beryl,

$Be_3Al_2Si_6O_{18}$. Beryllium oxide and hydroxide are slightly soluble; the concentrations of noncomplex-bound Be^{2+} in equilibrium at pH is $0.001-1$ $\mu g/l$. Complex hydroxide ions can raise the solubility a little, particularly at higher pH values. At low pH beryllium ions are easily adsorbed onto clay minerals. Extremely low concentrations (< 0.001 $\mu g/l$) of this very poisonous element are to be expected in groundwater because of the low solubility of its compounds (Hem 1970). Durfor & Becker (1964) found detectable beryllium in potable water in one case only (0.75 $\mu g/l$).

Strontium.   Strontium is a common element in igneous rocks, at 368 mg/kg (Table 64), in which it at times replaces calcium and potassium in silicate minerals, especially in granitic and syenitic rocks (Rankama & Sahama 1960). Strontianite ($SrCO_3$) and celestine ($SrSO_4$) occur in sediments. In spite of some strontium enrichment (617 mg/kg) in precipitates, the Sr/Ca ratio in most limestones is less than $1:1000$ (Hem 1970) (Table 68).

Table 68   Occurrence of Strontium in the Earth's Crust and in Seawater (After Turekian & Kulp 1956, and Turekian 1969)

	Ca (%oo)	Sr (%oo)	$\dfrac{\%oo\ Sr}{\%oo\ Ca} \times 1000$
Basalts (245)	71	0.465	6.5
Granites (175)			
0.1–1.0% Ca	6	0.100	16.7
1.0–5.0% Ca	19	0.440	23.0
Sedimentary rocks			
Limestones (150)	400	0.610	1.5
Shales (69)	50	0.300	6.0
Calcareous deep sea sediments	350	2.075	5.9
Seawater	0.411	0.0081	19.7

The solubility product of $SrSO_4$ of about 2.8×10^{-7} at 25°C permits at a SO_4^{2-} concentration of 10 mg SO_4^{2-}/l a maximum content of 244 mg Sr^{2+}/l; at 100 mg SO_4^{2-}/l only 24 mg Sr^{2+}/l. For equal pH values the solubility of $SrCO_3$ is rather less than that of $CaCO_3$: at ph 8 when the concentration of hydrogen carbonate is about 61 mg/l the strontium activity would be about 28 mg/l (Hem 1970).

The solubility limits of $SrSO_4$ and $SrCO_3$ are not reached in most groundwaters: in general the Sr^{2+} content is of the order of $0.01-1.0$ mg/l. For example Pfeilsticker (1937) has recorded values between 0.01 and 5 mg/l in spring waters in the neighborhood of Stuttgart. Groundwater in celestine-bearing Upper Silurian limestones and dolomites and in glacial deposits that are derived from these rocks in Champaign County, Ohio, contain up to 30 mg Sr^{2+}/l (Feulner & Hubble 1960). Values up to 52 mg Sr^{2+}/l have been reported by Skougstad & Barker (1960), Skougstad & Horr (1960), and Hem (1970) (Table 65, analysis 11) in fresh groundwaters in Iowa, Kansas, Minnesota, Nevada, North Dakota, Ohio, and Wisconsin. In mineral waters however, higher values have been found: 1400 mg Sr^{2+}/l in an oil field brine from the Zechstein Hauptdolomit in the borehole at Langensalza (Krejci-Graf 1963a), 3480 mg Sr^{2+}/l in the highly mineralized water from a borehole at Midland, Michigan (Table 65, analysis 4).

The behavior of strontium in the ground has been investigated in connection with the occurrence of its radionuclide, the occurrence and distribution of which are much dependent on sorption and ion exchange processes (cf. Section 2.2).

Barium. Barium has an average distribution of 595 mg/kg in igneous rocks (Table 64), and is more abundant than strontium in this type of rock. Its low geochemical mobility shows up in comparison with strontium as higher amounts in resistates and in the significantly lower content in seawater. The low solubility of $BaSO_4$ is probably the reason for this; $BaSO_4$ has a solubility product of 1.08×10^{-10} (Table 11). At sulfate concentrations of 10 mg/l and 100 mg/l the corresponding equilibrium concentrations of Ba^{2+} would be 0.14 and 0.014 mg/l, respectively. Besides this, barium is strongly adsorbed by metal hydroxides and oxides (Hem 1970). For public water supply undertakings in the United States, which rely on surface water and groundwater, a median value of 0.043 mg Ba^{2+}/l has been reported (Durfor & Becker (1964).

In spring waters in the immediate vicinity of Stuttgart, Pfeilsticker (1937) observed barium values between 0.05 and 2 mg/l. The ratio of barium to strontium, which varied between 0.08 and 7.0 in this area, showed a dependence on the type of aquifer (Table 69). The differences are so marked that they could be used as criteria for the origin of the groundwater.

Table 69   Ratio of Barium to Strontium in Spring Waters near Stuttgart (after Pfeilsticker 1937)

	Lias α	Keuper	Muschelkalk	Bunter Sandstone
	0.15	0.44	0.82	7.0
	0.25	0.83	1.1	6.4
	0.14	0.43	1.2	4.6
	0.08	0.72		4.1
		1.1		5.8
		1.0		
		3.8		
		1.5		
Average	0.16	1.2	1.0	5.6

In sulfate-poor or sulfate-free brines higher barium concentrations have been observed. Krejci-Graf (1963a) has reported values up to 51 mg Ba^{2+}/l at Kala, USSR, and Michel et al. (1974) values up to 2032 mg Ba^{2+}/kg in the Ruhr mining district.

Barium concentrations of over 50 mg/l in mineralized groundwaters in the Upper Cretaceous Woodbine Formation of the East Texas Basin have been correlated by Brooks (1960) with localized sedimentary barium enrichment.

Aluminum. Aluminum is the third most common element in igneous rocks at 79,500 mg/kg (Table 64). It is a constituent of many silicates, for example feldspar, feldspathoids, mica, and many amphiboles. In silicate lattices aluminum can substitute for silicon in fourfold coordination, and in magnesium and iron sites in sixfold coordination. In sedimentary rocks aluminum occurs as gibbsite $[Al(OH)_3]$, less often as cryolite $[Na_3AlF_6]$, and in alum, but most of all as a component of clay minerals. Its low geochemical mobility is clear from the low aluminum content of seawater (0.001 mg/kg). In aqueous solution the cation Al^{3+} is predominant at pH values below 4.0, probably in the hydrate with six water molecules; with increasing pH it appears that one of the water molecules becomes an OH^- ion, and at pH 4.5–6.5 polymerization starts, forming aluminum hydroxide complexes of various sizes, which form minute crystals with time. Above pH 7 the anion $Al(OH)_4^-$ is the most abundant dissolved form. The polymerization of aluminum hydroxide is accelerated by the presence of silica, which at sufficient concentration leads to rapid precipitation of poorly crystalline clay minerals (Hem 1970).

Aluminum forms strong complexes with fluoride ions, particularly AlF^{2+}, and AlF_2^+ in waters with a few tenths to a few milligrams F^-/l. The sulfate complex $AlSO_4^+$ can be predominant in acid solutions with a high sulfate content. Soluble phosphate complexes also occur (Hem 1970).

Generally the aluminum content in groundwater amounts to only a few hundredths or tenths of a milligram per litre; values above 1 mg/l, such as 28 mg/l in the previously mentioned manganese-rich water (Table 65, analysis 7), are rare. In groundwaters or acid mine waters with pH values below 4.0, higher values – for example 64 mg Al/l in an acid groundwater (pH 3.5) from Tertiary marine argillaceous rocks at Monticello, Drew County, Arkansas (Scott & Barker 1962) – even concentrations of several hundreds or several thousands of milligrams of aluminum per litre occur.

Vanadium. Vanadium is often present in magmas as a minor element (Table 64), mainly as a minor constituent in magnetite, pyroxene (especially aegirine), amphibole, and biotite. It is concentrated in sulfide deposits, mainly as patronite (VS_4), in oxidized sulfide ores in form of various vanadates, in muscovite, and in oxide deposits in sandstones mainly as montroseite $[(V,Fe)O \cdot OH]$ (Wedepohl 1978).

Vanadium occurs in solutions as anionic and cationic species, V^{3+}, V^{4+}, and V^{5+}. At nearly neutral pH, under moderately reducing conditions in which at least 0.5 mg Fe^{2+}/l remains in solution, the V^{3+} and V^{4+} species are soluble only up to approximately 1 μg/l (Hem 1970). Vanadium is possibly removed from solution by adsorption, for example at freshly precipitated ferric iron oxide (Krauskopf 1956$_b$).

In oxygenated groundwaters measurable vanadium values can be found, and concentrations as high as 70 μg/l have been observed (Durfor & Becker 1964, Hem 1970). In Japanese hot springs concentrations up to 247 μg V/l are reported, and in acid hot spring water contents up to 330 μg V/l (Wedepohl 1978).

Chromium. Among the various oxidation states of chromium, Cr^{3+} and Cr^{6+} are stable in water. The solubility of chromium is relatively low at pH 8–9.5 and in weakly reducing conditions (< 0.5 μg Cr/l). The anion species $Cr_2O_7^{2-}$ and CrO_4^{2-} appear to be relatively stable. In strongly reducing conditions below pH 8 the complex cation $CrOH^{2+}$ is the dominant species (Hem 1970).

Chromium is relatively well distributed. Its median value in United States public water supplies, which use groundwater as well as surface water, is 0.43 μg/l (Durfor & Becker 1964). Higher values are however rare. Silvey (1967) gives seven examples of California groundwaters and spring waters with value up to 21 μg/l. Chromate and dichromate contents are chiefly found in groundwaters polluted by man (cf. Section 2.6).

Molybdenum. Molybdenum occurs in different valence states and forms polymeric hydroxide anions; of these the anionic molybdate species are probably predominant in natural water (Hem 1970).

Molybdenum accumulates in some plants and toxic concentrations can arise.

The concentrations in groundwater are generally a few milligrams per litre. Durfor & Becker (1964) give an average value for United States public water supplies of 1.4 μg Mo/l. Regional averages of 3 μg/l occur. Silvey (1967) found in 48 spring waters in California an average content of 2.2 μg/l, and in 52 groundwaters an average 1.2 μg/l. Sugawara et al. (1961) mention molybdenum concentrations between 0.8 and 22.9 μg/l in Japanese thermal waters. Peak values reported from the USSR lie between 10 μgl/l and 10 mg/l (Vinogradov 1957).

Cobalt. Cobalt is quite abundant in igneous rocks; most of the highest concentrations occur in ultrabasic rocks and in company with other ores. Traces of cobalt are necessary as a nutrient in plants and animals. The element occurs especially in the divalent form. The solubility of $Co(OH)_2$ ($K_{SP} = 10^{-15.2}$ at 25°C) is similar to that of $Fe(OH)_2$ (Table 11), but that of $CoCO_3$ ($K_{SP} = 10^{-12.84}$ at 25°C) and CoS ($K_{SP} = 10^{-22.5}$ at 25°C) are lower than the respective iron compounds ($K_{SP} = 2.5 \times 10^{-11}$ at 20°C and $K_{SP} = 3.7 \times 10^{-19}$ at 18°C, respectively) (Wedepohl 1978, D'Ans-Lax 1967). At pH 8 in the presence of approximately 100 mg HCO_3^-/l, approximately 6 $\mu g/l$ Co^{2+} is soluble. The solubility can however increase through formation of ion complexes. Cobalt is absorbed by hydrous iron and manganese oxides, which together with the low solubility explains why cobalt usually occurs only in trace amounts, often below the level of analytical detection. Average concentrations of 20 $\mu g/l$ were recorded from mineralized zones in the southern Urals (Hem 1970).

Nickel. Nickel is comparatively abundant and occurs preferentially in igneous rocks and in association with other ores. It occurs as Ni^{2+} ion in water and is absorbed by hydrous iron and manganese oxides. The solubility of $Ni(OH)_2$ is almost equal ($K_{SP} = 10^{-15.2}$ at 25°C), that of $NiCO_3$ ($K_{SP} = 10^{-6.87}$ at 25°C) and of NiS ($K_{SP} = 10^{-20.7}$ at 25°C); and much higher than the corresponding cobalt compounds (Wedepohl 1978). Nickel apparently occurs in higher concentrations than cobalt in groundwater; for example in the mineralized zone of the southern Urals already mentioned the reported average is 40 $\mu g/l$ (Hem 1970). Concentrations of 4–40 mg Ni^+/l are reported by Krejci-Graf (1963a) for Venezuelan oil field waters. In mine waters increased nickel values have likewise been observed. For mixed water in the upper levels of the Friedrich Mine at Niederhövels, Siegerland (pH 1.9), Heyl (1954) records 1.5 mg Ni^{2+}/l; in a water sample from the 580 m level the concentration was 0.10 mg/l (pH 7.7), and in an acid water (pH 1.9) in the 500 m level of the Wolf Mine at Herdorf even 319.3 mg Ni^{2+}/l.

Copper. Copper is a relatively common element in the lithosphere (Table 64). It forms numerous mineral species, particularly sulfides, oxides, and hydroxycarbonates. In equilibrium with oxidized ores it is soluble at pH 7 in oxygenated water to 64 $\mu g/l$ and at pH 8 6.4 $\mu g/l$. In reducing conditions in the presence of reduced sulfur species the solubility is even lower. Under oxidizing conditions copper sulfides are oxidized to soluble sulfate. Copper is necessary as a trace element in plant and animal metabolism and finds a wide use as a raw material and in herbicides. The copper content in fresh groundwater is generally very low. In the region of copper deposits, as the literature on geochemical prospecting shows, the copper content is markedly increased; for example in parts of the southern Urals there is an average of 200 $\mu g/l$ (Hem 1970). In North Rhine–Westphalia some mineral waters have copper concentrations up to 194 $\mu g/l$ (Westerkotten) (Fricke & Werner 1957). In $CaCl_2$-bearing oil field brines in the Hauptdolomit of the Southern Harz region the average is 17 μg Cu/l (Herrmann 1961). In alkaline oil field brines in Azerbaijan up to 176 $\mu g/l$ was recorded, and in alkali chloride brines up to 344 $\mu g/l$ (Krejci-Graf 1963a). High concentrations occur in acid mine waters and in the runoff from tailings tips and launders (in ore dressing), for example 312 mg/l in water from the Burra-Burra Mine, Ducktown, Tennessee (Table 65, analysis 10) (Hem 1970), 1154 mg/l in water from the 500 m level of the Wolf Mine at Herdorf, Siegerland (Heyl 1954), and

45,633 mg/l in water from Mountain View Mine, Butte, Montana (Smirnow 1954).

Silver.   Geochemically rare silver (Table 64) is soluble in oxygenated water only between 0.1 and 10 μg/l. Its scarcity explains the low median value of 0.23 μg/l in average potable water in America, obtained from groundwater and surface water (Durfor & Becker 1964, Hem 1970).

Zinc.   Zinc is a relatively common element (Table 64). It can replace iron and manganese in silicate lattices, and it forms the mineral sphalerite (zinc blende), ZnS (Wedepohl 1953).

Zinc is a universally used metal, hence occurs in waste. It is therefore found in the vicinity of metallurgical works, in flue gases, in mine tailings, and in wastes from processing plants.

At pH $\leq$ 7 zinc occurs in aqueous solution as Zn^{2+} ion, whereas at pH 11.5 $Zn(OH)_3^-$ and $Zn\ OH_4^{2-}$, respectively, are present. The solubility is controlled mainly by ZnS ($K_{SP} = 1.1 \times 10^{-24}$) and $ZnCO_3$ ($K_{SP} = 6 \times 10^{-11}$) (Table 11) (Hem 1972). The solubility of zinc hydroxide has its minimum at pH 9.5, therefore Hem (1972) concluded that hydroxide solubility control is not attained until pH 9.3 is reached. This author further suggested that for total dissolved silica concentration of 10^{-3} and 10^{-4} M the zinc silicate willemite (Zn_2SiO_4) may be the least soluble species between pH 7.5 and 10.0. There is evidence that zinc coprecipitation with calcium as carbonate and phosphate (apatite) (Wedepohl 1978), complexation by organic material of low solubility (Hem 1972), and adsorption on clay minerals (Krauskopf 1956) and on manganese or iron oxide and hydroxide (Jenne 1968) have some importance.

The zinc content of fresh groundwater appears to be generally far below 10 μg/l. In 1424 groundwater samples in Pleistocene glacial outwash sediments in Schleswig-Holstein, West Germany, Neumayr (1979) found an average concentration of 63 μg Zn^{2+}/l and a maximum value of 630 Zn^{2+} μg/l, whereas groundwater samples from Miocene mica-bearing fine sands of the same area contained up to 980 μg Zn^{2+}/l (average 511 μg Zn^{2+}/l). By help of selective extraction methods it could be shown that in the Pleistocene glacial outwash the total zinc content (average of 180 sediment samples, 16 mg Zn/kg) is bound in the solid phase in carbonates (2%), oxides (45%), hydroxides (48%), and silicates (5%), whereas in the Miocene mica-bearing fine sands the total zinc content (average of 25 sediment samples, 59 mg ZN/kg) is bound mainly in carbonates (38%), sulfides (14%), oxides (16%), hydroxides (23%), and silicates (7%).

The actual zinc value is seldom limited by the solubility of zinc itself but apparently by its availability underground, so that it is useful for geochemical prospecting (Kennedy 1956). Downstream of known mineral deposits in the Western Harz, zinc values in spring water ranging from 3400 μg/l to undetectable amounts have been observed; in regions without known mineralization values from undetectable to 220 μg/l have been measured, possibly caused by flue gases from smelting works (Nowak & Preul 1971).

In North Rhine–Westphalia mineral waters with high zinc concentrations have sometimes been found (Medard, 1.9 mg/l, Westernkotten 1.15 mg/l) (Fricke & Werner 1957). One very high value of 177 mg Zn^{2+}/l was reported by Van Everdingen (1970) from an acid spring water (pH 2.5) at Ochre Hill in the Kootenay National Park, British Columbia. The high heavy metal content and the acid pH values (2.5–3.5) in adjacent springs is explained by the oxidation of sulfide ores in the Cambrian rocks of this area. Van Everdingen & Banner (1971) have determined experimentally that on contact with calcium carbonate the pH value rises into the normal range (7.0–7.6) and the zinc content decreases considerably (in the example given, by about 82.8%).

Acid mine waters can exhibit a considerable zinc content, for example 90.8 mg Zn^{2+}/l in water from the 500 m level of the Wolf Mine at Herdorf, Siegerland (Heyl 1954), 345 mg/l in the water from Victor Mine, Joplin District, Missouri (Emmons 1917), and 2412 mg Zn^{2+}/l in water from Alabama Coon Mine, Joplin, Missouri (Smirnow 1954). These high values are caused by the oxidation of the zinc sulfide ore (Section 2.3).

Cadmium. The geochemically scarce element cadmium (Table 64) occurs in association with lead and zinc ores. Here it is observed to replace other elements in their minerals, especially zinc. Usually 0.1–0.5 wt % Cd is present in sphalerite (ZnS) and smithsonite ($ZnCO_3$). The cadmium minerals greenockite (CdS), monteponite (CdO), and otavite ($CdCO_3$) originate mainly from weathering of cadmium-bearing zinc minerals (Wedepohl 1978).

At pH < 8 cadmium occurs in aqueous solutions as Cd^{2+} ion, whereas at higher pH $Cd(OH)_2(aq)$ and $Cd(OH)_3^-$ are present. The solubility is controlled mainly by CdS ($K_{SP} = 3.6 \times 10^{-29}$) and $CdCO_3$ ($K_{SP} = 2.5 \times 10^{-14}$). Cadmium forms complexes with chloride, sulfate, and chlorohydroxide ions. Chloride complexes [e.g., Cd(OH) Cl(aq)] may be important in water when they have Cl^- contents above 350 mg/l (Hem 1972).

Cadmium has been observed in concentrations up to 20 μg/l in spring waters and up to 71 μg/l in groundwaters in California (Silvey 1967). In groundwater in Pleistocene glacial outwash sediments in Schleswig-Holstein, West Germany, Neumayr (1979) measured in 1424 samples an average concentration of 1.8 μg Cd^{2+}/l and a maximum value of 13 μg Cd^{2+}/l, whereas 120 groundwater samples from sand and silt layers in Holocene marshland sediments of the same area contained up to 21 μg Cd^{2+}/l (average 4.6 μg Cd^{2+}/l). By help of the scanning electron microscope and the electron microprobe it could be proved that greenockite, otavite, and cadmium-containing sphalerite are present in the sediments.

High values are occasionally detected in mine waters, for example 41.1 mg Cd/l in an acid water from the 170 m level in the St. Lawrence Mine, Butte, Montana (Smirnow 1954).

Mercury. Mercury is generally undetectable in groundwater. Even in thermal waters it lies below the usual limits of detection, although found in associated deposits of sinter as HgS (White et al. 1963). Mercury can possibly enter into groundwater through man-made pollution, particularly as a constituent of pesticides (p. 157), and in areas of mining, mineral dressing, and general industrial waste (Hem 1970).

Germanium. This rare element (Table 64) is detectable in trace amounts in groundwater. Pentcheva (1967b) gives concentration of the order of 0.1 μg/l. In 19 California oil field brines Silvey (1967) found an average content of 14 μg/l. With higher contents (up to 50 μg/l) in California spring waters Silvey (1967) believed that there was a late magmatic influence. For nitrogenous thermal waters in Bulgaria, Pentcheva (1967b) gives values between 0.08 and 130 μg/l (average 29 μg/l).

Lead. Lead is distributed through rocks in the Earth's crust at low concentrations (Table 64), sometimes replacing potassium in silicates, particularly in feldspars, and in phosphates, but also in mineral deposits in specific minerals, galena (PbS), cerussite ($PbCO_3$), and anglesite ($PbSO_4$) (Wedepohl 1956).

Lead is used in insecticides and in high octane gasoline; it therefore can enter the ground through human activities, mostly by way of air pollution via incinerators, motor vehicle exhaust gases, gases from metal smelting, and lead-containing pesticides and fertilizers, as well as in solid and liquid waste, especially tailings from mineral processing, mine workings, and smelting works (Matthess 1972b).

Because of the low solubility of the compounds of lead and its very low geochemical mobility, groundwater mostly shows amounts of the order of a few micrograms per litre to 20 μg/l.

Higher lead contents are observed in the vicinity of mineral deposits (Kennedy 1956). Nowak & Preul (1971) found lead concentrations up to 1300 μg/l in springs in the Upper Harz ore district, but in unmineralized regions in which the influence of flue gases could not be ruled out the values were only 230 μg/l or less. The lead content of 85 mineral waters in North Rhine–Westphalia lay between < 5 and 52 μg/l (Fricke & Werner 1957); Krejci-Graf (1963a) mentions a lead content of 380 μg/l in an oil field water from the Zechstein Hauptdolomit (borehole at Langensalza). As an average value for oil field waters in the Hauptdolomit of the southern Harz region Herrmann (1961) gives 39 μg/l.

A very high lead content of 1.2 mg Pb^{2+}/l is also reported by Van Everdingen (1970) in an acid spring water (pH 2.5) at Ochre Hill, Paint Pot Springs, Kootenay National Park, British Columbia, a region of sulfide ore deposits in Cambrian rocks. Van Everdingen & Banner (1971) have stated that the pH value rises at the contact with calcium carbonate rock to the normal range of 7.0–7.6 and that the lead content is then considerably reduced – in the example given, by 93.3%.

Man-made factors can locally raise lead values to the above-mentioned orders of magnitude, particularly in areas of soils affected by smelting gases and mine and smelting works tips.

Boron. Boron is a trace element in igneous rocks (Table 64, 7.5 mg/kg), but in sandstone and in argillaceous rocks it is enriched (90 and 194 mg/kg, respectively), because of the resistance to weathering of boron-containing tourmaline. Boron is also present in biotite and amphiboles in small amounts. Volcanic gases are a further source of boron, containing orthoboric acid (H_3BO_3), or boron halogenides such as boron trifluoride (BF_3), so that groundwater in volcanic areas and many hot springs in volcanic regions exhibit appreciable concentrations of boron. Finally the element occurs in many terrestrial evaporite deposits in southeast California and Nevada, even in extractable amounts, in particular as colemanite ($Ca_2B_6O_{11} \cdot 5H_2O$) and kernite ($Na_2B_4O_7 \cdot 4H_2O$). Boron is an essential plant nutrient, but it can have a damaging effect in too high concentrations. Tetraborate is an ingredient in washing powder and ends up in sewage and industrial waste.

Little is known about boron in aqueous solution. The simplest assumption is that orthoboric acid species are most likely to occur, which are dissociated to an appreciable extent only in the pH range above 8.2. Beyond that it seems that complex solute species occur (Hem 1970). Fresenius & Fuchs (1930) and Fresenius & Quentin (1969) state that the boron content in mineral waters is in the form of undissociated metaboric acid (HBO_2), and only in alkaline waters can metaborate ion (BO_2^-) be expected.

Boron in groundwater mostly occurs at concentrations of less than 1 mg/l of the order of a few hundredths to tenths of a milligram per litre. High values, above 10 or even 100 mg/l, are known from oil field brines and thermal waters in volcanic regions. Krejci-Graf (1963a) gave values of up to 234 mg B/l for Russian oil field brines (Ishimbay, Cape Apsheronski), and Hem (1970) quotes a thermal water (77°C) at Sulphur Bank, Lake County, California, as showing 660 mg B/l. Probably boron would be helpful in water provenance studies: the B/Cl ratio of seawater and surface water is approximately 0.0002, of oil field brines 0.02, and of volcanic thermal waters 0.1 (Davis & DeWiest 1967).

Phosphorus. Phosphorus is an important constituent of the lithosphere (Table 64) and occurs chiefly in apatite. On weathering, the liberated phosphate is largely adsorbed on clay minerals. The various forms of phosphate fertilizers and liquid and solid waste provide a major source of phosphate in groundwater.

Phosphate plays a role in plant and animal metabolism and thus occurs in their waste products. The current use of sodium phosphate in detergents increases the supply of phosphorus in natural waters. Sodium polyphosphates, which are often used in water treatment to prevent precipitation of $CaCO_3$ and $Fe(OH)_3$, are unstable and eventually break down to orthophosphate. Phosphoric acid (H_3PO_4) occurs in the ion forms PO_4^{3-}, HPO_4^{2-}, $H_2PO_4^-$, and $H_3PO_4(aq)$. At about pH 6, $H_2PO_4^-$ ions are present almost exclusively, and at pH 8 the dominant species is HPO_4^{2-} in aqueous solution. Phosphate ions form complexes with many other solutes in natural waters (Hem 1970).

The strong bond of phosphate with clay minerals and metal hydroxides, particularly iron hydroxide, as well as its use in the biological cycle, are the causes of the low concentration of phosphate in groundwater (tenths or hundredths of a milligram per litre).

Arsenic. Arsenic is present in the lithosphere as the sulfide and as the arsenate (AsO_4^{3-}) in mineral deposits. Other sources of arsenic include waste tips and arsenical insecticides and herbicides.

In aqueous solutions the $HAsO_4^{2-}$ ion is predominant at pH values above 7.2; below this pH the ion $H_2AsO_4^-$ is found. In reducing conditions the form $HAsO_2(aq)$ is present (Hem 1970, 1977).

Under the prevailing pH conditions in groundwater the solubility of calcium and magnesium arsenates leads to concentrations of several milligrams of arsenic per litre. The solubility is however lowered in the presence of trace elements, for example at 65 $\mu g/l$ of copper it decreases to a few tenths of a milligram per litre. Beyond that arsenate is precipitated as Fe(II)-arsenate ($K_{SP} = 5.7 \times 10^{-21}$) (Chukhlantev 1956) or as Mn(II)-arsenate ($K_{SP} = 2 \times 10^{-29}$) (Meites 1963), coprecipitated and absorbed with $MnO_2(aq)$, $Fe(OH)_3$, and other substances with active surfaces. Gulens et al. (1979) indicated that As(V) and As(III) species form complexes with Fe(III) in solution, with the Fe(III)-As(III) complex being more soluble than the Fe(III)-As(V) complex.

In groundwater the arsenic concentrations mostly lie below 0.1 mg/l (Höll 1972). Locally, geological or human influences can cause higher values, particularly in mineral, oil field, and thermal waters; for example the content in a thermal water well at Steamboat Springs, Washoe County, Nevada, is 4 mg As/l (Hem 1970), and 17 mg As/l was reported in the Max Spring at Bad Dürkheim; Krejci-Graf (1963a) noted concentrations up to 32.3 mg/l. High concentrations also occur in fresh water, for example 1.3 mg As/l in a well with 363 mg/l total dissolved solids (Hem 1970) in Lane County, Oregon.

Man-made arsenic contents occur particularly in runoffs from tips and arsenic-containing wastewaters from chemical and metallurgical works (Höll 1972).

Selenium. Selenium is a very scarce element in the Earth's crust (Table 70). Its geochemical behavior very closely resembles that of sulfur. In aerated water SeO_3^{2-} is stable at pH values above 6.6; under reducing conditions the equilibrium species is elemental selenium and has presumably low solubility (Hem 1970).

In general selenium in groundwater occurs in concentrations of a few to tens of micrograms per litre; locally however significantly higher concentrations do occur (Table 70).

Very high values have been recorded in groundwater and in the agricultural drainage in the outcrop areas of the selenium-bearing Upper Cretaceous marine argillites in parts of the western United States, for example 1.98 mg Se/l in a 10,900 mg/l total dissolved solids in drainage water in Mack Mesa County, Colorado (Hem 1970). High selenium concentrations (up to 400 μg/l) were reported by D'Yachkova (1962) from the neighborhood of pyrite deposits in the Urals.

Table 70 Selenium Concentrations in Different Groundwaters (After Quentin & Feiler 1968)

	Concentration (μg/l)
France, groundwater (Department of Loire and Seine)	2–35
France, mineral springs	35–200
Argentine, groundwater (Cordoba Province)	48–67
United States, spring waters in soils with extremely high selenium content	70–400
Utah, springs	To 1600
Wyoming, springs	To 9500
Bad Gastein, mixtures of thermal waters	0.36

Fluorine.   Fluorine, with average distribution of 715 mg/kg (Table 64), occurs much more abundantly than chlorine in igneous rocks, but is very much scarcer in the Earth's crust, including the hydrosphere. Fluorite (CaF_2) and apatite ($Ca_5[F,Cl](PO_4)_3]$) are constituents of igneous and sedimentary rocks; amphiboles and micas can contain fluorine substituting for hydroxyl groups because the ionic radii are similar and the electric charges are the same.

The fluoride ion can be bound to mineral surfaces in the place of hydroxyl ions. However, with increasing pH it can be displaced from this bond by hydroxyl ions.

Fluorine is a biologically important constituent of teeth and bones and is often present in industrial flue gases (Section 2.6).

Fluorine forms F^- ions in water, which can form strongly soluble complexes with aluminum, beryllium, and ferric iron, and in the presence of boron, mixed fluoride-hydroxide complexes. At pH below 3.5 the HF^0 form may occur. The fluoride content in the presence of Ca^{2+} ions is presumably controlled by the solubility product of fluorite, CaF_2 ($10^{-10.57}$ at 25°C). For an activity of 40 mg Ca^{2+}/l the equilibrium activity of F^- would be 3.2 mg/l, but the concentration can be rather higher in the presence of other dissolved constituents; the significantly higher F^- contents in groundwaters are however associated with calcium deficiency. The mineral cryolite (Na_3AlF_6) is insoluble and will hardly affect the solubility because the aluminum content required for this is generally not present in groundwater (Hem 1970).

In most fresh water the fluorine content is below 1 mg/l. The fluorine background value of waterworks in western Germany (purified water) lies between 0.1 and 0.2 mg F^-/l (Gad & Fürstenau 1954). Quentin (1957) (Table 71) also notes maximum values of 0.50 mg F^-/l.

Table 71 Fluoride Content of Bavarian Drinking, Ground-, and Spring Water (After Quentin 1957)

Number of samples	Fluoride content (mg F^-/l)
10	0–0.05
31	0.05–0.10
12	0.10–0.25
2	0.25–0.50

In a low calcium groundwater (5.5 mg Ca^{2+}/l) from a Quaternary valley deposit in Cochise County, Arizona, 32 mg F^-/l was recorded (total dissolved solids = 389 mg/l). A groundwater in South Africa with 67 mg F^-/l has been reported (Hem 1970).

In 30 Bavarian mineral springs fluoride contents between 0.021 and 17.40 mg/kg were found (of which 9 springs had more than 1 mg/l), the highest values in Bad Wiessee, Upper Bavaria (Quentin 1957).

The concentration of 22 mg F^-/l in a basic thermal water (pH 9.2; 50°C) in Owyhee County, Idaho (Hem 1970), is possibly attributable to the displacement of adsorbed fluoride ions by OH^- ions. Very high fluoride concentrations occur in volcanic thermal waters: 806 mg F^-/l in a hot spring in the crater on White Island, Bay of Plenty, New Zealand (White et al. 1963).

Bromine. Bromine is significantly scarcer than the geochemically similar chlorine (Table 64). Its content in seawater at 67.3 mg/kg is considerably higher than in igneous rocks. As with chlorine, the bulk is attributed to volcanic emanations (Behne 1953). In natural waters it occurs as the Br^- ion. It is found in trace amounts in rainwater, at a higher Br^-/Cl^- ratio than in seawater, apparently because of the small size of its aerosol particles (Hem 1970).

Possibly the bromide ion is preferentially concentrated by osmosis, for it possesses a greater ionic radius than Cl^-. A further possibility for the cause of the enrichment is concentration by evaporation after reaching NaCl saturation (Hem 1970).

Little is known about the bromine content of normal groundwater. Presumably its concentrations lie considerably below 0.1 mg/l. High bromide contents (0.4–6.4 mg Br^-/l) have been recorded in the coastal plain of North Carolina in groundwaters associated with occurrences of phosphorite (Brown 1958).

The bromide ion is a characteristic constituent of oil field waters (Krejci-Graf 1963a, 1963b): the highly mineralized water from Midland, Michigan (Table 65, analysis 4) contains 3720 mg Br^-/l. Similar high values are known from Russian and Brazilian oil field waters. A very high value of 6500 mg Br^-/l has been measured in the weakly brackish oil field water at Filipesti, Romania (Krejci-Graf 1963a).

Iodine. Iodine is a scarce element in both the lithosphere and the hydrosphere (Table 64). It is however biologically concentrated in plants and animals. In marine clayey sediments, which contain many organic substances, iodine is liberated on the decomposition of these materials into the pore solutions. The concentration so brought about rises from the I^-/Cl^- ratio of 0.000003 in seawater compared with about 0.001, the usual ratio in oil field brines (Davis & De Wiest 1967).

Iodide (I^-) that is tagged with ^{131}I inverts in small amounts to higher oxidation states (e.g., I^+) in karst waters, though not to the same extent as in surface waters. This phenomenon has not yet been observed in other groundwaters. From the retardation of the movement of ^{131}I compared with that of tritium or uranium in tracer tests, it may be concluded that ^{131}I is absorbed below the surface to some small extent and is consequently elutriated (Institut für Radiohydrometrie 1966, 1968). The anion adsorption in young Pleistocene river terraces in Germany is however very low. The average sorption capacity lies between 2 and 5% of the total ^{131}I injected (9.129×10^5 Bq in 150 ml of $0.01N$ KI solution and 50 g sediment) (Weisflog 1968). Iodide ions can be coprecipitated with iron and manganese and aluminum hydroxides (Sugawara et al. 1958).

The iodine content in most fresh water should lie below the limit of detection. Dean (1963) gives values between 0.7 and 14.8 μg/l for New Zealand groundwaters. Higher iodine contents (0.2–2 mg l) were found by Brown (1958) in the coastal plain of North Carolina in groundwater associated with phosphorite deposits. Higher values are known in oil field waters, for example 48 mg/l in the highly mineralized water in Midland County, Michigan (Table 65, analysis 4), up to 77 mg I^-/l in the Vienna Basin, and 50–80 mg I^-/l at Tchusowskaja in the western Urals (Krejci-Graf 1963a).

Uranium. Natural uranium includes numerous isotopes, among which [238]U is dominant. This radionuclide decays with a half-life of 4.5×10^9 years via a series of intermediate products to the stable isotope [206]Pb. The solubility of reduced U^{4+} compounds is low. The more soluble forms are those in higher oxidation states, such as the uranyl ion UO_2^{2+}, which builds up anionic forms at higher pH values, or finally uranyl complexes with carbonate and sulfate.

As shown in the following equations

$$HCO_3^- \rightleftharpoons H^+ + CO_3^{2-} \tag{105}$$

$$UO_2^{2+} + 2CO_3^{2-} + 2H_2O \rightleftharpoons UO_2(CO_3)_2(H_2O)_2^{2-} \tag{106}$$

$$UO_2^{2+} + 3CO_3^{2-} \rightleftharpoons UO_2(CO_3)_3^{4-} \tag{107}$$

an increase in hydrogen carbonate displaces the equilibrium in favor of the formation of stable, soluble carbonate complexes. Uranium in groundwater originates in uranyl compounds or other uranium species present in the aquifer or in the unsaturated zone (Barker & Scott 1958). The nonionized complex UO_2SO_4 can likewise occur in appreciable amounts in groundwater (McKelvey et al. 1955, Tokarev & Shcherbakov 1956).

Uranium is widespread in groundwater in low concentrations in differing amounts, with values that depend on local variations of Eh, pH, and temperature (Garrels & Christ, 1965). Generally the uranium concentration lies between 0.05 and 10.0 μg/l. The uranium contents of 67 random samples in the United States ranged from less than 0.1 to 22 μg/l (median value 1.6 μ/l) (Barker & Scott 1958). The median value of 565 samples from potable water obtained chiefly from groundwater was about 1.5 μg/l (Scott & Barker 1962). Locally higher values have been described from fluviatile deposits with intercalated tuffs in the Ogalalla Series at Llano Estacado, Texas and New Mexico, with contents up to 12 μg/l (median value 6.2 μg/l) (Barker & Scott 1958). Groundwater from uranium-rich rocks can give higher values, for example 120 μg/l in a groundwater in the Permian Rush Springs Sandstone at Cement, Oklahoma, which contains uranium-bearing asphaltic bituminous sandstone beds (Scott & Barker 1962), and 90 mg/l in a groundwater in the USSR (Tokarev & Shcherbakov 1956).

Radium. Of the four radium isotopes occurring in nature, ^{223}Ra, ^{224}Ra, ^{226}Ra, and ^{228}Ra, the only ones of importance are ^{226}Ra (half-life 1620 years), a daughter product of ^{238}U, and, in considerably smaller amounts, ^{228}Ra, a daughter product of ^{232}Th. Radium behaves as an alkaline earth metal like barium. The solubility of the sulfate is rather lower than that of $BaSO_4$ (Table 11) (Hem 1970).

In general, radioactive equilibrium between parent and daughter product should be established after 10 times the half-life of the daughter. In the case of ^{238}U and ^{226}Ra this means 16,200 years. If this time is not available — say because of flow of the water, or when geochemical processes are effective in separating parent and daughter from each other — then no radioactive equilibrium can come about. Groundwaters are often in a state of radioactive disequilibrium, for example the measured Ra/U ratios in the groundwater of the Ogalalla Series, Llano Estacado (< 0.3 up to 7 mBq ^{226}Ra/μg ^{238}U), are different from the equilibrium ratio 12.6 mBq ^{226}Ra/μg ^{238}U. The ^{226}Ra deficit is attributed by Barker & Scott (1958) to adsorption onto clays and other ion exchanges. Groundwater generally has a ^{226}Ra content of less than 37 mBq/l. Higher values occasionally occur, for example 181 mBq/l in England (Kenny et al. 1966), and values up to 3700 mBq/l in Western Illinois, Kentucky, New Mexico, and Wisconsin (Scott & Barker 1961, 1962, Emrich & Lucas 1963). In many cases the radium content increases along the subsurface flow path of the groundwater. For example the content in Cambrian, Ordovician, and Silurian-Devonian aquifers in the northern part of Illinois rises from less than 37 mBq/l in recharge zones to 740–3700 mBq/l in the direction of the groundwater movement (Emrich & Lucas 1963), although by variable amounts. This is obviously the result of variations in the distribution in the aquifer of the parent isotope, uranium, and different contact surfaces and contact times, because the low values occur preferentially in areas of water extraction, with faster water circulation and reinforced inflow from groundwater recharge areas. A very high value of 720 pCi/l (26.6 Bq/l) was recorded in a brine in Midland County, Michigan (Table 65, analysis 4). Even higher values of the order of several hundreds of becquerels per litre are possibly explainable at times as combined values of radium and radon (Hem 1970).

Other Trace Elements. Little is known about the occurrences of other trace elements in groundwater. A survey of the occurrence of trace elements is given by Pentcheva (1967b) for nitrogenous thermal waters, Na–Cl waters, and other groundwater in Bulgaria (Table 72).

Table 72   Occurrence of Trace Elements in Bulgarian Groundwaters (Pentcheva 1967b)

	Occurrence (mg/l)		
	Average	Mininum	Maximum
Tungsten			
Nitrogenous thermal water	130	9×10^{-2}	460
Others	$< 10^{-1}$	$< 10^{-1}$	8×10^{-1}
Cesium			
Nitrogenous thermal water	43	1	500
Na-Cl water	1.8	< 5	30
Others	< 5	< 5	1
Gallium			
Nitrogenous thermal water	33	1×10^{-2}	100
Na-Cl water	$< 10^{-2}$	$< 10^{-2}$	8×10^{-2}
Others	$< 10^{-2}$	$< 10^{-2}$	9×10^{-1}

Cesium is reported furthermore in small quantities in oil field brines (up to 2 mg/kg) (Collins 1963), in warm springs ($<$ 0.1–0.55 mg/kg), and in thermal waters of New Zealand (between 0.02 and 4.7 mg/kg) (Ellis & Mahon 1964, Golding & Speer 1961). Cesium released by weathering or derived from atmospheric precipitation is rapidly and strongly absorbed by solid soil materials, especially by clays (Davis 1963, Jenne & Wahlberg 1968). Very small mobility of cesium was observed in field studies of radioisotope migration (Ames & Rai 1978; see Section 2.2).

Gallium values between 4.9 and 13 μg/l have been reported by Silvey (1967) in five spring waters from the San Jacinto Mountains, Riverside County, California.

Titanium should also be mentioned as one of the more abundant elements in the Earth's crust; the concentrations should be over 0.1 μg/l in some groundwaters. Silvey (1967) gives concentrations between 0.70 and 12 μg/l in spring waters, and 2.1 μg/l as the average of 14 samples from groundwater in California.

Bismuth is noticeably scarce. It was found in only one Californian spring by Silvey (1967) at 0.7 μg/l, and in three samples of ground (0.80–2.7 μg/l).

Tin has been detected in oil field brines in the borehole at Langensalza (Zechstein Hauptdolomit) at a concentration of 120 μg Sn/l (Krejci-Graf 1963a). Hermann (1961) gives an average tin content of 14 μg/l for oil field waters in the Hauptdolomit of the southern Harz district.

Cerium is relatively immobile in natural aquatic systems because of the low solubility of its phosphate and hydroxide $[Ce^{3+}][OH^-]^3 = 1.5 \times 10^{-20}$ (Vickery 1953) and its adsorption on ground materials. On the other hand cerium can be transported in form of ions, carbonate complexes, and perhaps organic complexes or colloids. Thus a higher content of HCO_3^- causes a higher solubility (Balashov et al. 1964). Field observations of the migration of liquid high level nuclear wastes showed that cerium was more mobile than plutonium, but less than ruthenium (Ames & Rai 1978). The single cerium value of a natural groundwater (0.5 μg Ce/kg) from a dug well in crystalline schists of the Virginia Piedmont area west of Washington, D.C. (Robinson et al. 1958), indicates its scarcity in natural systems.

The rare element ruthenium has not been reported in natural groundwater, but it has been found in groundwaters contaminated by nuclear fission products. Ruthenium exhibits several oxidation states varying from Ru(II) to Ru(VIII), with Ru(III) and Ru(IV) the most common oxidation states in aqueous solutions. In an alkaline medium Ru(IV) forms the insoluble hydrated oxide $RuO_2 \cdot nH_2O$, whereas in acid solution ruthenium may form cationic, anionic, and neutral complexes. Field observations indicate a high mobility of ruthenium (Ames & Rai 1978; see Section 2.2).

Noble Gases.   The noble gases helium and argon are constituents of the atmosphere (5.24×10^{-4} and 0.934 vol %, respectively). Both, as well as radioactive radon, are nuclear decay products of radioactive elements — helium and radon from the elements uranium and thorium, argon from the radionuclide ^{40}K. The other noble gases — neon, krypton, and xenon — are less abundant than argon (1.82×10^{-3}, 1.14×10^{-3} and 8.7×10^{-6} vol %, respectively).

The elements may possibly be concentrated in deep groundwater bodies with slow water movement. In shallow groundwater an equilibrium concentration comes about dependent on pressure and temperature conditions in the ground air–water phase boundary. In cases of confined groundwater flow, the concentrations of noble gases in the groundwater can be used as a measurement of the pressure and temperature conditions at the time of the groundwater recharge. The smallest subsequent changes in the groundwater zone naturally affect the noble gases, such as helium and argon. In this way Sugisaki (1961) was able to demonstrate by measurement of argon content the temperature at which groundwater is recharged from the Makita River; he discovered that summer infiltrated water had a lower argon content, and winter infiltrated water a higher argon content. This method was applied to measurement of groundwater flow velocities (Fig. 56).

On the basis of general geochemical distribution the He/Ar ratio of about 0.0005 in surface waters should increase to 1.0 in brines in regions of natural gas occurrences (Anderson & Hinson 1951; Junge 1963, Zartman et al. 1961).

Spectacular ^{3}He and ^{4}He enrichments of magmatic origin were observed in the hot brines in the bottom of the Red Sea. They are supersaturated with respect to atmospheric solubility equilibrium by factors of 3200 for ^{3}He and 370 for ^{4}He (Lupton 1977).

Noble gases occur in springs and hydrothermal areas on continents in about the same proportions as they are present in the atmosphere but significantly below saturation for reasonable recharge temperatures (Mazor & Wasserburg 1965, Gunter & Musgrave 1966, Mazor 1972, Mazor & Fournier 1973). Noble gas concentrations equivalent to 3% up to 64% of the saturated values are reported by Mazor & Fournier (1973).

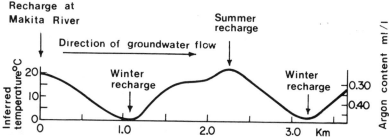

Fig. 56 Argon in groundwater in the vicinity of Takada, Japan. The profile is parallel with the groundwater flow direction so that the distance between temperature minima represents the flow path during 1 year (after Sugisaki 1961).

The most important radon isotope in groundwater is ^{222}Rn, a daughter product of ^{226}Ra; this noble gas has a half-life of 3.82 days. The other 11 radon isotopes are scarcer and have significantly shorter half-lives, thus ^{220}Rn (thoron), a daughter product of thorium, has a half-life of less than 1 minute. Radon that originates in the groundwater zone dissolves in the water there; there its concentration depends on the rate of production, the radioactivity decay rate, and the possibility of its escape into the atmosphere. With very slow flow velocities in deeper parts of the aquifer the dissolved radon can approach radioactive equilibrium with the radium content of the aquifer. In normal groundwaters one can consider the radon content to range from less than 3.7 Bq/l to approximately 1110 Bq/l; the higher concentrations are to be found more often in regions of granitic rocks. Higher radon concentrations of the order of 1850–16,200 Bq/l are found in uranium deposits, in deep fracture zones, and in thermal springs (Belin 1959, Jourain 1960, Mazor 1962). Even higher values have been recorded in Maine and New Hampshire, the highest value of 41,810 Bq/l in a well in the Fitchburg Pluton, New Hampshire (Smith et al. 1961, Kenny et al. 1966).

Radon escapes rapidly from the surface water. Locally raised concentrations in river water can, according to Rogers (1958), be used for the purposes of locating groundwater inflow into surface water.

3.2.2.1.3 *Organic Substances.* Organic substances are dissolved in groundwater originating either from the active soil zone or from aquifer materials. The substances include humic acids, pectins, hydrocarbons, simple fatty acids and their salts, and tannic acids and their degradation products (phlobaphene) (Csajaghy 1960). Polycyclic aromatics, which can be carcinogenic, are detectable in natural groundwater at concentrations of 1–10 ng/l down to depths of 50 m (Borneff & Kunte 1965, Borneff 1967, Kruse 1965). Polycyclic hydrocarbons occur as natural products of metabolism but also in man-made pollution. As slightly soluble substances they can reach the water table in appreciable amounts only under circumstances in which bacteria and viruses would have reached the groundwater (Borneff & Knerr 1960). Organic substances are produced partly by Actinomycetes and blue-green algae, which give rise to earthy odors that contain the dimethyl-substituted, saturated, two-ring *tert*-alcohol geosmin ($C_{12}H_{22}O$) as a major constituent (Medsker et al. 1968).

Hydrocarbons occur under anaerobic conditions, particularly in oil field waters, especially methane, ethane, and other low members of the methane series. Besides the already mentioned materials, one finds water-soluble organic acids like fatty acids and naphthenic acids (0.1–30 meq/l), phenols (content up to approximately 4 mg/l), aromatic oxygen compounds, esters, and asphaltic compounds that contain oxygen, nitrogen, and sulfur (Smith & Sutton Bowman 1930, Shvets 1964, Davis 1967, Al'Tovskii et al. 1961, 1962, White 1957b). Furthermore, one should mention that amino acids are present in concentrations of about 20–160 μg/l, not as free forms, but as

components of humic acids, organic "heteropolycondensates" from qui-
nones, phenols, and amino compounds as basic structural units (Degens et al.
1964, Degens & Chilingar 1967).

Krejci-Graf (1963a) presumed that metal-free porphyrin derivatives,
phthalates, sugar, and carotene are present in oil field brines; the last probably gives the water a rose color. The increase in the amount of dissolved
organic substances when water comes into contact with petroleum deposits is
clearly seen in an example in the Grozny-Daghestan region, USSR (Fig.
57). In recharge zone A, the Chernye Mountains, the organically bound carbon content of the groundwater amounted to 3.4 mg/l, in the vicinity of the
petroleum deposit (region B) 6.6–34.7 mg/l, and in the emergence zone C the
thermal springs of the chain of hills of Peredonye, because of microbial
breakdown, on average only 3.4 mg/l (Al'Tovskii et al. 1961).

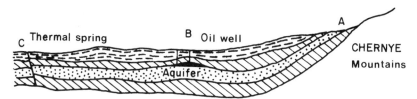

Fig. 57   Aquifer in a petroleum deposit (after Al'Tovskii et al. 1961): A, recharge area; B, oil-
bearing zone; C, emergence area.

In western Turkmenistan and eastern Georgia, USSR, Al'Tovskii et al.
(1962) found that groundwaters from oil-free beds contained on average 6.9
mg/l organic carbon, 1.86 mg/l organic nitrogen, and no naphthenic acids.
Waters in the vicinity of the oil-water contact were rich in organic carbon
(24.7 mg/l) with very low content of organic nitrogen: C/N ratio = 28. The
naphthenic acid content reached 900 mg/l.

Methane and ethane occur occasionally in thermal waters in volcanic
regions; these substances possibly originate from sedimentary rocks or from
organic substances that have come into contact with hot lava (White 1957b).
The dissolved organic substances serve as energy and carbon sources for the
microorganisms living in the groundwater and are therefore broken down
particularly rapidly in the presence of dissolved oxygen (Section 2.5). Except for a humus-rich subsurface, the adsorption of dissolved organic substances is generally negligible. The adsorption of surfactants in humus-rich
soil (Gottschaldt & Winter 1967) and the adsorption of nonionized aliphatic
compounds by calcium-montmorillonite are examples (Hoffmann & Brindley 1960).

Dissolved organic substances are often the cause of water coloring, and in
the most minute amounts they act as odorivectors and taste carriers. In very
many cases the dissolved organic substances are attributable to man-made

pollutants that enter the groundwater with polluted river water, by pollution with petroleum products, or as a result of disposal of solid, liquid, and gaseous wastes.

Odorivectors, which for example bring about peculiar odors in the bank-filtered river water at Coblenz, Bad Godesberg, and Düsseldorf, are caused by a neutral oily mixture extractable by carbon tetrachloride. By thin layer chromatography it was determined that oxygen- and chlorine-containing (ethers and alcohols of chlorinated aromatics) and oxygen- and chlorine-free hydrocarbons (hydroaromatics) from the sesquiterpene group are the chief odorivectors. Furthermore, paraffins and elemental sulfur are detectable (Koppe 1966).

The organic matter in solution is mostly determined as bulk values, named according to the method of extraction and determination. The two chief criteria of the amount of organic material present are the permanganate value and the chemical oxygen demand (COD), as measurements of the oxygen necessary to remove different groups of oxidizable constituents.

Permanganate value: 1 g $KMnO_4$ corresponds to an oxygen demand of 0.253 g O_2

Chemical oxygen demand: 1 g $K_2Cr_2O_7$ corresponds to an oxygen demand of 0.167 g O_2

The determination of organic carbon (i.e., TOC or DOC), which embraces all dissolved organic matter, is a better test (Kempf 1970).

3.2.2.2   Insoluble Constituents

3.2.2.2.1 *Suspended Matter.* Suspensions are particularly noticeable in karst water, and to a lesser extent in groundwater in fissured rocks, because turbulent flow velocities allow the movement of large particles. The commonest are suspensions of clay and fine silt. Organic detritus (seeds, etc.) also may be transported into fracture systems.

In granular aquifers the tortuosity of flow paths generally hinders the transport of suspensions. Suspended matter can occur in wells when ferric hydroxide precipitates form, or when clay, silt, or fine sand is carried into the well from its immediate surroundings as a result of very high inflow velocities.

Clay and silt suspensions of this kind may be detrimental to groundwater quality. For example the water from the coarse Pleistocene gravels and sands in the Upper Rhine Valley at Freiburg im Breisgau and Emmendingen were found to be unsatisfactory for use in photographic laboratories because fine sand and silt (grain size $\leq$ 0.1 mm) were carried along into the water supply, which in turn spoiled the emulsion coatings. This trouble was eliminated by the use of large diameter well screens, which cut down the flow rates (Braun et al. 1953, Truelsen 1965).

Davis & De Wiest (1967) have shown that deposition in water supply mains of a concentration of 5 mg/l of suspended clay, silt, and fine sand can cause considerable trouble and can also lead to land subsidence in the vicinity of wells and damage to the wells themselves. With a content of 5 mg/l some 1300 kg of material is extracted within a month from below ground with a yield of 100 l · s^{-1}. Concentrations above 500 mg/l hardly ever occur.

3.2.2.2.2 *Organisms in Groundwater.* Groundwater is a natural habitat for numerous microorganisms. The occurrence and distribution of these has been fully described in Section 2.5.

Microorganisms that belong to other habitats are occasionally carried below the surface, but they cannot survive for long because the conditions do not meet their demands. To this group belong most pathogenic microbes and the indicators of fecal pollution (*Escherichia coli*) (cf. Section 2.5).

The conditions in central Europe are such that a period of 50–70 days underground is sufficient to eliminate the microbes that enter groundwater with domestic sewage (Knorr 1966).

Laboratory studies have shown that viruses carried below the surface are adsorbed onto clayey and silty matter. The sorption process is influenced by the pH value of the water-rock system, chiefly in consequence of the amphoteric nature of the protein sheath, round viruses. At pH values below 7–7.5 the sorption is quick and highly effective. At higher pH values the efficiency of sorption falls rapidly as a result of ionization of carboxyl groups of the virus proteins and increasing negative charge on the soil particles. With increasing cation concentration in the water this detrimental effect is partly or completely neutralized, so that the sorption is favored. In general sorption increases with increasing clay or silt content and increasing ion exchange capacity; however, good virus adsorption has also been found in materials that because of grain size composition and the exchange capacity would seem to be unsatisfactory. Hence in porous media there is little danger of viruses entering the groundwater (Drewrey & Eliassen 1968).

Antagonistic and antibiotic microorganisms seem to have different effects on viruses that do enter with the percolate. Smaller viruses are generally less affected than larger viruses. The greatest effect is to be expected in the region of the highly active humus-rich soil layer (Schinzel 1968).

The possibility of entry and propagation of viruses and other microorganisms through the groundwater depends on the subsurface conditions. An undisturbed soil cover, and fine-grained, highly adsorbent ground and aquifer materials prevent groundwater contamination. Fissured rock and karst aquifers, and very coarse outwash sandy-gravelly deposits with high groundwater flow velocities are particularly susceptible to contamination, especially when an extensive soil or subsoil cover is lacking. Thus infectious Hepatitis Epidemica viruses from subsurface percolation points have been found to move up to 200 m through the groundwater body (Schinzel 1968). No data on the conditions in the aquifer concerned are given. The same holds for poliomyelitis viruses in London, spread through polluted spring and groundwater as reported by Carlson (1965), infectious Hepatitis Epidemica viruses at Mount Allison University in Sackville, Canada, and other examples in the United States, Czechoslovakia, and East Germany. The hepatitis epidemic at Split, Yugoslavia, caused by infected karst water (Carlson 1965) suggests that unfavorable underground conditions were present also at the other places. The outbreak of a typhus epidemic at Glion, Switzerland, in 1945 was caused by groundwater infected with percolating sewage (Schinzel 1968).

The transport of bacteria and viruses in a groundwater system may be treated like that of other contaminants using the general transport equation (Bear 1972), including terms for the biological elimination of these organisms in the groundwater milieu and for adsorption (Keswick & Gerba, in press, Meinick et al. 1978, Matthess & Pekdeger 1981).

The occurrence of higher organisms in karst groundwater is well known. The most famous example is the 20–30 cm long, blind, colorless cave newt (*Proteus anguinus*), a tailed amphibian found in the Yugoslavian karst. Less well known is the adapted fauna in the smaller pores of permeable sands and gravels, which include representatives of turbellaria, nematodes, oligochetes, copepods, and halacarids. The characteristic organisms of these zones are highly adapted to a life in tiny water passages and therefore have elongated wormy shapes. Because the elongation of the tiny water passages depends on the grain size distribution of the sediments, the faunal elements can be classified as a series with differing space requirements related to the grain size of their habitats, and these can be seen very well in slow sand filters. The characteristic forms include *Parastenocaris*, *Bathynella*, *Parabathynella*. The topic has been covered by Husmann (1966a, 1966b, 1968).

In sediments with an average grain diameter of about 0.6 mm investigators have observed, for example, harpacticoids *Phyllognathopus vigueri, Epactophanes richardi, Parastenocaris fontinalis bora,* and *Nitrocrella reducta;* water mites *Lobohalacarus weberi quadriporus;* and cyclopoids *Graetiriella unisetigera.* The other cyclopoids require a more spacious environment. In filter sands of medium grain size ($\approx$ 1.5 mm) *Paracyclops fimbriatus,* a species in the Diacyclops-languidoides group, a Cladoceres-type *Alona guttata,* and species of the genus *Candona,* together with chironomid larvae, oligochetes, and nematodes were identified (Husmann 1968). Groundwater in sands and gravels is the preferred environment of *Phyllognathopus, Epactophanes, Parastenocaris,* and *Nitrocrella reducta* (Husmann 1968).

In addition to the normal groundwater fauna of oligochetes and turbellarians, Crustacea, *Nyphargus,* protozoans, and Rotifera which have been reported by Kooijmans (1966) in wells 20–60 m deep; other organisms may have entered the groundwater during well construction and adapted themselves to the conditions there.

In addition to the generally phylogenetic ancient forms mentioned, which are completely adapted to the subsurface conditions, still other species occur in underground areas. Thus Schwoerbel (1961) has identified 14 species of water mites (Hydrachnellae) of the genus *Atractides* in several mountain streams in southwest Germany, in the water-filled interstices between the boulders of the stream bed and along the banks, which are ecologically positioned between the surface water and groundwater. The organisms show a progressive adaptation to this environment below the stream bed, which is named in ecology the hyporheic region. Seven species are restricted to differing surface waters with an inherent endemic hyporheic *Atractides* fauna. This indicates that these species are so young phylogenetically that they have not yet been able to spread.

3.3 INFLUENCE OF AQUIFER MATERIALS ON GROUNDWATER QUALITY

Groundwater is subject to diagenetic changes during its passage through the aquifer because of dissolution, hydrolysis, precipitation, adsorption, ion exchange, oxidation and reduction, and microbiological metabolism (Chapter 2). The properties of the aquifer are more clearly shown the more soluble are the rock-forming minerals, the more quickly their breakdown proceeds, and the greater the contact surface and the contact time between rock and water. Ion exchange requires the presence of exchangers; microbial metabolism a supply of organic matter. Groundwaters of similar age from similar rock masses in the same region have generally conformable quality. As a result of differences in the quality of the recharge water and differing influences on the subsurface water passages, various periods of stay underground, and diagenetic processes of different types, waters from one rock type can exhibit greater or less degrees of variations in quality. Variations can also occur among multiple aquifers within the same rock mass; these can be brought about through the causes mentioned, and through the different confining strata and the effects of geochemically distinct superimposed waters.

The connection between aquifer and groundwater quality is sufficiently narrow for hydrochemical maps to be made by consideration of the petrographic properties of the rock, so that the groundwaters can be classified according to genetic types (Gerb 1953).

Igneous and Metamorphic Rocks. Igneous rocks can be divided into plutonic (deep) and volcanic (surface) types. Most igneous and metamorphic rocks form dense masses that are impermeable except through fractures (faults and joints). Volcanic rocks are partly highly porous, so that some make good aquifers.

For igneous and metamorphic rocks of the same mineralogical composition the same principles hold during weathering. Volcanic rocks weather more easily because they are more porous and possess rather more surface internally for reactions, and to some extent they contain noncrystalline glassy matter (Hem 1970).

In general groundwaters contain only small amounts of dissolved solids when they are in crystalline rocks because of the slow weathering of silicates. Arid regions are an exception; here evaporation can lead to high concentrations of dissolved solids (Section 2.1).

High silica content is characteristic of groundwater in crystalline rocks (Hem 1970). This mostly amounts 25–55 mg/kg; groundwaters in quartzitic rocks, argillites, and phyllites however generally have less than 30 mg/kg silica (Davis & De Wiest 1967). An appreciable amount of the silica freed during silicate weathering combines with aluminum to form clay minerals (Section 2.1).

The alkalis and alkaline earths are liberated by weathering of silicates; however their ratios in groundwater are generally different from those in the original rocks because of the different degrees of resistance to weathering of the minerals and different fixation of the weathering products (Table 64). Potassium especially is incorporated in the weathering products. The iron liberated by weathering of pyroxenes, micas, amphiboles, and pyrite goes into solution, although some of it is precipitated as amorphous iron hydroxides. High dissolved iron concentrations are due to reducing conditions (Section 3.2.2.1.1). A really high iron content of 5.1 mg/l has been reported by White et al. (1963) in groundwater from the Waterloo Gabbro, Maryland.

The SO_4^{2-} and Cl^- contents do not generally exceed 2 meq/l, except in arid regions. Sulfate ions are formed by oxidation of sulfides, especially pyrite. The low Cl^- content derives from the low chlorine content of the rock and from atmospheric input.

Most of the biogenic carbon dioxide takes part in the hydrolysis of silicates and often is present as H_2CO_3. The pH value in such cases is low, but it is shifted with the escape of CO_2, and with dissolution of secondary carbonates the water can become alkaline.

Groundwater in Granite, Rhyolite, Gneiss, and Similar Rocks. These rocks contain quartz, orthoclase, sodium-rich feldspar, biotite, and muscovite as the major minerals. The slow breakdown of these rocks leads to low solute concentrations (mostly considerably below 300 mg/l total dissolved solids). Relatively low silica content and alkali concentrations, mostly below 2 meq/l with sodium dominance over potassium, are characteristic (Table 65, analyses 12–16). Higher alkali concentrations, for example 11.23 meq/l in groundwater in granite in the Transvaal (White et al. 1963), occur only in dry regions. With exception of hornblende- and augiterich rocks, alkaline earths are lower in concentration than the alkalis. Thus calcium is predominant over magnesium, as in the parent rock (cf. Table 73) (see further discussion in Garrels 1967, Paces 1972, 1973).

In hornblende-granites, diorites, and similar rocks calcium-rich plagioclase and hornblende are the dominant minerals, while the mica and – in the case of diorite – quartz diminishes. Groundwaters in these rocks contain characteristically more Ca^{2+} and more dissolved silica.

Table 73   Ratios of Sodium to Potassium and Calcium to Magnesium in Crystalline Rocks and Spring Waters (After Dittrich 1903a, 1903b)

		Granite		Granodiorite	
		Spring water (Löwenbrunnen)	Fresh water	Weathered rocks	Spring water
Na_2O/K_2O	1:1	1:0.2	1:0.93	1:4.2	1:0.22
CaO/MgO	1:0.83	1:0.33	1:0.73	1:1.67	1:0.28

Groundwater in Gabbro, Basalt, and Similar Crystalline Rocks. These rocks contain more calcium- and magnesium-bearing minerals. The groundwaters contain more silica than the water described above, very often a few tens of milligrams of SiO_2 per litre, with a frequency distribution maximum between 20 and 40 mg/l. Furthermore, alkaline earths are dominant over alkalis; however, except in dry regions, not more than 3 meq Ca^{2+}/l and 2 meq Mg^{2+}/l are met. The sodium content does not generally exceed 2 meq/l, and residues after evaporation 400 mg/l (Table 65, analyses 17–20).

Comparison of slightly acid groundwaters from granite, gneiss, mica-schists, and rhyolites with slightly alkaline waters from gabbro, diorite, hornblende-gneiss, and andesite (LeGrand 1958) shows clearly that groundwaters in metamorphic rocks have chemical qualities similar to those in igneous rocks. Water from mica-schists are alkali rich, those from amphibolites on the other hand are richer in alkaline earths (Table 74).

Table 74   Comparison Between Water from Granitic and Dioritic Rocks in North Carolina (After LeGrand 1958)

	Granitic rocks (mg/l)				Dioritic rocks (mg/l)			
	Median	Average	Minimum	Maximum	Median	Average	Minimum	Maximum
SiO_2	30	28	9.9	39	32	34	7.1	60
$Fe^{2+(3+)}$	0.2	1.2	0.01	8.7	0.2	0.6	n.d.	5.7
Ca^{2+}	5	5	1.8	12	38	49	13	174
Mg^{2+}	2	2	0.6	4.5	12	12	2.6	40
$Na^+ + K^+$	7	7	1.3	14	11	14	3.3	35
HCO_3^-	34	35	12	80	127	137	50	304
SO_4^{2-}	2	4	0.1	23	17	44	0.1	391
Cl^-	2	4	1 1	17	14	28	1.4	204
F^-	0.1	0.1	n.d.	0.9	0.1	0.2	n.d.	1.8
NO_3^-	0.9	2.9	n.d.	14	1.3	4.3	n.d.	30
Total dissolved solids	71	75	25	123	233	269	106	696
pH	6.5	–	5.8	7	7.1	–	6.5	8.2
Number of analyses		29				23		

The low grade metamorphic slates, especially phyllites and clay slates, provide greater surfaces of contact for groundwater and are permeated more slowly. Hence the total dissolved solids concentration and the concentrations of the individual substances are generally higher than in igneous rocks and high grade metamorphites. Higher SO_4^{2-} concentrations and acid pH values occur in pyritous slates.

Groundwaters in metamorphosed limestones and dolomites (marbles) do not differ from those in unaltered limestones and dolomites and are therefore treated together with them.

Groundwater in Sedimentary Rocks. The four most important sedimentary rocks, the sandstones and other psephitic and psammitic hard rocks; clays, marls, and silty rocks; carbonates; and gypsum, anhydrite, halite, and potassium salts give a characteristic imprint to the groundwater because of their soluble constituents. These differences can be blurred through the occurrence of important constituents such as carbonates, gypsum, anhydrite, and other salts as cementing material or intercalated lenses. Thus gypsum and anhydrite often contain NaCl, Na_2SO_4, and $MgSO_4$. There are peculiar features in groundwaters that have been in contact with caustobiolites.

Groundwater in Sandstones and Other Psephitic-Psammitic Hard Rocks. Sandstones show in addition to their joint system permeability a variable matrix permeability. The effective pore spaces present offer extensive contact surfaces and cause relatively low permeability, hence a long period of contact between water and rock (cf. Section 2.1).

The character of the dissolved solids in sandstone groundwaters therefore depends on the material in the sandstone. Pure siliceous material without soluble cement leads to groundwaters with very low total dissolved solids. The CO_2, controlled by its partial pressure in the ground air, can become aggressive because of the lack of bases in the rock and can bring about low pH values (5-6). The low dissolved solids value originates essentially from other sources, such as atmospheric precipitation, especially Na^+, Cl^-, and SO_4^{2-}, so that these values can even exceed the Ca^{2+} and Mg^{2+} and hydrogen carbonate content (Table 65, analyses 21-23).

Groundwater in sandstones with soluble cementing materials may show appreciable amounts of SO_4^{2-}, Cl^-, Na^+, Mg^{2+}, and Ca^{2+}, even more than groundwater in carbonate rocks, particularly in hot, arid countries.

Groundwater in Unconsolidated Aquifer Materials. Unconsolidated sand and gravel form permeable porous aquifers that can differ greatly in their petrological properties, in which the sedimentation controls (marine, fluviatile, lacustrine, glacial, or aeolian) and the properties of the detrital material play an important part. The large surface of contact between sediments and groundwater, and the long duration of stay underground may in many cases allow the establishment of equilibrium between rock and groundwater properties. The residence time underground depends on the groundwater flow system (terrace deposits alongside surface water channels, deposits in large zones of depression like tectonic grabens or coastal plains, dunes, etc.). The water quality is often affected by ingress of sea or flood waters or by man-made causes. Depletion of O_2 brings about reducing conditions, leading to an increase in the iron and manganese content. The most irregular occurrence of this kind of trouble is caused by small pockets of organic matter, which bring about low *Eh* and pH values. The groundwater quality in such rock formations can thus be very variable.

When unconsolidated beds contain lime or gypsum, a common feature with detrital sedimentary rocks in general, the water shows clearly increased alkaline earth and hydrogen carbonate contents (typically 50–200 mg Ca^{2+}/l, 10–50 mg Mg^{2+}/l, 100–250 mg HCO_3^-/l) and higher sulfate content (50–300 mg SO_4^{2-}/l). If on the other hand the sediments have been derived from metamorphic or igneous rocks, then they contain far fewer soluble substances, and the groundwater shows a low dissolved solids content. The sulfate and chloride contents are then essentially low (each > 50 mg/l), whereas the silica content is typically higher, especially in the presence of volcanic matter.

Most shallow groundwaters in such aquifers in dry regions show marked variability, particularly an increase in concentrations, because in these porous sediments the capillary rise, and thus the evaporation are greatly increased.

Groundwater in Carbonate Rocks. The solution of carbonate rocks depends on the content of free dissolved carbon dioxide, which in groundwater amounts mostly to between 20 and 50 mg/l and is explained by the increased partial pressure in the ground air (Sections 1.2.1, 1.2.3.2.1, and 3.1.2). The variations in the free dissolved CO_2 are also related to the variations between 100 and 200 mg/l of bound carbon dioxide. Hence the groundwater cannot dissolve much more than 150–300 mg/l of limestone (Schoeller 1962). The velocity of the solution process, hence the liberation of other salts, depends on the petrological properties of the limestone concerned.

The commonest carbonate rocks, limestones and dolomites, are of extraordinarily diverse origin — inorganically precipitated calcareous mud, coquina, biological waste, carbonate sand, and coral reef masses and their debris. The primary porosity and permeability can be very variable, ranging from practically impermeable, dense, pore-free rocks to extremely permeable sediments rich in voids. The primary permeability in crystalline limestones, marbles, and fine-grained calcilutites is practically negligible ($\leq 10^{-8}$ m $\cdot$ s^{-1}), whereas in porous limestones, particularly in poorly cemented coarse breccias and oolites, it can be considerable ($\sim 10^{-2}$ m $\cdot$ s^{-1}).

Dolomitization involves some increase in porosity because the change of calcite into dolomite involves a 13% volumetric reduction in the rock. On the other hand the diagenetic processes mechanical compaction, cementation, solution and crystallization, and the forming of authigenic minerals collectively tend to reduce the porosity and the permeability.

Carbonate rocks contain other water-soluble salts (Table 75). The $CaSO_4$ fraction comes chiefly from interlaminated lenses of this sulfate, while the other salts chiefly or totally form intracrystalline liquid inclusions that in dolomites are smaller but more numerous than in limestones.

In carbonate rocks the permeability of the whole rock mass is considerably greater than that of the rock itself, and groundwater movement takes place chiefly through joints and larger fissures in the body of the rock. The surface of contact between water and rock is thus limited. The soluble salts in the carbonates, chlorides, and sulfides are released only as the mass of the rock is dissolved away at the water channel interfaces. Porous or weathered carbonate rock can however be leached more easily. The chemical quality of the groundwater reflects the chemical composition of the limestones. Groundwater in limestones contains more alkaline earth and carbonate ions, relative to alkalis, chloride, and sulfate.

Table 75 Concentrations of Soluble Salts in Limestones and Dolomites (After Lamar & Schrode 1953).

	Average concentrations (wt % of rock)	
	Limestone	Dolomite
Na^+	0.008	0.013
K^+	0.003	0.004
Ca^{2+}	0.017	0.004
Mg^{2+}	0.003	0.026
HCO_3^-	0.014	0.051
SO_4^{2-}	0.038	0.028
Cl^-	0.012	0.045
Σ	0.093	0.172

The low solubility of carbonates and the limited contact surface explain the small content of dissolved solids. The pH of the groundwater from carbonate rocks is generally above 7.

For dense limestones the greater part of the residue after evaporation consists of HCO_3^- and Ca^{2+} ions; SO_4^{2-} and Cl^- ions are present only in very small amounts. The water from fine-grained limestones shows slight increases in the SO_4^{2-} and Cl^- contents, whereas the water from porous limestones has increased HCO_3^- and Ca^{2+} content. (Table 65, analyses 24–29).

Groundwaters in dolomite rocks are subject to the same diagenetic processes as those in other limestones. The equivalent ratio Mg/Ca in the groundwater is rather less than in the rock, as a result of the preferential solution of calcium carbonate, and this can lead to magnesium enrichment in the remaining rock mass, that is, to further dolomitization (Gorup-Besanetz 1872, Bär 1924, 1932, Keilhack 1935).

Groundwater in Clayey, Marly, and Silty Rocks. The porosity of fine-grained clayey, marly, and silty rock formations decreases significantly as a consequence of the mechanical pressure forces, which in turn depend on the depth of the overburden. Unconsolidated fine-grained marine muds exhibit porosities between 50 and 90% (Von Engelhardt 1960). In general the greatest decrease in porosity occurs because of onset of diagenetic changes approximately as far down as 500 m. In the zone between 1000 and 3500 m the porosity decreases proportionally to the depth. At 4000 m depth porosities below 1% have been measured. During the consolidation process considerable amounts of water are squeezed out. In the case of marine muds these waters possess a high content of dissolved mineral matter, and this can lead to mineralization of the groundwater.

Since water movement is very slow in these rocks of poor permeability, there is a long period of contact; and because of the small size of the pores, there is a high surface of contact between rock material and water. By adsorption and ion exchange salts are incorporated into these rocks, particularly chlorides and sulfates. A fraction of these substances has been locked in during deposition within voids and cannot be easily leached out because of the low water velocity.

Groundwater from clay-silt-marl deposits is therefore very often rich in dissolved solids – up to several grams per litre – which is particularly attributable to the high SO_4^{2-} and Cl^- content. The high values of SO_4^{2-} are associated with high Ca^{2+} and Mg^{2+} content, and high Cl^- with high Na^+ content (Schoeller 1962). Cation exchange is a characteristic feature of these rocks (Section 2.2).

The silica content is higher than in other waters. Groundwater in argillaceous rocks has comparatively higher iron and fluoride content, and a relatively low pH (Davis & De Wiest 1967).

Groundwater from Gypsum-. Anhydrite-, and Salt-Bearing Rocks. In these rock formations the permeability through joints and solution channels is the only effective means of control of flow rates. The high solubility of gypsum- and anhydrite-bearing rocks leads to the formation of karst phenomena with master water courses underground, while salt rocks in humid regions are quickly dissolved away. Groundwater from gypsum and anhydrite deposits shows high SO_4^{2-} content in combination with Ca^{2+} or Mg^{2+}, which is also contained in gypsum. The solutions are often saturated with respect to $CaSO_4$ so that only magnesium can go into solution.

During the solution of salt-bearing rocks very high levels of SO_4^{2-}, Cl^-, Ca^{2+}, Mg^{2+}, and Na^+ are reached, so that the total dissolved solids can exceed 200 g/l.

Because the solubility of $CaSO_4$ increases in the presence of NaCl, Na-Cl waters exhibit a lower equivalent ratio Mg/Ca than do mineral-poor groundwaters.

Groundwater in Contact with Caustobiolites. Caustobiolites such as coal, lignite, peat, and petroliferous hydrocarbons alter the groundwater quality by consuming oxygen, which leads to a reducing environment (Table 76). Coal and lignite beds can act as aquifers because of their jointing, peat deposits because of their inherent porosity. Oily hydrocarbons come into contact with the groundwater at the oil-water and gas-water phase boundaries. Typical constituents of the groundwater under these reducing conditions include H_2S, ammonia, iron(II) and manganese(II) ions, and raised CO_2 contents. Furthermore, hydrocarbons, organic acids, carbohydrates, and other organic compounds are found in these groundwaters, particularly in oil field waters (Section 3.2.2.1.3). Common constituents of oil field waters also include phosphate, ammonia, iodide, bromide, as reported by Krejci-Graf (1963a), who describes them as biophilic elements and substances having a connection with petroleum formation.

Table 76   Groundwater in Contact with Peat, Lignite, and Coal (After Schoeller 1962)

	Peat, France (mg/l)		Lignite, Dakota (mg/l)			Coal, France (mg/l)		
	1	2	3	4	5	6	7	8
Mn^{2+}	—	—	—	—	—	0.32	—	—
Fe^{2+}	—	7	0.10	0.8	0.58	0.008	1.8	—
Ca^{2+}	87	7	16	7.6	30	30	82	9
Mg^{2+}	24	8	6.6	4.3	16	11	32	4
Na^+	—	13	585	430	515	—	(60)	(322)
Cl^-	—	20	36	6.0	134	10.3	18	92
SO_4^{2-}	tr.	16	11	400	100	28	35	164
$(HCO_3^- + CO_3^{2-})$	187	95	753	319	584	—	234	255
NO_3^-	—	—	3	2	2	4.4	18	17.5
SiO_2	—	—	—	12	14	32	11.5	—

According to Palmer (1924) oil field waters can be grouped into sulfate-bearing groundwater (type 1), brines (type 2), and alkali carbonate waters (type 3). For waters of type 3, which are to be regarded as mixtures of the other two types or transitional between them, the formation of hydrogen sulfide by sulfate-reducing bacteria in the presence of hydrocarbons is typical.

Table 77 Changes in Groundwater Quality (Samples 1, 2, 5, 7) Caused by Contact with Hydrocarbons (Samples 3, 4, 6, 8, 9) (After Rogers 1917 and Schoeller 1962)[a]

Sample Nos.	1	2	3	4	5	6	7	8	9
Fe mg/l	–	–	–	–	–	–	–	–	0.1
Ca^{2+} mg/l	303	193	27	75	11	26	13	20	10
Mg^{2+} mg/l	112	121	40	44	9.6	2.9	1.2	8.6	1.2
$Na^+ + K^+$ mg/l	804	933	717	2872	1059	2809	51	296	1550
Cl^- mg/l	404	606	332	2961	1125	2920	35	58	1418
SO_4^{2-} mg/l	2181	1673	170	23	38	8.7	12	7.5	0
CO_3^{2-} mg/l	76	282	686	1435	443	1226	52	384	–
HCO_3^- mg/l	–	–	–	–	–	–	–	–	1708
SiO_2 mg/l	54	249	95	67	12	17	23	52	–
Σ mg/l	3498	4076	2070	7533	2699.3	7009.6	192.2	826.1	3821.3
rSO_4^{2-}/rCl^-	40.0	2.04	0.39	0.006	0.03	0.002	0.25	0.10	0
rMg^{2+}/rCa^{2+}	0.62	1.04	2.47	0.98	1.45	0.19	0.15	0.71	0.20
$\dfrac{rCl^- - Na^+}{Cl^-}$	−2.07	−1.38	−2.44	−0.49	−0.45	−0.49	−1.24	−6.27	−0.25
H_2S mg/l	–	–	104	–	–	–	–	–	–

[a] CO_3^{2-}: Total HCO_3^- and CO_3^{2+} reported as CO_3^{2-}. SiO_2: Inclusive of suspended matter.

The resulting carbon dioxide reacts with the excess cations to form carbonates. Hence almost insoluble alkaline earth carbonates are precipitated. The end result of the reaction after mixture of a sulfate-bearing water of type 1 with chloride-rich, hydrocarbon-bearing water (type 2) is accordingly an alkali carbonate water (type 3), which can possess an excess of hydrogen sulfide if the H_2S cannot escape, oxidize to sulfur, or be precipitated as insoluble sulfides. One important feature therefore is a relatively high sulfate content. Sulfate-free, highly saline oil field waters of type 2 on the other hand show no active sulfate reduction (Davis 1967).

Most oil field waters do not have the chemical composition of seawater not only because they are subject to diagenetic changes. In many cases they have taken part in the hydrologic cycle and possibly have been completely replaced since the time of the formation of the deposit. The salt content, which is often higher than that of seawater, was taken up by the water originally in the sediment and entered the subsurface zones with migration of highly concentrated waters.

Table 77 shows the changes through microbiological sulfate reduction and ion exchange that are displayed in oil field groundwaters. Hence the estimate of an ion exchange under discussion is made difficult if the sulfate reduction has taken place and the equivalent ratio $(Cl^- - Na^+)/(SO_4^{2-} + CO_3^{2-})$ is to be used.

Waters 1–4 originate from different horizons in the Caolinga oil field in California; the boreholes are spaced at most 360 m round a central point. Water 1 comes from a 300 m well in strata 240–300 m above petroliferous sands; water 2 apparently comes from groundwater-bearing strata at 330 m depth. The upper limit of the petroliferous sands lies at 478 m depth. The hydrogen sulfide-bearing water (3) was encountered at 404 m in the zone below asphalt-bearing sands (351–361 m) above the petroleum-bearing beds starting at 478 m. Water 4, from 540 m depth, is a typical oil field brine, and the associated oil occurrence is found at 514–533 m. Waters 5 and 6 come from the same borehole. Water 5, from 414 m depth, above a gas-bearing horizon, was unaffected; water 6, from 584 m, may also have been affected by an oil-bearing bed below via the overlying gas-bearing bed just mentioned. The brines from the oil field at Kern River, California, originate from an unaffected well (7); the other wells tap oil field waters.

When caustobiolites occur close to the surface, the abundant oxygen available oxidizes the enriched sulfide minerals so that there is an increase in sulfate in the groundwater. For example in a pyritic peaty soil in which concrete mains were laid, SO_4^{2-} contents up to 1254 mg/l in the groundwater were recorded (Bömer 1905). The pyrite content in carbonaceous strata, which come into contact with air during mining activities, is also subject to oxidation, so that the reducing effects of the organic matter apparently take on a subordinate role (Matthess 1961).

Chapter 4 Classification and
Assessment of Groundwater

4.1 REPORTING AND HANDLING OF PHYSICAL AND CHEMICAL DATA; METHODS OF ILLUSTRATING ANALYSES

The geological, geochemical, hydrochemical, and hydrological literature contains abundant reports of chemical groundwater analyses. There is also a wealth of unpublished groundwater analyses, held by chemical, geological, medical, and hydrological institutions, water supply authorities, industrial undertakings, and mining companies. In many cases however only partial analyses are available. Furthermore these data may be of questionable value — because of improper sampling, preservation, and analytical methods. Information about the way in which the samples were obtained has often been omitted — sampling depth, well depth, type of well construction (diameter, filter packs), multiple aquifers, and so on. With spring water and shallow groundwater, single analyses and temperature measurements are often insufficient for the proper assessment of the conditions in the aquifer; in these cases a series of tests must be undertaken.

The methods of conducting water analyses are laid down in the relevant instruction manuals and handbooks (e.g., Brown et al. 1970, Rand et al. 1975, Skougstad et al. 1970, U.S. Environmental Protection Agency 1974; in Germany, Eichelsdörfer et al. 1969, Haberer 1969b, Fresenius & Quentin 1969, Höll 1972). Beyond this no further mention is made of the current analytical methods because rapid developments in analytical chemistry tend to make the details out of date. A consideration of these methods would be beyond the scope of this geochemical and hydrological textbook and the reader is referred to the relevant handbooks.

Further information on microbiology and general biological research methods can be obtained from the manuals and handbooks by Reploh (1969), Dombrowski (1969), Lüdemann (1969, 1970), Bringmann (1969), and Peter (1970).

4.1.1. Numerical Presentation of Analyses

Because the dissolved solids in water are almost completely dissociated – only a few form undissociated molecules – their content in ionic form
is reported as milligrams per litre (mg/l) or milligrams per kilogram (mg/kg)
of solution. The statement "mg/kg" corresponds to the formerly used standard "ppm" (parts per million – weight of dissolved substance per million
parts by weight of the solvent). For trace elements the concentration is given
as micrograms per litre or per kilogram (μg/l or μg/kg) because this is more
practical. The values given in mg/kg (formerly ppm) and in mg/l are practically equal in dilute waters. In hydrogeological practice therefore, with
waters having solution concentrations up to about 7000 mg/l, one can omit
the conversion and simply replace the mg/l unit by mg/kg; not until higher
concentrations are reached is the conversion given by equation 108 necessary.

$$mg/kg = \frac{mg/l}{\text{density of the solution}} \qquad (108)$$

In English-speaking countries the following units may still be met:

$$1 \text{ grain per U.S. gallon} = 17.12 \text{ mg/l}$$
$$1 \text{ grain per Imperial gallon} = 14.3 \text{ mg/l}$$

In earlier publications the analyses were often reported in terms of salts,
which were calculated by more or less empirical rules. This procedure is
however to be deprecated, because a given content of sodium, calcium, sulfate, and carbonate ions originates by solution not only of $CaSO_4$ and
Na_2CO_3 but also of Na_2SO_4 and $CaCO_3$. Any decision about which of these
possibilities is the more likely depends on the geochemical history of the
groundwater, particularly the type of aquifer.

The conversion of the data from old salt content tables and reports into
ion form can be carried out by the following method:

For the compound A_nB_m:

$$mg/kg \text{ ion } A = (mg/kg \ A_nB_m) \cdot \frac{\text{nuclear atomic weight A}}{\text{molecular weight } A_nB_m}$$

$$mg/kg \text{ ion } B = (mg/kg \ A_nB_m) \cdot \frac{\text{molecular atomic weight B}}{\text{molecular weight } A_nB_m}$$

If ion B is a molecule formed of more than one element, such as SO_4^{2-},
then the molecular weight of the ion is used. For example in the calculation
for water with 54 mg $MgSO_4$/l, the SO_4^{2-} ion content is found as follows,
using the atomic weights of $Mg = 24.312$, $S = 32.064$, and $O = 15.9994$:

$$54 \text{ mg/l} \times \frac{1 \times (32.064 + 4 \times 15.9994)}{24.312 + (32.064 + 4 \times 15.9994)} = 43.1 \text{ mg SO}_4^{2-}/\text{l}$$

Calculations on and comparisons of analyses used in geochemical studies are made easier by converting the results into milliequivalents per kilogram (meq/kg = epm = equivalent parts per million). A milligram equivalent, or milliequivalent for short, of an element or ion species is that weight of the element or ion which is equal to the equivalent weight in milligrams.

Statements in terms of milliequivalents per litre (meq/l) as proposed by Stabler (1911) can be designated by the letter r in front of the chemical symbol. Concentrations in meq/kg or meq/l are obtained by dividing the ionic concentration in mg/kg or mg/l, respectively, by the equivalent quantities of the ion concerned. The equivalent quantity is the atomic weight, or, for polyatomic ions, the molecular weight divided by the valence of the ion. The significance of this unit in groundwater chemistry is derived from SI definitions by defining an equivalent part. The following examples illustrate the procedure.

(a) Conversion of 57 mg Ca^{2+} in meq/l
    Atomic weight Ca = 40.08
    Valence = 2
    Equivalent quantity $\dfrac{40.08}{2} = 20.04$

    $57 \text{ mg Ca}^{2+}/\text{l} = \dfrac{57}{20.04} = 2.84 \text{ meq/l}$

(b) Conversion of 154 mg HCO_3^-/l in meq/l
    Atomic weight H = 1.00
    Atomic weight C = 12.011
    Atomic weight O = 15.9994
    Molecular weight $HCO_3^- = 61.017$
    Valence = 1
    Equivalent quantity = 61.017

    $154 \text{ mg HCO}_3^-/\text{l} = \dfrac{154}{61.017} = 2.52 \text{ meq/l}$

Table 78 gives the equivalent quantities of the most important ions.

Table 78   Equivalent Quantities of the Most Important Ions

Na^+	22.9898	Fe^{2+}	27.9235	NO_3^-	62.005
K^+	39.102	Mn^{2+}	27.469	SO_4^{2-}	48.031
Ca^{2+}	20.04	CO_3^{2-}	30.005	Cl^-	35.453
Mg^{2+}	12.156	HCO_3^-	61.017		

In a complete analysis the number of the negatively and the positively charged equivalents must be equal. Therefore the sum of the equivalents of the cations and of the anions can be used to evaluate the accuracy of the analysis and the content of the other dissolved constituents present, but not yet determined. The completeness of the analyses can be checked by comparison of the sums of cation and anions, thus:

Cations	mg/l	meq/l	Anions	mg/l	meq/l
Na^+	8.26	0.3593	HCO_3^-	357.5	5.86
K^+	1.17	0.0299	Cl^-	12.8	0.3610
					6.22
Ca^{2+}	84.3	4.207			
Mg^{2+}	25.5	2.098			
		6.69			

Because the cation and anion totals are not equal, either the analysis is incomplete or some values are inaccurate. The anion deficit is probably caused by the presence of sulfate and nitrate ions, which in the case above were not determined. The value for calcium of older analyses may include any strontium present.

For estimation of the percentage error (i.e., "ion-balance error") the usual method is the following:

$$e = \frac{rc - ra}{rc + ra} \times 100$$

in which rc represents the cation sum and ra the anion sum (in meq). Any value of $e > 5\%$ suggests an error in the analysis or the calculations; for example one element might not have been determined. In general the value of e should be less than 2%. However Hem (1970) points out that errors exceeding 5% are sometimes unavoidable when the total of cations and anions is less than 5 meq/l. Naturally, with this procedure colloidal or suspended matter such as Al_2O_3, Fe_2O_3, and part of the SiO_2 is not included.

When the data for the most important ions HCO_3^-, Cl^-, SO_4^{2-}, and NO_3^-, as well as the figure for the total hardness are available, the alkali content can be estimated, possibly given as sodium, in the following way:

Cations	meq/l	°German hardness	Anions	mg/l	meq/l
Total hardness ($Ca^{2+} + Mg^{2+}$)	1.643	4.6	HCO_3^-	97.7	1.600
			Cl^-	8.5	0.240
			NO_3^-	0.72	0.012
			SO_4^{2-}	12	0.250
	1.64				2.10

The difference between total anions and total cations of 0.46 meq/l can now be accounted for by the alkalis Na^+ and K^+: the alkalis are commonly reported as sodium. To this end one multiplies by the equivalent weight of sodium (22.9898) and obtains the value (0.46 × 22.9898 = 10.5 mg/l). This estimate is inaccurate because all the errors of the analysis of the ions measured will affect the result, and undetermined ions can be present in the water. Finally, the H^+ or OH^- content should be considered in the calculation of ion totals, especially when pH values are markedly different from neutral, because they make up an important part of the total ion content: for example in a strongly basic water from the Aqua de Ney spring near Mount Shasta, California (Feth et al. 1961), with pH 11.6, the OH^- content is 68 mg/kg (3.98 meq/kg) [Table 65, analysis 8: 1430 mg/l, 84.08 meq/l)]. In the acid thermal water from Sulphur Spring, New Mexico (Hem 1970), with pH 1.9 the H^+ content reaches 12.7 mg/kg (12.6 meq/kg).

When making assessments a careful selection of suitable sampling methods is very important. In contrast to springs, where sampling usually involves no difficulties, although preservation does, with wells and boreholes some postulates have to be met if the natural conditions in the aquifer are to be investigated. A net of observation wells is drilled in the area, having diameters to suit the various pieces of extraction and measuring apparatus. A minimum diameter of 150 mm has proved satisfactory, although smaller submersible pumps and measuring devices are available.

If extraction is not intended to be horizon specific, open buckets and weighted flasks can be used. For horizon-defined extraction a modified type of bucket apparatus – for example a type of Ruttner baler used in oceanographic work (Fig. 58) and semiautomatic groundwater samplers (Andersen 1978) may be used. Closely spaced samples of interstitial water in unconsolidated material can be obtained from the unsaturated and groundwater zone with help of a special coring technique (Patterson et al. 1978). The interstitial water can be squeezed out by displacement with an immiscible fluid (Patterson et al. 1978) or gas, or by centrifuge extraction (Edmunds & Bath 1976). A pumped water sample gives a representative average for a zone within the aquifer. When choosing and operating a pump one should remember that it depends on the pumping interval and the pumped volume whether more distant groundwater or water from other groundwater aquifers is drawn in.

If a stratigraphically separated groundwater system exists, inflatable packers can be used that permit extraction of water from each of the individual aquifers in the system. However, there may be inflow of other groundwaters along the permeable beds in the borehole or via gravel packs.

For the appraisal of physical, chemical, microbiological, or biological results obtained for the water from the wells, changes in the water quality need to be considered, which occur as a result of pumping activities: aeration of the aquifer in the cone of depression during pumping, degassing, and

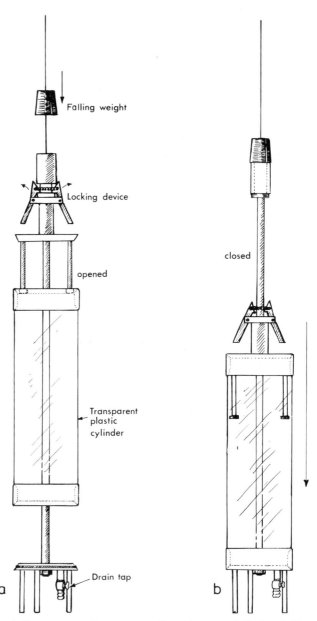

Fig. 58 Ruttner baling apparatus for water sampling: (*a*) open and (*b*) closed. (For supplies see oceanographic suppliers' trade catalogs.)

precipitation within the well or in its immediate vicinity. These effects may be caused by the well itself, and by the pumping installation. Degassing can be avoided by the use of pressure pumps, but suction pumps hardly ever provide perfect samples for gas analysis. Peristaltic pumps may be used to obtain samples from shallow aquifers.

Finally, in wells that do not completely penetrate the aquifer there can be inflow of water from greater depths and distances caused by the drawdown in the well. Representative samples for the whole aquifer unit are seldom or never obtained in such wells. Particularly troublesome in this respect is the case of a groundwater body that has distinct stratification of water quality (especially in respect of salt content) (Richter & Wager 1969).

The results of the sampling are assembled into a report, which includes the ion content in milligrams or milliequivalents per litre, and is correlated with the sum of the milliequivalents of the cations and anions in the relative unit meq %. The content of an ion species should not be referred to the sum of the cations and anions given in the relative unit mg %. Suspended matter and nonionized dissolved solids such as silica (SiO_2), and certain organic substances, are not reported in ion form, nor in gram equivalents, but in millimoles or milligrams per litre.

Table 79 shows an example of a report on the important properties. In many cases such a table can be expanded to include additional substances. In addition, there may be reports on the hydrogeological origin of the water sample (type of aquifer, groundwater body), and notes on any peculiarities in the sampling procedure.

4.1.2 Graphical Presentation of Analyses

Graphical representations of analytical results are intended to simplify comparison and evaluation of different analyses. A number of different methods can be used to this end. They can be classified into three main groups:

1   Pictorial diagrams, in which the values of the analyses of a water sample are represented (e.g., on a map).
2   Multivariate diagrams, in which several analyses can be directly and quickly compared.
3   Diagrams combined with maps or cross sections.

The results of analyses can be reported as absolute values or as percentages. When absolute values are neglected, percentage reports may conceal correlations or suggest incorrect ones, thereby leading to incorrect interpretations.

Table 79   Water Analysis Report (Example Given by Scharpff 1972)

Descriptive details
Date of sampling: July 26, 1968
Sampling point: spillway at well
Location: Sheet 5818 Friedberg, R 34 79 87   H 55 74 57
Cloudiness: clear   Odor: no foreign odors, no peculiarities
Color: colorless   Taste: mildly mineralized
Physical and chemical measurements
Specific electrical conductance S/cm: n. m. (not measured)
pH   6.9   Temperature: 11.5°C
Density at 20°C: 0.9987 g $\cdot$ cm^{-3}

Cations	mg/l	meq/l	meq %
Na$^+$	8.26	0.3593	5.35
K$^+$	1.17	0.0299	0.45
NH$_4^+$	0.12	0.0080	0.10
Ca^{2+}	84.3	4.207	62.69
Mg^{2+}	25.5	2.098	31.26
Fe$_{total}^a$	0.24	0.0086	0.13
Mn^{2+}	0.03	0.0011	0.02
		6.711	100.00

Anions	mg/l	meq/l	meq %
SO$_4^{2-}$	22.2	0.4622	6.87
Cl$^-$	12.8	0.3610	5.37
NO$_3^-$	2.61	0.0421	0.62
HCO$_3^-$	357.5	5.86	87.13
HPO$_4^{2-}$	0.02	0.0004	0.01
		6.726	100.00

Undissociated substances
Silicic acid (meta) (H$_2$SiO$_3$)   13.5 mg/l   0.173 mmole/l
Other constituents
Residue after evaporation 350.5 mg/l
Residue after ignition n.m. (not measured)
Loss on ignition   n.m.
m-value   5.68
p-value   –
Total hardness   17.7° German hardness
Carbonate hardness   16.4° German hardness
Noncarbonate hardness   1.3° German hardness
Gaseous substances
Free carbon dioxide (CO$_2$)   77.3 mg/l
Associated carbon dioxide content   28.4 mg/l
Aggressive carbon dioxide defined by Heyer   n.m.

Table 79 *(Continued)*

Aggressive carbon dioxide defined by Tillmans		48.9 mg/l
Hydrogen sulfide (H_2S)	n.d.	(not detectable)
Oxygen (O_2)	n.d.	
Permanganate value		
Unfiltered	n.m.	
Filtered	n.m.	

[a] Conversion into meq/l based on Fe^{2+}.

4.1.2.1 Pictorial Diagrams. Pictorial diagrams are suitable for the presentation of physical and chemical properties of a particular aquifer; the measured values are shown in columns (bars) or circular diagrams. The pictorial form (e.g., diagrams of representative groundwater) can be presented in cartographic form, to facilitate comparison of regional or facies variations in the water. It is however difficult, or even impossible, to represent the analyses of several groundwaters of quite different geochemical origins clearly on one diagram.

Bar Graphs. Single variable analysis values such as Cl^- or Na^+ content can be represented in a simple way by bars or columns, the length of which shows the concentration depending on a predetermined scale. On map diagrams various types of shading can be used to facilitate the visual comparison of the concentrations concerned (Fig. 59).

The commonest type of bar graph gives the fractional content of cations and anions in absolute values (meq/l or meq/kg) as fractions of the column length. The total length of the bar represents the total dissolved solids. By showing the cations and anions as mg % or meq % (based on the total of anions and cations, respectively), the bar graph can be plotted with the same length notwithstanding the different concentrations, or it can serve as a measure of the total concentration. Cations are conventionally shown on the left-hand side of the bar, anions on the right (Fig. 60). The sequential arrangement in the diagram varies with different authors: for example, in order of increasing magnitude, Ca^{2+}, Mg^{2+}, Na^+ and HCO_3^-, CO_3^{2-}, SO_4^{2-}, Cl^- (Collins 1923), or Ca^{2+}, Mg^{2+}, Na^+ including K^+, ($HCO_3^- + CO_3^{2-}$), SO_4^{2-}, and ($Cl^- + NO_3^-$) (Renick 1924a).

By use of equal scales for graphing milliequivalents per kilogram (e.g., 2.5 mm² for 1 meq/kg), a quick visual comparison of different groundwater samples is possible. As the unit of the height scale, 2.5 mm is suitable for weak concentrations, and $n \times 2.5$ mm ($n = 0.1, \ldots$) for strong concentrations. The resulting areas can be differentiated by colors or shading. Renick (1924a) also gave the depth of extraction of water in the form of a column.

The representation of analysis values in bar form following Preul (in Kar-

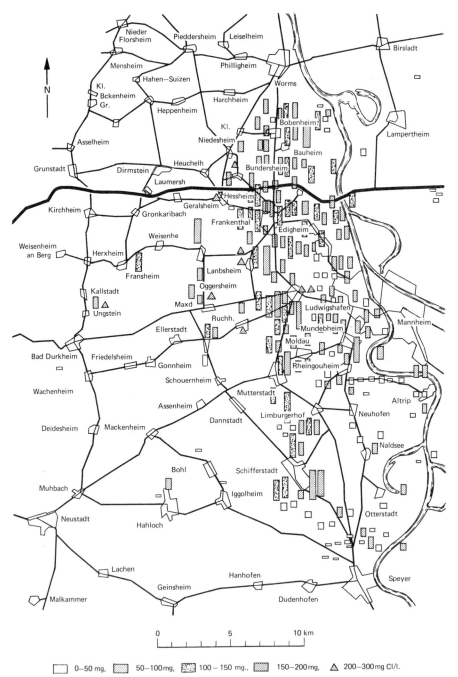

Nieder Florsheim
Pieddersheim
Leiselheim
Birsladt
Mensheim
Phillgheim
Hahen–Suizen
Worms
Kl. Bckenheim Gr.
Harchheim
Heppenheim
Bobenheim
Kl. Niedesheim
Lampertheim
Asselheim
Bauheim
Bundersheim
Grunstadt
Heuchelh
Dirmstein
Laumersh
Hessheim
Kirchheim
Geralsheim
Gronkaribach
Frankenthal
Edigheim
Weisenhe
Weisenheim an Berg
Herxheim
Fransheim
Lanbsheim
Kallstadt
Oggersheim
Maxd
Ludwigshafen
Ungstein
Mannheim
Ruchh.
Ellerstadt
Mundebheim
Moldau
Bad Durkheim
Friedelsheim
Rheingouheim
Gonnheim
Schouernheim
Wachenheim
Mutterstadt
Altrip
Assenheim
Limburgerhof
Neuhofen
Deidesheim
Mackenheim
Dannstadt
Naldsee
Bohl
Schifferstadt
Muhbach
Iggolheim
Otterstadt
Neustadt
Hahloch
Lachen
Hanhofen
Speyer
Geinsheim
Malkammer
Dudenhofen

0   5   10 km

☐ 0–50 mg, ▨ 50–100mg, ▦ 100 – 150 mg., ▨ 150–200mg, △ 200–300mg Cl⁻/l.

Fig. 59 Map of Cl⁻ concentration near Mannheim, Rhine. In the bar graphs 1 mm corresponds to 30 mg/1 (after Matthess 1958).

300

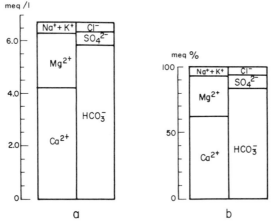

Fig. 60 Bar graph (based on water analysis data in Table 79).

renberg et al. 1958) is often used on maps. The basis is a square of 8 mm sides (the height of the bar can however be adapted to the map scale) (Fig. 61), which is divided into six equal vertical columns and subdivided by a horizontal line through the center. The six columns are allocated to the most important components of the analysis, for example SO_4^{2-}, Cl^-, HCO_3^-, NO_3^-, Fe^{2+}, and Mn^{2+} (Fig. 61a). For each ion a critical concentration is established according to the intended use of the water, and this is appended to the center line. For example the limiting values for drinking water may be used (see Table 82, Section 4.2.3.1). Each bar has its own scale. All bar fractions above the center line, which therefore represent undesirably high concentrations, are colored differently or uniformly red, or shaded in different ways (Fig. 61b).

The upper limit of values of the constituents is decided according to the concentrations observed in the groundwaters under comparison, so that all the concentration ranges of practical significance can be given in the individual bars. In the example considered the maximum values are as follows: 1000 mg SO_4^{2-}/l, 1000 mg Cl^-/l, 1000 mg HCO_3^-/l, 100 mg NO_3^-/l, 80 mg Fe^{2+}/l, and 10 mg Mn^{2+}/l.

Circular Diagrams. The results of the analysis of the ions concerned can be illustrated simply by a circle. Visual comparison between circular diagrams always requires that the surface area F of the circle (in mm²) be made proportional to the concentration (in mg/kg or mg/l). From there it follows that the radius of the circle $r = \sqrt{F/\pi}$. A special diagram developed by Carlé (1954) combines several analytical elements — for example total dissolved solids and Cl^- and HCO_3^- contents — into one statistical diagram in which the elements in the analysis are depicted in separate eccentric circles, as shown in Fig. 62.

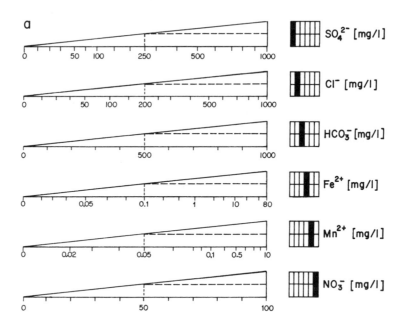

Fig. 61 Bar graphs illustrating selected analyses of water (after Preul, in Karrenberg et al. 1958). (*a*) The six vertical bars represent chemical values of a water sample. The scales are shown adjacent to the bars for quick reference. (*b*) Data from Table 79.

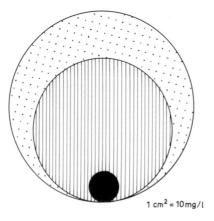

$1 \, cm^2 = 10 \, mg/l$

Total dissolved solids

Cl^-

HCO_3^-

Fig. 62 Circular diagrams for Cl^- and HCO_3^- and total dissolved solids concentrations. Data from Table 79.

302

The illustration of solution concentration by circles is even more versatile when the individual ions are represented by sectors, the angles of which are proportional to the concentration in meq % of the ion concerned (circular diagrams). In the form perfected by Udluft (1953, 1957) the circular diagram is suitable for illustrating comprehensive water analyses and chemical peculiarities (Fig. 63). The radius of the circle is calculated by the procedure given above based on the total dissolved solids in mg/kg. The cation values are put in the upper half-circle, the anion values in the lower. The sectors, which can be variously shaded, are proportional to the percentage fraction of the individual ions (in meq %). The smallest sector shown has an angle of 1^g (complete circle = 400 new degrees (grads) = 400^g), which corresponds to a concentration of 0.5 meq %. Smaller concentrations are shown as strokes having length proportional to the concentration and placed in the sectors of the related ions. The undissociated dissolved constituents, CO_2, HBO_2, and SiO_2, are shown as concentric circles with true area relationship. Ions whose concentrations exceed 20 meq % are underlined (Na^+ and Cl^- in Fig. 63); the convention likewise calls for the symbol for CO_2 to be underlined for contents above 1 g/kg (carbonic acid water).

Other conventions of the Udluft diagram not shown in Fig. 63 are as follows. Thermal springs are shown by circles added outside, each of which signifies 20°C. The radioactive constituents (radium and radon) are shown

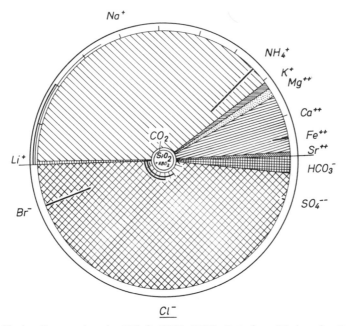

Fig. 63 Cicular diagram given by Udluft (1953, 1957). Data from Rheingrafen Spring, Bad Münster am Stein.

by concentric semicircles. Radium, the content of which is given in multiples of 10^{-7} mg/kg, is shown as a cation, upward in concentric circles that are subdivided at intervals of 20^g to represent increasing orders of magnitude from inside outward. The innermost line corresponds to one decimal place, the second corresponds to the figure before the decimal point, and so on. Radon is shown in a similar way round the exterior circle of the undissociated constituents from the center line downward.

There is an inherent inconsistency in Udluft's method in that the circle area refers to the concentration in mg/l while the angle of the sector is related to the fraction of individual ions in meq %. This defect can be easily eliminated by specifying the concentration of dissolved substances in meq/l, and the undissociated fraction in mmole/l in terms of area of the circle. The inner circles, which represent the undissociated substances, are thus smaller and sometimes are rather difficult to read. If the scales are suitably chosen, there is little change in the diagram.

Radial Diagrams.   In radial diagrams the concentrations (in meq/l, meq %, or mg %) are shown as radial lines from the center. Each value or group of values is shown as a ray. One drawback to these diagrams is the difficulty in representing low concentrations clearly; it is also difficult to amplify the information by including additional components.

In the hexagonal diagrams used by Tickel (1921, after Schoeller 1956) (Fig. 64) the concentrations of alkali ions ($Na^+ + K^+$), alkaline earth ions ($Ca^{2+} + Mg^{2+}$), carbonate species ($CO_3^{2-} + HCO_3^-$), sulfate ions (SO_4^{2-}), and chloride ions (Cl^-) are in meq % and the concentration given. The lines joining points form an irregular hexagon, the shape of which shows the overall

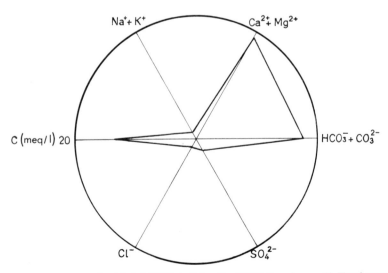

Fig. 64   Radial diagram after Tickel (1921, in Schoeller 1956). Ions in meq %. Total concentration in meq %/l. Data from Table 79.

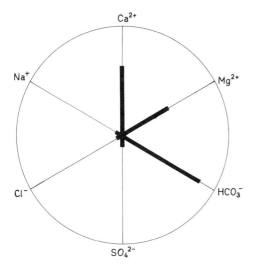

Fig. 65 Radial diagram after Dalmady (1927). Data from Table 79.

picture of the water analysis. In certain cases it is possible to show several water analyses at once.

The method is even applicable to incomplete analyses in which total hardness only is determined as cations. In a diagram given by Dalmady (1927) (Fig. 65) the ions Na^+, Ca^{2+}, Mg^{2+}, Cl^-, SO_4^{2-}, and HCO_3^- are shown in meq % on the six axes; in each case the axis length corresponds to 100 meq % of cation and anion totals, respectively. The sum of the dissolved constituents, the meq total, the temperature, and the amounts of free CO_2 and of minor constituents are given in figures.

The illustration may be varied, say by choice of other ions for representation, or other dimensions for the concentration results (meq/l, mg %). The components can also be represented as vectors with corresponding lengths.

In the four-axis diagram used by Girard (1935) the concentrations (in meq/l) of the ion pairs Ca^{2+} and HCO_3^-, SO_4^{2-} and Mg^{2+}, and Cl^- and Na^+ are shown on single axes (Fig. 66). The cations are connected by solid lines, the anions by dashed lines. Instead of meq/l the concentrations can also be given in mg/l, or meq %. In the four-axis diagram used by Frey (1953) the alkaline earth carbonates are shown in an upward direction, alkali sulfates to the left, chloride downward, and alkali carbonates to the right (in meq %).

The 16-axis diagram given by Maucha (1932) shows the concentrations in meq % of the anions and cations, starting with a regular 16-sided polygon constructed to give an area of 200 mm², so that the 16 sectors have areas of 12.5 mm² and an angle of 22.5° at the center (Fig. 67). The axial lengths for this area work out to 8.082 mm.

The dividing line AE separates the polygon of Fig. 67 into right and left halves, which are subdivided into four sectors (ion fields) by the axes $A–E$, $B–F$, $C–G$, and $D–H$. The anions (HCO_3^-, CO_3^{2-}, SO_4^{2-}, and Cl^-) are plotted to the left, and the cations (K^+, Na^+, Ca^{2+}, and Mg^{2+}) to the right. Hence a

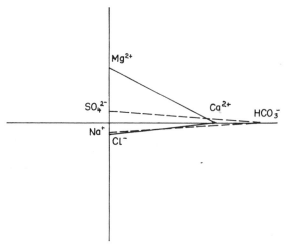

Fig. 66 Diagram after Girard (1935); data from Table 79. The left-hand vectors for H⁺ and CO₃²⁻ are not used, because the concentrations are too low.

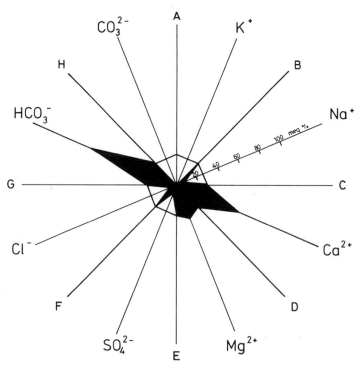

Fig. 67 Radial diagram after Maucha (1932). Data from Table 79.

length l is marked from the center along the axis of the ion field concerned. The point obtained is joined to the adjacent corners of the polygon. Thus one obtains two congruent triangles. The length l is chosen so that the area of the two triangles (in mm²) corresponds to the ion fraction (in meq %) of the anions and cations, respectively. For a sulfate concentration of 7.45 meq % the area of the triangle should be 74.5 mm². Hence the length l is:

$$l = \frac{74.5}{8.082 \sin 22.5°} = \frac{74.5}{8.082 \times 0.38268} = 24.1 \text{ mm}$$

This time-consuming procedure can be simplified if the lengths of the axes are made equal to $8.082 \times 3.093 = 25$ mm, because then the percentage values can be plotted directly (Schoeller 1962).

4.1.2.2 Multivariate Diagrams. Multivariate diagrams bring together the results of analyses of different water samples and permit direct comparison of various groundwater types and genetic and classification statements. By the use of different symbols, it is possible to distinguish waters from stratigraphically different aquifers or from different depths or of different evolutionary status. The diagrams can be classified as trilinear, square, rectangle, combination, and parallel scale diagrams.

Trilinear Diagrams. Concentrations of ions in meq % or − less often − in mg % of cations and anions can be illustrated in trilinear diagrams, which permit the representation of the various cations and anions (shown by different symbols) within a single triangle (Fig. 68), or in two separate triangles

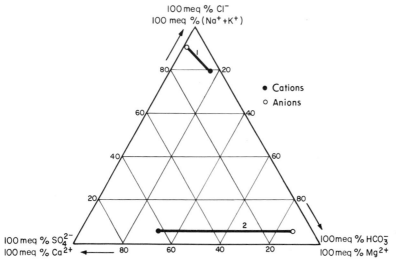

Fig. 68 Trilinear diagram showing cations and anions. Correlated cations and anions can be joined by straight lines; 1, seawater; 2, analysis in Table 79.

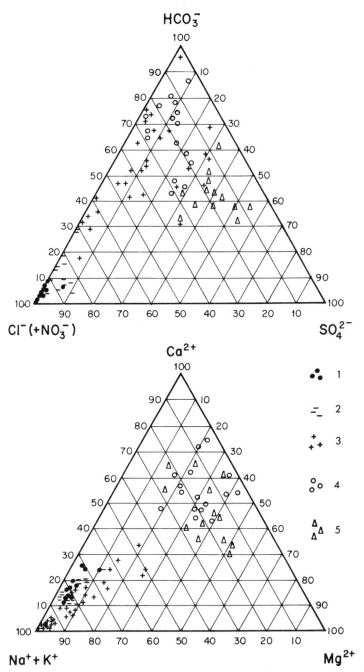

Fig. 69 Trilinear diagrams with separately plotted cations and anions for groundwater from the Mainz Basin and its surroundings (after Schwille 1953a). 1, Na-Cl water from a Tertiary aquifer; 2, Na-Cl water from Permian aquifers in the border area of the eastern part of the Mainz Basin; 3, NaHCO₃-containing water from the Mainz Basin and its surroundings; 4, water from the predominantly calcareous Tertiary formations and the Pleistocene of the western Mainz Basin; 5, waters from the predominantly marly Tertiary beds, also from the Pleistocene marly catchment area in the western Mainz Basin.

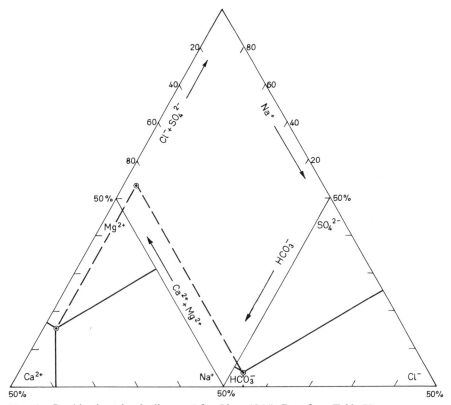

Fig. 70 Combination triangle diagram (after Piper 1944). Data from Table 79.

(Fig. 69), and if necessary in a combination of triangles to make rhombohe-
dral, rectangular, or triangular fields to record the analysis of a water sample
(Figs. 70 and 71).

In the diagram given by Durov (in Chilingar 1956a) the results are plotted
by points that are located by the Cartesian coordinates for the anion and the
cation on two trilinear diagrams (Fig. 71). This procedure can lead to ambi-
guity in that waters of different compositions and concentrations may be
defined by the same point.

Equilateral triangles are generally used, with each side referring to one ion
or ion group, for example SO_4^{2-}, Cl^-, CO_3^{2-}, HCO_3^-, Na^+, Mg^{2+} and Ca^{2+}.
For the combined representation of cations and anions it is advisable to ar-
range the ions Cl^- and Na^+, ($HCO_3^- + CO_3^{2-}$) and Ca^{2+}, and SO_4^{2-} and Mg^{2+}
on one side at a time (Schoeller 1962).

The hardness triangle devised by Schwille (1957) is a special construction
that compares total hardness, carbonate hardness, and noncarbonate
hardness. The lengths of the sides of the triangle are determined by the
highest values of the hardness to be plotted. The plotting of the points or the

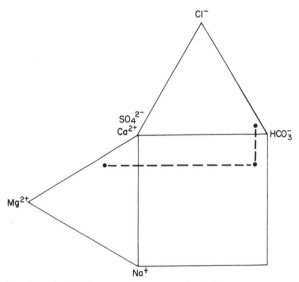

Fig. 71 Combination triangle diagram (after Durov, in Chilingar 1956a). The values are given in meq % of the cation and anion totals (= 100% respectively). Data from Table 79.

reading of the values relating to them proceeds in the appropriate directions, which in Fig. 70 are shown by arrows.

The concentration can be given in degrees of hardness or in meq/l. In Fig. 72 the hardness triangle is combined with two diagrams with rectangular coordinates, in which the associated SO_4^{2-} and Cl^- values are plotted. With the reproduction of a large number of hardness values in the least space, the basic triangle is divided by a corresponding number of lines into triangular elements, the lengths of whose sides for example represent $\frac{1}{4}°$, $\frac{1}{2}°$, or $1°$ (German hardness). The points are plotted in the small triangles using a sharp pencil and as little space as possible, added up, and the number so obtained once more inserted in the small triangle. From this base, lines of equal frequency can be constructed. The areas between two adjacent lines are colored or hatched. Furthermore, the lines can be drawn so as to enclose, in the manner of contours, any predetermined percentage of the marked points. In Fig. 73 for example the thicker line encloses about 66% of all the plotted points (solid and hatched areas). Only 33% of all values lie in the surrounding areas shown by large dots. With widely spaced values this is often the only clear way of illustrating the results.

Hardness triangles are almost indispensable for the preparation of maps of groundwater hardness. When these diagrams are used, the mean hardness of the water related to a definite stratum or to a geologically or geographically delineated area must be determined, and thus a model can be constructed that will accommodate all the relevant variables.

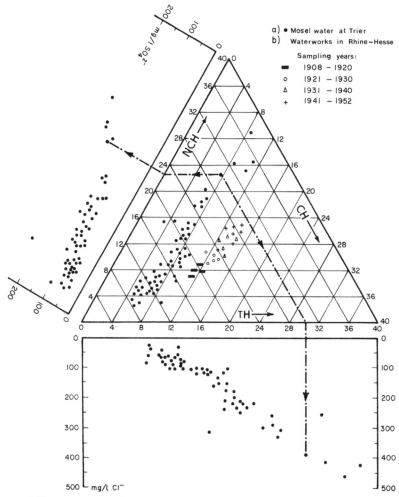

Fig. 72 Trilinear hardness diagram from Schwille (1957) combined with two diagrams with rectangular coordinates; TH, total hardness; CH, carbonate hardness; NCH, noncarbonate hardness. (*a*) Mosel water in 1951. (*b*) Well water from a waterworks in Rhine-Hesse, 1908–1952.

Trilinear diagrams not only permit comparison between numerous waters, the recognition of singularities, and the range of the values, but also show the effect of mixing between waters, because mixtures of two different waters will lie on a straight line. Furthermore, from the convergence of two sets of values along two straight lines toward a common point in the diagram, one can infer that there is a common genetic source for some ions, or a mixture. The addition or removal of solution constituents by ion exchange, precipitation, or solution of salts can also be recognized.

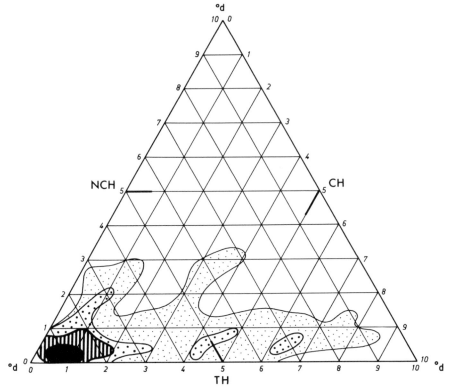

Fig. 73 Hardness triangle of waters from clay slates, greywackes, and quartzites in the Trier District; TH, total hardness; CH, carbonate hardness; NCH, noncarbonate hardness (after Schwille 1957).

Four-Coordinate Diagrams.   Each corner in a four-coordinate diagram is allocated to one component, which is stated in meq % or mg % of the total dissolved constituents, or in cation or anion totals. One can therefore enter four groups from the analysis results, for example ($Na^+ + K^+$), ($Ca^{2+} + Mg^{2+}$), ($Cl^- + SO_4^{2-}$), and ($HCO_3^- + CO_3^{2-}$) in meq %, or for only two ion groups and with them the point representing the analysis. In the four-coordinate diagram given by Langelier & Ludwig (1942) the percentage fractions of alkalis (Na^+) are plotted on the ordinates against the alkaline earths ($Ca^{2+} + Mg^{2+}$), and on the abscissas those of the noncarbonate anions ($SO_4^{2-} + Cl^-$) against the carbonate ions ($HCO_3^- + CO_3^{2-}$) (Fig. 74). Hence the genetically important ratios Ca^{2+}/Mg^{2+} and SO_4^{2-}/Cl^- cannot be represented. The diagram given by Tolstikhine (1939) (in Chilingar 1957) differs from the above only in the direction of the axes.

As the example in Fig. 75 shows, these diagrams also lend themselves to elaboration according to size and number of points through the addition of different symbols.

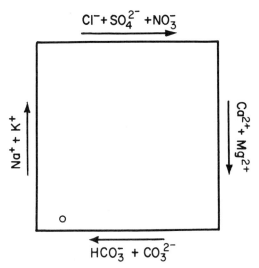

Fig. 74 Four-coordinate diagram after Langelier & Ludwig (1942). Data from Table 79.

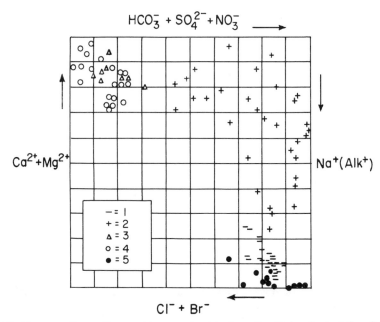

Fig. 75 Four-coordinate diagram for groundwaters from the Mainz Basin (after Schwille 1955). 1, Na-Cl water from Permian strata; 2, NaHCO₃-containing water; 3 and 4, water from marly and calcareous Tertiary and Pleistocene formations; 5, Na-Cl water from Tertiary strata.

313

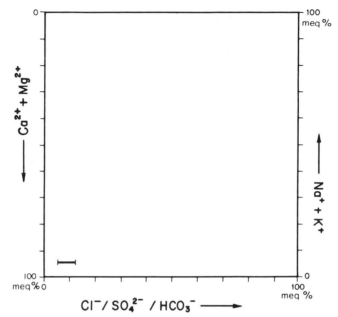

Fig. 76 Four-coordinate diagram after Käss (1967). Data from Table 79.

In an additional approach developed by Käss (1967), the alkaline earths are compared with the alkalis on the vertical axes, while the anions chloride, sulfate, and hydrogen carbonate appear on the horizontal axes (Fig. 76). The analysis is then presented not by a point, but by a line, the length of which corresponds to the percentage sulfate content; the distance of the line from the sides shows the chloride and hydrogen carbonate contents.

Two-Coordinate Diagrams.   In two-coordinate diagrams the relationships between different variables are illustrated, for example the connection between the total dissolved constituents and the Cl^- content (Davis & De Wiest 1967), or that of the alkalis/alkaline earths ratio as functions of depth (Anrich 1969). If the values spread over several tens percent, the use of logarithmic graph paper is recommended.

Parallel Scale Diagrams.   This group is defined by parallel scales on the drawing, from left to right ("horizontal") or from top to bottom ("vertical"). The scales are usually logarithmic for concentration values in meq/l or mg/l, or linear for meq % or mg %. Pictorial diagrams of this kind can accommodate several analyses. By the use of the linear scales for the axes, which are used to plot anions (Cl^-, SO_4^{2-}, CO_3^{2-}), cations (Na^+, K^+, Mg^{2+}, Ca^{2+}), and the residue after evaporation, it is difficult and often impossible to represent on one diagram very different concentrations in water, using the results (as meq/l or mg/l). In this case the percentage values (meq %, mg %) must be used, despite all the disadvantages involved.

Logarithmic scales have the advantage of avoiding calculations when the values are plotted. Furthermore they permit the use of concentration directly in mg/l or meq/l and allow one to distinguish the ratios of the components with respect to one another. In this type of diagram it is possible to include additional ions or components by a corresponding increase in the number of axes used in the diagram.

Horizontal Scale Diagrams. In the *German Spa Handbook* the water analyses are plotted as bars (Hintz & Grünhut 1907). The upper bar shows from left to right Na^+, Ca^{2+}, and Mg^{2+}; the lower bar shows Cl^-, SO_4^{2-}, and HCO_3^- (in meq/kg). In addition, the total concentration (in mg/kg) is shown as a solid line above the bars and the free CO_2 content is inserted as an extension to the anion bar.

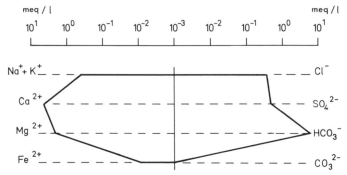

Fig. 77 Stiff diagram (after Stiff 1951). The scale of the Na-Cl axis is 1–100 linear in the original, that of the other axes 1–10; for higher concentrations the scales of all axes can be raised by the factor of 10 or can be replaced by a logarithmic scale. Different scales can be used for individual ions, and these can be indicated against the ion concerned. Data from Table 79.

In Stiff (1951) diagrams (e.g., Fig. 77) the analysis values are plotted along horizontal axes, usually three or four, which are separated from each other by equal distances and are divided by a vertical center line. Cations are shown on one side, usually the left (e.g., Na^+, Ca^{2+}, Mg^{2+}, Fe^{2+}) and anions on the opposite side (e.g., Cl^-, SO_4^{2-}, HCO_3^-, CO_3^{2-}). Joining the points produces a polygon with a shape that is characteristic of the particular water. In some cases a picture of several analyses can be built up. The adaptability of the Stiff diagram is very great because of the possibility of increasing the number of axes and choosing other units (mg/l, mg %, meq %).

Vertical Scale Diagrams. Several forms of vertical scale diagram are in use, however only the semilogarithmic diagrams of Schoeller (1935, 1938)

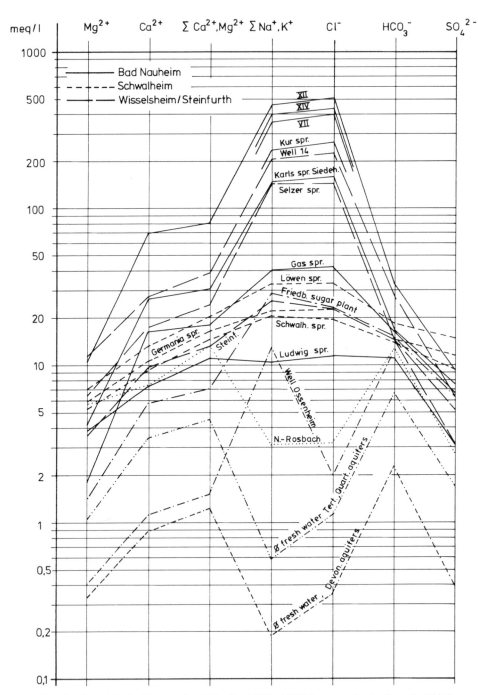

Fig. 78 Semilogarithmic diagram after Schoeller (1935, 1938) for groundwater from the Friedberg area, Upper Hesse (after Scharpff 1972).

316

are widely used. The concentration of ions (in meq/l) is plotted in the sequence Ca^{2+}, Mg^{2+}, K^+, Cl^-, SO_4^{2-}, HCO_3^- + CO_3^{2-} on the vertical axes using logarithmic scales. There are equal distances between individual axes for each ion. Schoeller restricts himself to the constituents named, but other important ones, such as B, Br^-, K^+ (by itself), Fe^{2+}, NO_3^-, or combinations of elements can be accommodated in the same way (Fig. 78). The points so obtained are joined together by straight lines. The parallel directions of the connecting lines from one point to the next indicate the same ionic ratios at different concentrations.

For an approximate estimate of the relevant state of saturation Schoeller (1938, 1956, 1962) has introduced nomograms for simplified calculations for different solubility products ([Ca][SO$_4$]), calcium carbonate–carbon dioxide equilibrium) by consideration of the effective physical-chemical quantities such as pH, temperature, and ionic strength. Schoeller considers the NaCl concentration by itself to be the ionic strength.

With the use of logarithmic scales there is the possibility of determining graphically the milliequivalents of the ions from the usual statements about weight in the analysis reports. This is especially feasible when the values are given in terms of oxides or salts, as was formerly the practice. Simple conversion charts can be made up by using strips of logarithmic paper assembled in the manner of a slide rule, showing equivalent values of the various ions, compounds, and elements. Plotting of the analytical values is then a matter of simple logarithmic division. For example, to plot 180 mg Ca^{2+}/l, the mark Ca = 20 on the strip is placed against the value 1 on the logarithmic scale of the nomogram and the point 180 on the strip transferred to the log scale of the diagram. Figure 78 reproduces water analyses of the topographical map (1:25,000 sheet) of the Friedberg area (Scharpff 1972).

Hem (1970) shifts the scales for the individual ions opposite the base line for the appropriate equivalent weight, so that the weights can be plotted directly in mg/l (Fig. 79). This is particularly advantageous for the plotting of numerous analyses; however, Schoeller's method is preferable for calculation.

4.1.2.3 Plotting Data on Maps.

With cartographic reproduction of chemical data one has to ensure that waters from interrelated groundwater bodies are compared with each other. Localized high chloride values can for example be caused by pollution of the shallow groundwater body, or by mineralized water, which has been encountered in deep wells. The simplest form of water quality map is a representation of individual values or analyses, if necessary with the help of the diagrams described above (Figs. 60–62). The representation of results of analysis on maps by lines of equal concentration or of equal ionic ratios (Fig. 80) is widely used. However the construction of such maps requires fundamental information on the hydrogeological conditions (groundwater flow direction, arrangement of hydrostratigraphic units in a multiaquifer system) and the relevant geochemical factors. Illustrations

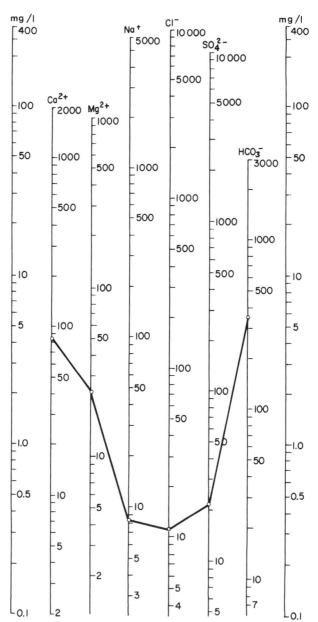

Fig. 79   Semilogarithmic diagram from Hem (1959). Data from Table 79.

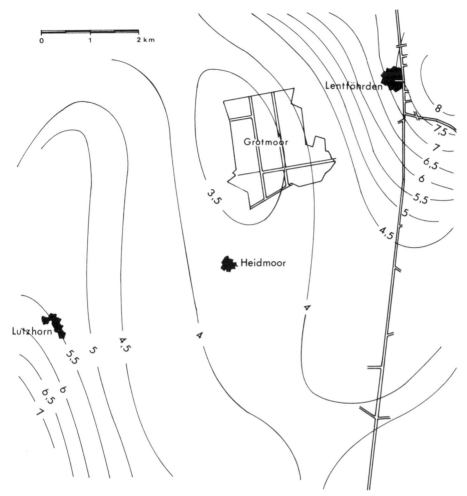

Fig. 80 Cartographic representation with Fourier trend surface contours of equal concentrations. Data from Kaltenkirchen airport project (after Schulz 1977). Lines of equal ionic ratios can also be used in the same way; see for example Hölting (1966).

of this type make it possible to recognize the changes in concentrations of a substance within an aquifer.

Another method is the cartographic representation of the distribution of groundwater types, to which certain distinguishing chemical features can be ascribed (Gerb 1953, 1958).

For the determination of characteristics and for the demarcation of individual areas, as many sampling points as possible should be the aim, from which water samples are taken and tested for pH, Eh, Ca^{2+}, Mg^{2+}, Na^+, K^+, Cl^-, SO_4^{2-}, HCO_3^- and NO_3^-. When sampling points – springs, exposed groundwater in ponds or lakes, observation boreholes, or wells – are not available in sufficient number in the area under investigation, it can prove difficult to achieve a picture of sufficient accuracy about the groundwater quality of a region. In such cases other methods can sometimes be employed. Measurements on the surface with geoelectrical equipment (resistivity measurements) can separate saline water occurrences from fresh water. Electrical resistivity shows a very distinct change at the boundary between fresh and saline water. Determining the boundary between fresh and saline water presents difficulty when clays and saline waters are adjacent in horizontal or vertical directions because both have similar low values of electrical resistivity. For further details see Flathe et al. (1961) and Section 3.1.4.

Because under natural conditions in dry periods the dissolved solids in river water originate almost exclusively from the groundwater, the nature of the flow rate of the river under consideration can be useful for drawing conclusions about the groundwater quality in the catchment area (Section 3.1.3).

In regions with aquifers in only slightly soluble rock material (e.g., silicate rocks) information on the quality of the precipitation and on the evaporation fraction can be used to predict the probable composition of the groundwater (Section 3.2.2.1.1). In areas subject to human influence, however, this procedure is to be used only in exceptional cases. Insufficient knowledge about the degree of evaporation is also a cause of difficulty here.

4.1.3 Data Processing and Storage

In the description and illustration of chemical groundwater conditions one should note that the groundwater chemistry is a dynamic system for which a single determination is not necessarily representative of the conditions prevailing over a period. Besides these variations in time in the groundwater quality at a single place, there can be changes over short distances within the same aquifer.

It is therefore useful to employ modern electronic computing and data processing methods to simplify the study of these variations. The procedures offer the possibility of making statistical and thermodynamic interpretations after the appropriate encoding of the raw data. Suitable computer programs are available for ratio and correlation coefficient calculations.

Punched cards offer the simplest method of storing the results from the sampling points, allowing rapid organization of the information and its retrieval. Edge-punched cards and indexed punch cards are satisfactory because they can be sorted by the punchholes without special devices.

More advanced techniques need special devices. The data processing machines use data carriers adapted for mechanized reading. For this purpose punch cards and handwritten and document-reader schedules are used. Furthermore, the data can be stored on magnetic tapes, films, discs, and cores. Further data processing uses electronic computer and storage units. Computer programs are available for different purposes — for example FOR-

TRAN programs for the plotting of diagrams and Piper diagrams (Morgan & McNellis 1969, McNellis & Morgan 1969, Walger & Schulz 1976) — or for the calculation of the chemical equilibrium of the groundwater (Plummer et al. 1976). The widely used FORTRAN IV computer program WATEQF models the speciation of inorganic and complex species for a given water analysis (Plummer et al. 1976). The application of computers to groundwater studies is continually in development and the reader is advised to consult special textbooks, manufacturers' instruction manuals, and vendor information sheets (e.g., *IBM News*).

4.2 GROUNDWATER CLASSIFICATION

For groundwater classification purposes genetic, chemical, and hydrological viewpoints are the major items of geological and geochemical interest. In water supply studies the physical and chemical properties are incorporated into classifications for assessment of possible applications and exploitation of the water.

4.2.1 Classification on Basis of Origin

Subsurface water is composed of cyclic water and juvenile water. Cyclic water forms part of the hydrologic cycle, arises from infiltration of atmospheric water into the Earth's surface, and after some time emerges at the surface. To this group belong waters in the unsaturated zone (soil water, percolate, or infiltration water) and in the saturated zone (groundwater). Groundwater can also move, as flow theory shows, below sea level, and in closed basins below the level of lakes and surface drainage channels lying below sea level (Dead Sea, −394 m; Death Valley, California, −86 m; Qattara Depression in Egypt, −137 m). Visible evidence for this is in the localized or broad areas of welling up of mineralized groundwater, which may discharge into the main river drainage system, although in small amounts. The flow velocity decreases with increasing depth. It amounts, for example, in the Angara-Lena Basin (USSR) at a depth of 1000–3000 m to 1–9 cm per year according to Pinneker (1976). The water below these reference levels may be referred to as deep-lying water (Weithofer 1933) or "profound" water, also often, and misleadingly, as dead or stagnant water. Some of these bodies of water are cut off from the hydrologic cycle by practically impermeable rocks and in the strict sense can be designated as stagnating waters. The name "fossil water" for deep-lying water should be avoided, because the term "fossil" in geological usage denotes material from pre-Holocene times; in this case the impression is given that fossil deep water has originated in the Pleistocene period or earlier. A valid proof is possible only when there is available an error-free ^{14}C dating (Section 3.2.2.1.1), or dating has been accomplished by other methods, for example potassium-argon or helium-argon

(Pinneker 1967). In this connection Degens & Chilingar (1967) define as "fossil" the groundwaters that have been syngenetically incorporated in ancient sediments. "Deep static water," the alternative concept proposed by Weithofer, should be avoided because deep water in the strict sense is only occasionally static (stagnant). The lower boundary of deep water forms a zone in the Earth's crust, down to which open spaces can exist in the body of the rock.

The assumptions in this respect vary widely (10–50 km), but ought to be nearer the lower value of 10 km given by Van Hise (1904). As noted by Smith (1953) water can act as a solventlike liquid even at pressures and temperatures above the critical temperature prevailing at the greater depths in the Earth's crust, and this property may be of great geochemical importance.

Deep-lying water is often strongly mineralized, and several different explanations have been proposed for this phenomenon.

1. Long-term leaching of aquifer materials as a consequence of the long residence time. This may bring about chemical equilibrium between water and rock. Moreover, a small amount of soluble salts in the rock and a very small amount of chemically vulnerable minerals are sufficient to give the water high total dissolved solids and for it to approach saturation (Section 2.1).

2. Solution of occurrences of salt, by which the resulting highly concentrated brines are caused by their very high density to lie beneath less dense groundwater, until finally the deepest groundwater-bearing rock formations are filled with brines. The proved possibility of brine migration explains the occurrence of highly mineralized groundwaters in regions in which no salt deposits are known.

3. Connate water, which has been trapped in the pores of rocks during their deposit as sediments and during diagenesis has undergone chemical changes there that have led to a rise in the concentration. Degens & Chilingar (1967) limited the concept of connate water to marine water alone, departing from the definition of the original author (Lane 1908). A strict application of the term "connate" excludes migration out of the rock group, even out of the original pores. A wider definition to include water that has been pressed out of the original clay or lime mud sediment by compaction during diagenesis (compression water, after Schoeller 1962) is therefore misleading. The same statement is valid for the definition that even includes water that has moved several kilometres (White 1957b), the only requirement being that the water be out of the hydrologic cycle for at least an appreciable part of a geological period. The various definitions of the concept and the insufficient opportunities of observation often lead to arguments about whether a particular occurrence of deep water is indeed connate.

"Connate water" in the narrower sense presupposes the preservation of some porosity in the rock and a low permeability. The lower the permeability, the better can the original water be retained. A marl with high porosity and low permeability is therefore more likely to contain original water than a sand or sandstone. It is critical for the retention of connate water that the occurrence form part of a closed hydraulic system, in which no water movement can occur because of absence of any potential gradient or because of partial or complete confinement by practically impermeable rock. Thus the similar hydraulic prerequisites for the retention of petroleum occurrences are reasons for referring to associated oil field brines as connate waters.

The changes from the original composition of the pore water include partial or total disappearance of sulfate (biological reduction in the presence of organic matter, precipitation of gypsum or anhydrite when the solubility product has been exceeded), raising of the hydrogen carbonate–carbonate ion content (as a result of biogenic CO_2 production), and falling Mg^{2+}/Ca^{2+} ratio (because of base exchange, whereby most calcium is displaced by sodium, and because of dolomitization processes) (Degens & Chilingar 1967).

One real problem in the explanation of the quality of supposed connate waters is the phenomenon they often display — a much higher total ionic concentration than that of the original liquid, assumed to be seawater. One possible explanation is that there has been an increase in the concentration by degassing processes (Mills & Wells 1919); there may have been a filter press action during diagenesis of the parent sediment (Fairbridge 1967). With the upward movement of the original water, any intercalated clay beds could act not only as geochemical filters or as natural chromatographs, but also as semipermeable membranes obeying the laws of osmosis (Section 2.2).

Concerning the pressing of water out of clays it can be experimentally established that on moderate initial compaction (at a pressure of 6.895 bar) the squeezed-out water possesses an increase in dissolved solids compared with the original pore solution. With increasing pressure (up to 13,790 bar) pressed-out water shows an exponential decrease in the total electrolyte content, also a corresponding decrease of the most important cations and anions (Fig. 81). Hence there results a lower solution content in the pore water remaining in the clays compared with the pore water in the sandstones interbedded with the clay beds (Von Engelhardt 1960, Degens & Chilingar 1967).

Total dissolved solids contents below the seawater concentration in an occurrence of presumed connate water are explainable either by nonmarine origin or by dilution with hydrologic cycle water.

Because other deep-lying waters, for example oil field waters, are subject to similar geochemical processes (sulfate reduction, base exchange), some convergence of water quality is to be expected, which an undoubted separation of true "connate" waters from other deep-lying waters does not make possible, even when many resemble the quality of seawater (cf. Hahn 1972).

One example of this problem is the discussion about the origin of the brines in the Ruhr area, which are interpreted either as connate water or as leached-out brines from the Zechstein saline deposits lying to the north (Puchelt 1964, Kühn 1964, Michel & Rüller 1964, Käss 1964, Dombrowski 1964, Michel 1965). In any discussion of this question the hydrogeological

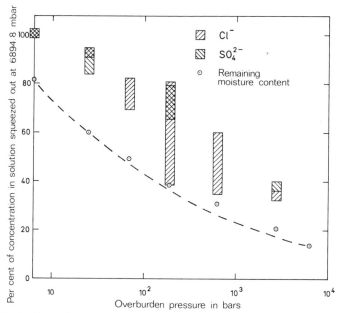

Fig. 81 Dissolved constituents in waters that have been pressed out of montmorillonite saturated with seawater at different pressures. Remaining moisture shown as interrrupted line in percentage of the dry weight (after Degens & Chilingar 1967).

viewpoint has to be considered, in addition to the geochemical and pollen analysis arguments. These waters are undoubtedly "fossil"; in four saline water samples ^{14}C could hardly be detected, so that a model age of at least 24,000 years is indicated (Jacobshagen & Münnich 1964).

4. Metamorphic water. If one includes in the concept "connate water" some displacement of water from the pores, it is possible for changes to occur by expulsion of free pore water and water of crystallization by regional and contact metamorphism. Schoeller (1962) designated this water as "regenerated" water, and White (1957b) called it "metamorphic" water. According to White metamorphic water ranges from normal to moderately high thermal temperatures and probably contains much CO_2 or carbonate and boron, less Cl^- than seawater, and an equal or smaller D/H ratio and a greater $^{18}O/^{16}O$ ratio than seawater.

5. Marine water, invading seawater in the sense used by White (1957b) in the region of recent coastlines (Section 3.1.4). There is a possibility of infiltration by seawater during periodic invasions of the sea, as noted by Mazor (1968) in explanation of some of the mineral waters in the Jordan–Dead Sea sector of the Syrian–East African rift valley.

Water is subject to diagenetic changes in the aquifer, often showing a greater Ca^{2+} content and the same or approximately the same isotopic composition as that of seawater.

Juvenile water in the broad sense comes from magma, from which it is dislodged during crystallization of minerals. Juvenile water has never been in the atmosphere. This dewatering process can set free considerable amounts of water according to Sosman (1924): for an assumed water content of 5 wt % a magma body 1 km thick, with a 10 km^2 surface area, and having a density of 2.5 contains 1.25×10^9 m^3 water, which escapes over an assumed cooling period of 1 million years, at a rate of 1.25×10^3 m$^3 \cdot a^{-1} =$ approximately 0.04 l $\cdot$ s^{-1}. Magmatic water can, as stated by White (1957a, 1957b) be designated as plutonic water for a depth of the magma from a few kilometres, and as volcanic water for a shallow magma. Waters of this kind, mostly connected with volcanic activity, are strongly thermal and according to White (1957a, 1957b) and Weithofer (1933) show relatively high contents of Na^+, Li^+, Cl^-, F^-, SiO_2, B, S, Ba^{2+}, and CO_2, low contents of I^-, Br^-, Ca^{2+}, Mg^{2+}, and probably of nitrogen compounds, and a smaller D/H ratio and a higher $^{18}O/^{16}O$ ratio than seawater.

According to Suess (1909) juvenile water in the narrow sense originates through oxidation by atmospheric oxygen of hydrogen from the interior of the Earth. Because Suess also spoke of juvenile thermal water, he plainly assumed that this oxidation process occurs below the Earth's surface as well as in the atmosphere.

4.2.2   Classification on Basis of Dissolved Constituents

The simplest classification depends on the total concentration of dissolved constituents. Davis & De Wiest (1967) have proposed the following classification on this basis:

	Concentration of dissolved constituents (mg/kg)
Fresh water	0–1,000
Brackish water	1,000–10,000
Saline water	10,000–100,000
Brines	> 100,000

The classification above has the disadvantage that a salty taste is already noticeable below 1 g/kg. Richter & Wager (1969) therefore consider Cl^- contents above 350 mg Cl^-/l (corresponding to approximately 650 mg NaCl/l) as saline water. In mining terminology waters with more than 40 g/l, and in German mineral water terminology (Quentin 1969) waters with at least 14 g of dissolved salts (chiefly NaCl) per kilogram, are called brines.

A classification based on chemical composition is widely used, in which the percentage fraction of dissolved ions is reported either in mg % following the model of Clarke (1924), or in meq % following that proposed by Palmer (1911).

Classification is often simplified by the use of diagrams. One possibility here is the suggestion by Davis & De Wiest (1967), which involves the total

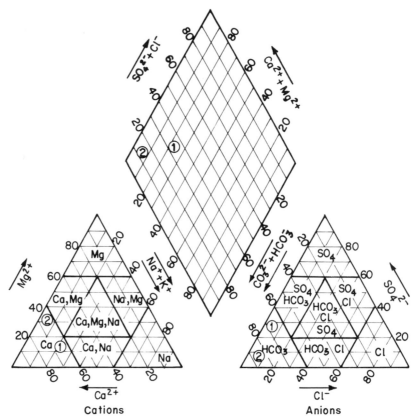

Fig. 82 Classification based on equivalent percentage (meq %) (after Davis & De Wiest 1967): Example based on data from Table 79: 1, seawater, 2, groundwater (Table 79).

dissolved solids (Fig. 82): point 1 represents an Na-Cl water and point 2 an Ca-HCO$_3$ water. With a total concentration of 528.25 mg/kg point 1 is considered to be fresh Ca-HCO$_3$ water.

Groundwaters can also be designated by numbers, based on the coordinates of a field in which the analysis point falls, for example in a composite trilinear diagram (Fig. 83) or in a four-coordinate diagram (Fig. 84).

Palmer (1911) takes the concepts of salinity and alkalinity as the basis of his classification, in which salinity is caused by nonhydrolyzed salts, and alkalinity by free alkaline bases, through the hydrolyzing effect of water on solutions of hydrogen carbonates or on solutions of salts of other weak acids.

All cations including hydrogen can contribute to salinity, but among anions only those of the strong acids can act in a similar way.

Because salinity depends on the combined effects of the equivalent values of anions and cations, its extent is limited only by the amounts of strong acids present, the value of which must be multiplied by 2. On the other hand complete alkalinity can be calculated by doubling the amount of cations that exceeds the content of strong acids.

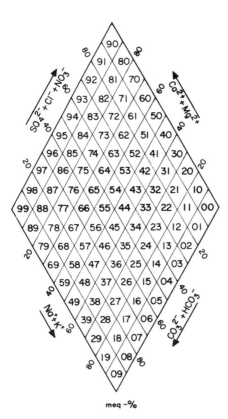

meq -%

Fig. 83 Water type diagram after Furtak & Langguth (1967). Example based on Table 79 falls in field 98.

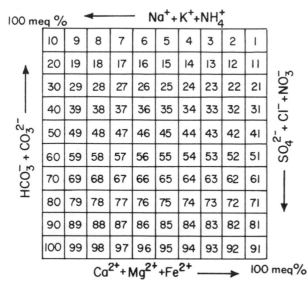

Fig. 84 Water type diagram after Tolstichin (according to Alekin 1962). Example based on Table 79 falls in field 11.

Thus ions can be grouped as follows:

a   Alkali ions ($Na^+ + K^+ + Li^+$) meq %

b   Alkaline earth ions ($Ca^{2+} + Mg^{2+}$) meq %

c   Hydrogen ion (H^+) meq %

d   Strong acids ($Cl^- + SO_4^{2-} + Br^- + NO_3^-$) meq %

e   Weak acids ($HCO_3^- + CO_3^{2-}$) meq %

Salinity and alkalinity can now be further classified:

1   Primary salinity (alkali salinity) is salinity that does not exceed twice the sum of the alkali ions.
2   Secondary salinity (noncarbonate hardness) is the excess of salinity over the primary salinity; it does not exceed twice the total of the alkaline earths.
3   Tertiary salinity (acidity) is the excess of salinity over the primary and secondary salinities.
4   Primary alkalinity is the excess of twice the total of alkalis over the salinity.
5   Secondary alkalinity is the excess of twice the sum of the alkaline earths over the secondary salinity.

Palmer has proposed five classes, as follows:

Class			Possible types of salinity or alkalinity
Class 1	$d < a$	2d	Primary salinity
		$2(a - d)$	Primary alkalinity
		2b	Secondary alkalinity
Class 2	$d = a$	2a or 2d	Primary salinity
		2b	Secondary alkalinity
Class 3	$a + b > d > a$	2a	Primary salinity
		$2(d - a)$	Secondary salinity
		$2(a + b - d)$	Secondary alkalinity
Class 4	$d = a + b$	2a	Primary salinity
		2b	Secondary salinity
Class 5	$d > a + b$	2a	Primary salinity
		2b	Secondary salinity
		$2(d - a - b)$	Tertiary salinity or acidity

Various graphical methods have since been developed from Palmer's classification.

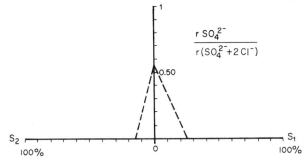

Fig. 85 Diagram of primary (S_1) and secondary (S_2) salinity and base exchange properties after Rogers (1917). Data from Table 79.

The bar graph used by Rogers (1917) shows from the bottom upward in the left-hand column Cl^-, SO_4^{2-}, CO_3^{2-}, and HCO_3^-, and in the right-hand column Na^+, Ca^+, Ca^{2+}, and Mg^{2+}. The center column gives the salinity and alkalinity based on Palmer's classification. A modified version of this graph appears in Fig. 86. The dissolved material is stated in meq % of the ion content. With equal scale values of 1 meq % a direct comparison of different diagrams is possible.

The three vector diagram used by Rogers (1917) (Fig. 85) show the ratios

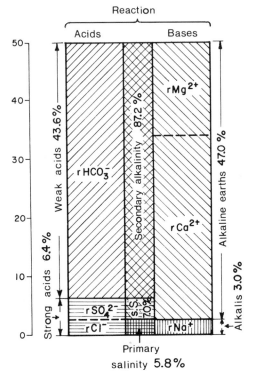

Fig. 86 Diagram of salinity and alkalinity after Rogers (1917). Data from Table 79.

of the equivalents $rSO_4^{2-}/(rSO_4^{2-} + rCl^-)$ in %, and on the other vectors the secondary alkalinity and primary salinity in %. Figure 85 shows only the ratio SO_4^{2-}/Cl^- and the base exchange.

Hill (1940, cf. Langelier & Ludwig 1942) has placed two equilateral triangles in contact with a rhombohedron as in Fig. 87. The cations and anions are plotted in each triangle. The triangles are subdivided into smaller triangles designated with Roman numerals as types and Arabic letters as subtypes. From the two points A, which represent the water, lines are drawn parallel to the base lines of the two triangles. The point of intersection of these lines A' is located in one of the type fields designated by Palmer (Fig. 87).

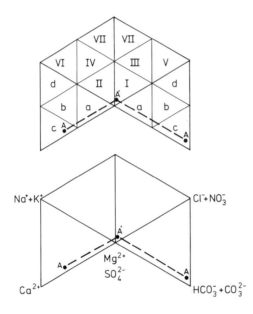

Fig. 87　Water type diagram after Hill (in Langelier & Ludwig 1942). Data from Table 79.

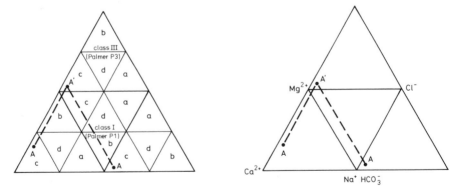

Fig. 88　Water type diagram after Langelier & Ludwig (1942). Data from Table 79.

Langelier & Ludwig (1942), departing from Hill's method, place the triangles with bases parallel (Fig. 88) and increase the sides to give a type triangle after Palmer. Through points A lines are drawn parallel to the sides of the large triangle, and the point of intersection A' lies in a type triangle. The conversion to percent values, and the fact that point A' corresponds to a whole series of points A is a disadvantage of this method.

Palmer's method groups together substances that could be of quite different origin and have been subject to different diagenetic processes.

Schoeller (1956) sought to avoid this in his classification by considering several viewpoints, given below in order of decreasing importance:

1   Differentiation according to Cl^- concentration.

Super chloride water $>$ 700 meq Cl^-/l.

Marine chloride water 700–420 meq/l. The Cl^- content varies about the average seawater content of 560 meq/l:
560 + 140 meq/l.

Strong chloride water 420–140 meq Cl^-/l.

Medium chloride water 140–40 meq Cl^-/l. A chloride content of 40 meq/l corresponds to the upper limit for normal human consumption.

Oligochloride water 40–15 meq Cl^-/l.

Normal chloride water 10 meq Cl^-/l.

2   Differentiation according to sulfate concentration.

Supersulfated water $>$ 58 meq SO_4^{2-}/l. 58 meq SO_4^{2-}/l is the concentration in sea water.

Sulfate water 58–24 meq SO_4^{2-}/l; 24 meq SO_4^{2-}/l is the upper limit for normal human consumption.

Oligosulfate water 24–6 meq SO_4^{2-}/l.

Normal sulfate water $<$ 6 meq SO_4^{2-}/l.

Finally, waters in this category can also be subdivided when they are approximately saturated with respect to $CaSO_4$, or close to saturation, so that the milliequivalent content of sulfate ion can be greater or less than that of calcium.

3   Differentiation according to concentration of hydrogen carbonate + carbonate ions.

Supercarbonated water 7 meq $(HCO_3^- + CO_3^{2-})$/l.

Normal carbonate water 2–7 meq $(HCO_3^- + CO_3^{2-})$/l.

Undercarbonate water $<$ 2 meq $(HCO_3^- + CO_3^{2-})$/l.

4   Indices of chloride-alkali unbalance or base exchange I_{BA}.

Positive index of base exchange:

$$rCl^- > rNa^+ \qquad \text{or} \qquad r(CO_3^{2-} + SO_4^{2-}) < r(Mg^{2+} + Ca^{2+})$$

equilibrium $rCl^- = rNa^+ \qquad$ or $\qquad r(CO_3^{2-} + SO_4^{2-}) = r(Mg^{2+} + Ca^{2+})$

$(I_{BA} = 0)$

Negative index of base exchange:

$$rCl^- < rNa^+ \qquad \text{or} \qquad r(CO_3^{2-} + SO_4^{2-}) > r(Mg^{2+} + Ca^{2+})$$

5   Differentiation on basis percentage anions and cations. The following six groups can be established on anion basis:

   (a) $rCl^- > rSO_4^{2-} > rCO_3^{2-}$ (b) $rCl^- > rCO_3^{2-} > rSO_4^{2-}$

   (c) $rSO_4^{2-} > rCl^- > rCO_3^{2-}$ (d) $rSO_4^{2-} > rCO_3^{2-} > rCl^-$

   (e) $rCO_3^{2-} > rCl^- > rSO_4^{2-}$ (f) $rCO_3^{2-} > rSO_4^{2-} > rCl^-$

which can be combined with the following six cation groups

   (a) $rNa^+ > rMg^{2+} > rCa^{2+}$ (b) $rNa^+ > rCa^{2+} > rMg^{2+}$

   (c) $rMg^{2+} > rNa^+ > rCa^{2+}$ (d) $rMg^{2+} > rCa^{2+} > rNa^+$

   (e) $rCa^{2+} > rNa^+ > rMg^{2+}$ (f) $rCa^{2+} > rMg^{2+} > rNa^+$

to give a total of 36 types.

The classification given by Souline (1948, in Schoeller 1962) distinguishes between fundamental types by emphasis on base exchange:

1   Na-SO$_4$ water: $r\dfrac{Na^+ - Cl^-}{SO_4^{2-}} < 1$

2   Na-HCO$_3$ water: $r\dfrac{Na^+ - Cl^-}{SO_4^{2-}} > 1$

3   Mg-Cl water: $r\dfrac{Cl^- - Na^+}{Mg^{2+}} < 1$

4   Ca-Cl water: $r\dfrac{Cl^- - Na^+}{Mg^{2+}} > 1$

which are subdivided as necessary according to anion ratios into the groups of hydrogen carbonate, sulfate, and chloride waters. Various classes are formed by combination of these factors:

Class A_2 with dominant secondary alkalinity.

Class S_2 with dominant secondary salinity.

Class S_1 with dominant primary salinity.

Class S_3 with tertiary salinity.

Class A_1 with dominant primary alkalinity.

Class A_3 with tertiary alkalinity.

The individual fundamental types can be collected into the possible classes given in Table 80.

Table 80   Groundwater Classification According to Souline (1948, in Schoeller 1962)

Classes for sodium sulfate type $r\dfrac{Na^+ - Cl^-}{SO_4^{2-}} < 1$	Classes for sodium hydrogen carbonate type $r\dfrac{Na^+ - Cl^-}{HCO_3^-} > 1$	Classes for magnesium chloride type $r\dfrac{Cl^- - Na^+}{Mg^{2+}} < 1$	Classes for calcium chloride type $r\dfrac{Cl^- - Na^+}{Mg^{2+}} > 1$
	Hydrogen carbonate water group		
A_2	A_2	A_2	A_2
	A_1		
	S_1		
	Sulfate water group		
S_2	S_1	S_2	S_2
S_1		S_1	S_1
	Chloride water group		
S_1	S_1	S	S_1
		S_2	S_2

Alekin's (1962) classification differentiates waters into three classes based on the dominant anions: hydrogen carbonate and carbonate waters ($HCO_3^- + CO_3^{2-}$), sulfate waters (SO_4^{2-}), and chloride waters (Cl^-). Each class is subdivided into three groups according to the dominant cation: Ca, Mg, and Na groups. Each group is broken down on a basis of ionic ratios (in meq/l) into three or four water types.

Type I $HCO_3^- > (Ca^{2+} + Mg^{2+})$.

Type II $HCO_3^- < (Ca^{2+} + Mg^{2+}) < (HCO_3^- + SO_4^{2-})$

Type III $(HCO_3^- + SO_4^{2-}) < (Ca^{2+} + Mg^{2+})$ or $Cl^- > Na^+$.

Type IV $HCO_3^- = 0$, which of course does not occur in the class of carbonate waters.

Each class is denoted by the chemical symbol for the anion concerned (C, S, Cl). To characterize the group, the chemical symbol of the anion is placed as a superscript to the right of the class symbol. The type membership is given by a Roman numeral subscript. Thus a water type in the hydrogen carbonate class, calcium group, of the second type is described by the symbol C_{II}^{Ca}.

In addition, the total dissolved solids (to an accuracy of up to 0.1 mg/kg) is added before this symbol, and before the group symbol the total hardness (to whole number accuracy in meq/l). If the second most abundant anion or cation is not more than 5 meq % below the concentration of the dominant ion, the symbol for this anion or cation is included in the index.

The standard classification used in Germany for mineral waters (Quentin 1969) gives the percentage fraction of the individual dissolved constituents. The ions used for nomenclature are those that make up at least 20 meq % of the cation or anion totals (= 100% for each). The dissolved constituents are thereby given in the sequence cations to anions, in decreasing concentration. Water with 71 meq % Na, 23 meq % Ca, 64 meq % Cl, and 29 meq % SO_4 is therefore designated as Na-Ca-Cl-SO_4 water. This classification has been introduced for mineral waters that contain either more than 1 g/kg total dissolved solids or at least 1000 mg/kg free dissolved CO_2 (carbonic acid water, medicinal water).

Natural salty waters that contain at least 14 g/kg dissolved salts (predominantly NaCl, 5.5 g = 240 meq Na^+ and 8.5 g = 240 meq Cl^-) are described as brines, and types with temperatures above 20°C are called thermal (Table 81). In the United States the term "thermal water" is particularly associated with volcanic activity.

Table 81   Common Types of Mineral and Medicinal Waters

Chloride waters	Sodium chloride waters (formerly muriatic water), calcium chloride water (formerly alkaline earth muriatic water)
Hydrogen carbonate waters	(Formerly alkaline water): sodium hydrogen carbonate water, calcium hydrogen carbonate water, magnesium hydrogen carbonate water
Sulfate waters	Sodium sulfate water (formerly saline water), magnesium sulfate water (formerly bitter water), calcium sulfate water (gypsum water)

Finally the description of mineral waters can be expanded by including the following limiting values by the addition of additional marks:

Iron-containing waters	10 mg/kg iron
Arsenical waters	0.7 mg/kg arsenic, corresponding to 1.3 mg/kg hydrogen arsenate ion
Iodine waters	1 mg/kg iodine
Sulfur waters	1 mg/kg titratable sulfur

Radon waters   66.6 kBq/l

The occurrence and quality of mineral and thermal waters is a matter for another textbook and is not discussed further here.

The classifications above are most appropriate for mineralized waters, but they are not suited to normal fresh groundwater. Because of the low mineral content, quite small variations in individual substances can lead to considerable changes in the percentage composition, so that by rigid application of this classification genetically related groundwaters can be allocated to different classes.

The application of meq % or mg % as the unit does not take into account the concentration, and it therefore causes waters of the most diverse concentrations to appear in the same class. This would be justified if the geochemical processes in the groundwater were limited to dilution and concentration. The numerous processes described above, and the various solubilities of the individual constituents, however, lead to the expectation that genetically related waters in the course of their movement underground will change their fractional compositions. Many processes can be recognized on the basis of changes in certain of the equivalent ratios, namely:

$$\frac{SO_4^{2-}}{Cl^-} \qquad \frac{Mg^{2+}}{Ca^{2+}} \qquad \frac{Cl^- - Na^+}{Cl^-} \qquad \frac{Cl^- - Na^+}{SO_4^{2-} + HCO_3^- + NO_3^-}$$

and the absolute contents of individual substances as well as the solubility limits.

Gerb (1953, 1958), and following him other authors (Fast & Sauer 1958, Thews 1972), have proposed for certain hydrogeologically uniformly structured regions groundwater types such as crystalline type groundwater or Alpine limestone type groundwater. Each of these types has definite chemical properties, which result from interpretation of the analyses of the water. A refinement in the classification hence brings in the separation of the normal type from the "reduced" type, which is recognized by oxygen deficiency and the occurrence of characteristic features such as iron(II) ions or ammonia (cf. Section 2.3). The procedure chosen by Gerb simplifies the cartographic representation of the distribution of water types on the basis of geological maps because the genetic correlation between aquifers and groundwaters stands out clearly.

4.2.3   Classification on Basis of Potential Use

Groundwaters can be classified with regard to their suitability for human use, for domestic, industrial, and agricultural purposes, and for recreational purposes. Statements about water quality offer an important base in this context. For many applications it is economically acceptable to process any naturally available water.

4.2.3.1 Potable Water. Water that is to be used for drinking must meet high standards of physical, chemical, and biological purity (DIN 2000-2001). The water should be appetizing – even considering its origin – clear, transparent, colorless, odorless, and of constant temperature. In central Europe the water temperature should be between 8 and 12°C, or as close to this range as possible. In regions with higher or lower average annual temperature other limits may have to be accepted.

For human consumption in a wide variety of lands the minimum acceptable quality of potable water has been very widely fixed. It should be fundamentally free from undesirable physical properties such as color, cloudiness, and objectionable odors and tastes. The odor and taste test is naturally a subjective one: many substances are in fact perceptible in the most minute amounts, for example chlorophenol present at a few micrograms per litre and free Cl_2 at a few tenths of a milligram per litre. For other substances the threshold of taste is considerably higher, for example for Cl^- ions 400 mg/l. The odor and taste thresholds of petroleum products vary from about 0.001 mg/l (gasoline, diesel fuel, heating oil) to about 10 mg/l (benzole). At present an overall "oil" concentration is specified as a limiting value. In spite of this, waters with values below 1 mg/l are to be rejected whenever the smell and taste of oily substances can be noticed. The sensitivity of the smell and taste organs is generally sufficient to give warning of the danger of toxic effects of petroleum products, because at toxic levels the water is unpalatable to both man and animals.

By health and bacteriological tests it should be ascertained that in 100 cm³ of potable water not more than 4 coliform bacteria (as indicators of fecal pollution) occur. The arithmetic mean of all samples must not exceed 1 per 100 ml (Committee on Environmental Improvement 1978). According to German standards the total bacteria count has to be less than 100 per cubic centimetre of water. Coliform bacilli are themselves not directly harmful; however, when they are present they indicate the possible presence of pathogenic microbes. Conversely the absence of coliform bacilli has a high probability of indicating the absence of pathogenic organisms, so that the problems of direct determination of the latter is usually avoidable. Possible fecal influence can also be detected by urochrome determination (Höll 1972, Lüdemann 1970).

Efficient methods of sterilization permit use of groundwater that because of hydrogeological conditions may contain harmful microorganisms. Since the methods used here have been thoroughly described in the literature (e.g., Beger 1966, Fair et al. 1971, Höll 1972, Holden 1970, Huisman 1972, Oehler 1969, Rubin 1974), the topic need not be pursued further here. The simplest method for the individual user of untreated water is to boil it.

The limiting chemical values given Table 82 follow the limits laid down by the European Health Organization (Müller 1971) and the requirements specified in the United States for potable water (U.S. Public Health Service

1962). The minimum standards take various viewpoints into consideration. Harmful physiological effects can occur when the limiting value is exceeded, for example with lead, fluoride, arsenic, sulfate, sodium, chloride, nitrate, selenium, and chromium. Odor, impaired taste, and damage (e.g., corrosion or encrusting in the mains) can arise from hydrogen sulfide, iron, and manganese. The stated limits for arsenic, barium, cadmium, chromium, cyanide, fluoride, lead, selenium, and silver should not be exceeded so far as is possible because of their known or assumed toxicity. Other substances, such as ammonia and nitrite, indicate possible pollution. Public water supply authorities very often treat water before it is supplied to the public. The treatment usually includes removal of iron by oxidation (aeration) and by filtering off flocculated iron hydroxide, or by base exchange; removal of manganese by oxidation or base exchange; removal of acidity by adding lime, (MgO), by neutralizing, or by anion exchange; removal of hardness by precipitation or exchange processes using inorganic cation exchangers or organic exchangers; the partial or total removal of salt, which is possible by exchangers, by electrodialysis, or by reverse osmosis (cf. Oehler 1969).

The standards of quality cannot be met in many places; climate, weight and age of persons, and individual eating and drinking habits are factors that modify the health requirements. Thus in many arid regions human beings all their lives consume groundwater with considerably more than 1000 mg/l total dissolved solids without any obvious harmful effects. In semiarid regions, according to Richter & Wager (1969), there are at times accepted upper limits for potable water of 2400 mg SO_4^{2-}/l and 2800 mg Cl^-/l. One should remember that the potable water forms only part of the intake of nutrients — some foodstuffs can on occasion contain harmful substances in higher doses. Special attention is necessary in respect of dissolved radioactive nuclides, because by intake into the human body with drinking water they can lead to a high radiation dose.

Radioactive substances change themselves spontaneously without any exterior influence and during the process emit radiation. For every radionuclide there are essentially three characteristic features that suffice for their identification: type and energy of the radiation emitted, and the half-life of the radionuclide.

The half-life of a radionuclide is that period during which half the radioactive atoms of the nuclide concerned have decayed. This varies for the different nuclides over a wide range from small fractions of a second to many billions of years.

The energy of corpuscular radiation is a measure of the velocity of the particles emitted; for electromagnetic radiation it is a measure of its penetrative power. The electron volt (eV) serves as the unit of measurement of the energy of radioactive radiation: 1 eV is the energy an electron or proton gains on moving between two points in an electric field between which there is a potential difference of 1 volt (practical units: 1 keV $= 10^3$ eV, 1 MeV $= 10^6$ eV).

Table 82 Limiting Values of Dissolved Constituents for Potable Water, Giving Maximal Permissible Concentration (mg/l)

Element	U.S.A. (U.S. Public Health Service 1962) 1962,	Europe (Müller 1971)	Germany F.R. (Trinkwasser-Verordnung 1975)
Sb^{3+}	0.05	–	–
$As^{3+(5+)}$	0.05	0.05	0.04
Ba^{2+}	1.0	–	–
Cd^{2+}	0.01	0.01	0.006
Cr^{6+}	0.05	0.05	0.05
CN^-	0.2	0.05	0.05
F^-	1.5 (see temperature data below)	1.5 (see temperature data below)	1.5 (see temperature data below)
H_2S	1.0	0.05	–
Pb^{2+}	0.05	0.1	0.04
Hg	0.002^a	–	–
Se	0.01	0.01	0.008
Ag^+	0.05	–	–
Zn^{2+}	5	5.0	2
Cu^{2+}	1	0.05	–
$Fe^{2+(3+)}$	0.3	0.1	–
Mn^{2+}	0.05	0.05	–
NH_4^+	–	0.05	–
Na^+	200	–	–

Ca^{2+}	200	—	—
Mg^{2+}	125	125 At more than 250 mg SO_4^{2-}/l only 30 mg Mg^{2+}/l	—
Total hardness	—	2-10 meq/l	—
HCO_3^-	500	—	—
Cl^-	250	200	—
SO_4^{2-}	250	250	240 (exception $CaSO_4$ bearing aquifers)
NO_2^-	—	Trace	—
NO_3^-	45	100 Less than 50 mg recommended	90
B	20	—	—
CO_2	—	0	—
O_2	—	at least 5	—
Alkyl benzene sulphonate (ABS)	0.5	—	—
Anionic detergents	—	0.2	—
CCl_4 extract	0.2	(0.2-0.5) provisional proposal	—
Polycyclic aromatic hydrocarbons		0.2×10^{-3} (after treatment)	0.00025 (total C)
Insecticides:			
Endrin	0.0002*	—	—

*Data from Committee of Environmental Improvement (1978).

339

Table 82 (*Continued*)

Element	U.S.A. (U.S. Public Health Service 1962) *Committee of Environmental Improvement 1978)	Europe (Müller 1971)	Germany F.R. (Trinkwasser-Verordnung 1975)
Lindane	0.004*	-	-
Methoxychlor	0.1*	-	-
Toxaphene	0.005*	-	-
Herbicides:			
2,4 D	0.1*	-	-
2,4,5-TP (Silvex)	0.01*	-	-
Phenolic compounds (as phenol)	0.001	0.001	-
Total dissolved solids	500	-	-

1) Corresponding to an average daily maximum air temperature of about 13°C. In this case the limits are temperature dependent (U.S. Publ. Health Service 1962).

°C	Upper limit mg/l	Optimum (value) mg/l	Lower limit mg/l
10 - 12.0	1.7	1.2	0.9
12.1 - 14.6	1.5	1.1	0.8
14.7 - 17.6	1.3	1.0	0.8
17.7 - 21.4	1.2	0.9	0.7
21.5 - 26.2	1.0	0.8	0.7
26.3 - 32.5	0.8	0.7	0.6

In the present context the most interesting types of radiation are α-, β-, and γ-rays. α-Rays consist of a double positively charged helium nucleus (α-particle = 2 protons + 2 neutrons) that has a high stability. α-Particles are emitted with great velocity (energy) from the atomic nucleus and occur almost always with radioactive elements of high atomic number. The maximum range of an α-particle with an energy of a few MeV in air is a few centimetres and in water only micrometres.

β-Particle radiation consists of electrons that are emitted from the nucleus at high velocity. Frequently positrons (β^+-rays) are also emitted, which behave exactly as β-rays apart from the effects of their positive instead of negative charge. The energy of the particle lies statistically divided between zero and a maximum value (continuous β-spectrum). In general a β-emitting radioactive substance is characterized by the observed maximum value of the energy. β-rays with radiation energy of a few keV to 1 MeV have a range in air of a few metres, but in water only a few millimetres.

γ-Rays are electromagnetic rays that are similar to the very energetic and penetrating X-rays. Their penetration ability is dependent on their energy. The most common energy levels of the emitted γ-quanta lie between 0.2 and 2 MeV. The intensity of γ-radiation decreases in exactly the same way as for visible light, as the inverse square of the distance from the source, which is highly significant in practical radiation protection.

During the transit through material of the forms of radiation mentioned above, interactions that involve an energy transfer occur between the radiation and the atoms of the material. Charged particles (α- and β-particles) are gradually stopped during transit by inelastic collisions with the atoms, and the latter become excited or ionized.

By the interaction of γ-radiation with the atoms of the material three different effects are possible: the photoelectric effect, the Compton effect, and ion pair production. The energy of the γ-quanta in these processes is either completely or only partly transferred to the atoms of the irradiated material.

For detailed information about the interaction processes of charged particles and photons with atoms of the irradiated material, the reader should consult the literature on atomic physics (e.g., Hendee 1970, Finkelnburg 1967).

The biological effect and thus the hazard of ionizing radiation depends mainly on the absorbed energy. It is necessary to determine how much of the radiated energy released by radioactive decay is absorbed by the biological material.

By the term "radiation doses" one understands the radiation energy absorbed by the material per unit mass. The unit of radiation dose is the gray (Gy)

$$1 \text{ Gy} = 1 \text{J/kg} \; (= 100 \text{ rad})$$

The concept of radiation dose holds for arbitrary ionizing radiation and is independent of the type of material involved. The dose per unit time (e.g., $Gy \cdot s^{-1}$) is known as the dose rate.

Besides the energy dose rate there is a further concept defined specially for X-rays and γ-rays: the ionizing dose. Its unit is the roentgen (R): 1 R is the amount of X- or γ-radiation that results in 258.4×10^6 C ionization per kilogram of air at 0°C and 1013.24 mbar. The ionizing dose rate here is given in R/s, R/min, R/h, and so on.

The connection between radiation dose and ionization dose depends on the medium. For air, water, and body tissue 1 R corresponds to approximately 0.01 Gy (1 R in air $= 8.77 \times 10^{-3}$ Gy, in water $= 9.4 \times 10^{-3}$ Gy).

The unit of radioactivity is the becquerel (Bq), the rate at which the atoms of a radioactive element are decaying: 1 Bq is a decay rate of 1 disintegration per second (dps). As a result of recent changes, and standardization of units, this unit replaced the curie (1 Ci $= 3.7 \times 10^{10}$ Bq).

Many analyses give the α- and β-activity per litre or millilitre, without any definition of the type of radionuclide (bulk activity). The radiation absorbed in the organism can give rise to biological damage, that is partly somatic, partly genetic. Somatic damage affects the radiated person; genetic damage, through damage to the genes, affects succeeding generations. The special danger from drinking water contaminated by radioactivity lies in the fact that radioactive substances enter the human body, can be absorbed there, and in some cases can be concentrated in certain organs and systems of organs. In this respect absorption, distribution, and excretion are different according to the chemical properties of the radionuclides. For example plutonium, radium, and strontium concentrate in bone, and iodine in the thyroid glands. Carbon and tritium on the other hand pass quite rapidly through the body.

For the assessment of the radiation absorbed by the body, besides the type and energy of the radiation, the physical half-life and the residence time of radionuclides (biological half-life) in the body must be considered. The biological effects of the different types of radiation are different for the same dose. Physiological factors must be considered when the effect of the radiation on the body is to be measured. Different body cells react with differential sensitivity toward radiation. This is taken into consideration by the quality factors (QF) — factors for the relative biological effect (RBE). The RBE, when multiplied by the radiation energy dose, gives the dose equivalent (DE). The unit is joules per kilogram (J/kg), formerly rem (1 J/kg $= 100$ rem), which is the absorption of an ionizing radiation by the body, which gives rise to the equivalent effect as the absorption of X- or γ-radiation: DE (Gy) $= D$ (Gy) $\times$ QF. Table 83 gives quality factors for different types of radiation.

Maximum permissible dose rates have been laid down for the protection of populations, and from these the maximum permissible concentrations

Table 83 Quality Factors (QF) for Different Types of Radiation (Aurand et al. 1927b)

Type of Radiation	QF
X-Rays, γ-rays	1
Electrons, positrons	1 for $E > 0.03$ MeV
	1.7 for $E < 0.03$ MeV
α-Rays	10
Slow neutrons	5
Fast neutrons	10

(MPC) for a large number of radionuclides in potable water have been determined. These are based on a daily water intake of 2.2 litres over 50 years. This means that a person could drink water containing these concentrations for a whole lifetime without exceeding the maximum permitted radiation. Differences in the physical and radiobiological behavior of the individual radionuclide account for the deviations of several orders of magnitude between values; for example the MPC for tritium is 11.1×10^7 Bq/l, and that for ^{90}Sr 14.8×10^4 Bq/l (Aurand et al. 1972b), because after a few weeks most of the tritium has left the body, whereas ^{90}Sr is stored in the bones. In this connection the total dose must be accounted for: this is the whole amount to which the person is exposed, including exposure to other foodstuffs as well as to general background radiation. Table 84 gives some hydrogeologically significant radionuclides and their maximum permissible values (see also Committee on Environmental Improvement 1978).

Without data on the effective radionuclides, radiation limits are set for total α-activity at 11.1 Bq/l and for total β-activity at 111 Bq/l (Müller 1971). In the ideal case water that is used for drinking or for production and preparation of foodstuffs should not contain any radioactive material, either in solution or in suspension. This is however hardly possible in practice because water always contains the natural radioisotope ^{40}K and very often the natural isotopes ^{14}C and tritium.

4.2.3.2 Water for Agriculture. Livestock and irrigation have different requirements in respect of water quality.

For livestock fundamentally the same requirements hold as for human consumption; however, animals can drink water with moderately high dissolved solids (about 10 g/l) when NaCl is the chief constituent, but the concentrations should not exceed 5000 mg/l as far as possible. For optimum cattle rearing the values should be less than those given in Table 85 (Hem 1970).

Table 84 Hydrogeologically Significant Natural and Artificial Radionuclides [a]

Element	Radioisotope mass number	Half-life	Radiation	Maximum permissible concentration above natural background in water (MPC) (1) Bq/l	(2) Bq/l	Maximal permissible radioactive ingestion by food or drinking water (3) (Bq)
Calcium	45	153 d	β^-	3.3×10^3	–	1.6×10^4
Carbon	14	5570 a	β^-	3.0×10^5	–	1.4×10^6
Cerium	144	285 d	β^-, γ	3.7×10^3	–	2.2×10^4
Cesium	135	2.0×10^6 a	β^-	3.7×10^4	–	2.0×10^5
	137	30 a	β^-, γ	7.4×10^3	7.4×10^4	2.6×10^4
Chlorine	36	2.5×10^5 a	β^-	3.0×10^4	–	1.0×10^5
Chromium	51	27.8 d	γ	7.4×10^5	–	2.6×10^6
Cobalt	57	267 d	γ	1.9×10^5	–	6.6×10^5
	60	5.24 a	β^-, γ	1.9×10^4	–	6.0×10^4
Hydrogen	3	12.4 a	β^-	1.1×10^6	1.1×10^5	5.8×10^6
Iodine	129	1.7×10^7 a	β^-, γ	1.5×10^2	2.2	3.3×10^2
	131	8.05 d	β^-, γ	7.4×10^2	1.1×10^1	1.8×10^3

Element	Mass no.	Half-life	Decay			
Plutonium	238	86.4 a	α, γ	1.9×10^3	–	9.0×10^3
	239	2.43×10^4 a	α, γ	1.9×10^3	–	7.8×10^3
	240	6580 a	γ	1.9×10^3	–	7.8×10^3
	242	3.8×10^5 a	γ	1.9×10^3	–	8.4×10^3
Radium	226	1620 a	α, γ	3.7	–	2.2×10^1
	228	6.7 a	β^-	1.1×10^1	–	4.9×10^1
Radon	222	3.8 d	α, γ	–	–	no limit
Rubidium	86	18.6 d	β^-, γ	2.6×10^4	–	4.2×10^4
	87	5×10^{10} a	β^-	3.7×10^4	–	no limit
Ruthenium	103	40 d	β^-, γ	3.0×10^4	–	1.4×10^5
	106	368 d	β^-, γ	3.7×10^3	–	2.2×10^4
Sodium	22	2.6 a	β^-, γ	1.5×10^4	–	5.3×10^4
Strontium	89	51 d	β^-	3.7×10^3	1.1×10^2	2.2×10^4
	90	28 a	β^-	3.7×10^1	1.1×10^1	7.2×10^2
Sulfur	35	87.9 d	β^-	2.2×10^4	–	1.1×10^5
Silicon	32	~650 a	β^-	–	–	3.4×10^5
Uranium[b]	235	7.1×10^8 a	α, γ	1.1×10^4	–	$6.6 \times 10^{32\,b}$
	238	4.5×10^9 a	α, γ	1.5×10^4	–	$1.0 \times 10^{32\,b}$
Zinc	65	245 d	β^+, γ'	3.7×10^4	–	1.7×10^5

[a] Data sources. (1), Strahlenschutzverordnung 1965, Fair & Parks 1973; (2) Committee on Environmental Improvement 1978; (3) Strahlenschutzverordnung 1976.

[b] Uranium: Because of the chemical toxicity of uranium, daily ingestion shall not be more than 150 mg/day, independent of radionuclide composition.

**Table 85   Upper Limits of Total
Dissolved Solids in Water for
Livestock Use**

Livestock	Upper limit (mg/l/
Poultry	2,860
Pigs	4,290
Horses	6,435
Dairy cattle	7,150
Beef cattle	10,100
Lambs (fattening)	12,900

Poisonous substances that occur particularly in polluted groundwaters can cause injury to livestock; however, it is often found that the poisoning has been caused by ingestion of plants, in which the poisons are enriched, and less often through water.

Assessment of the suitability of a groundwater for irrigation purposes requires consideration of the total dissolved solids, the concentration of any substance that may be toxic to plants, and the relative amount of certain of the constituents. There are however other factors to be considered.

During uptake of water through the plant roots some of the dissolved content is selectively absorbed for nutrition, but much is left behind, thus causing an increase in concentration near the surface, when large amounts of water are transpired. The soil water can become quite saline. This process is accentuated in hot arid climates. Substances of low solubility (e.g., $CaCO_3$) precipitate; the residue must however be removed by leaching, so that the osmotic gradient between the root sap and the soil water is maintained. This salinity problem depends on the quality of the irrigation water, the soil type and its structure, the local topography, the amount of irrigation water applied, the method of applying it, the local climate, particularly the amount of precipitation and its distribution in time, and finally the drainage system.

With deep-lying water tables and permeable surface strata there is a steady downward movement of salt when there is an excess of water, which guarantees steady percolation. With high groundwater levels this is not possible. Here the concentration of salts can easily come about by evaporation. In such cases suitable drawdown and drainage of the groundwater must be provided. The percolate has more dissolved solids than the inflowing water, but it can be recirculated several times, until finally the water has reached too high a concentration for further irrigation use. When drainage conditions are unsatisfactory (poor drainage, clayey soil), more stringent water quality requirements must be met.

The groundwater below the irrigated area is of course polluted by salts

(Eldridge 1963). The most severe pollution occurs when irrigation water is obtained from the groundwater body itself and is used in the vicinity of wells, because this practice accelerates the local hydrologic cycle. With higher groundwater demand the salt content is concentrated as in a closed basin (Hem 1970). The different sensitivities of plants to NaCl are shown in Table 86.

Table 86   Relative Tolerances of Crop Plants to NaCl (Richards 1954)

Low tolerance	Moderate tolerance	High tolerance
Pear	Grape	Date palm
Apple	Olive	Beetroot
Orange	Fig	Asparagus
Almond	Pomegranate	Spinach
Apricot	Tomato	Barley
Peach	Cabbage	Cotton
Lemon	Cauliflower	
Avocado	Lettuce	
Radish	Maize	
Celery	Carrot	
Lima bean	Onion	
Clover	Lucerne (alfalfa)	
	Wheat	
	Rye	
	Oats	
	Sunflower	

The tolerance of plants toward poor water can be increased by use of larger quantities, because when this is done local damaging salt concentrations may be avoided.

In certain cases, even when present in small amounts, some constituents can impair the suitability of water for irrigation. One of these constituents is boron, a plant nutrient that is occasionally added to fertilizers to combat boron deficiency in the soil. For some plants however excessive boron can be toxic. The various tolerances toward boron are given in Table 87.

Irrigation water that has a high sodium content can bring about a displacement of exchangeable cations Ca^{2+} and Mg^{2+} from the clay minerals of the soil, followed by the replacement of the cations by sodium. Sodium-saturated soils peptize and lose their permeability, so that their fertility and suitability for cultivation decreases. The process is reversible, however, and it can be counteracted by addition of suitable substances, commonly gypsum.

Table 87   Tolerance of Plants to Boron: Tolerance Increases in Each Column from Top to Bottom (Richards 1954)

Excellent water:	< 0.33 mg/l	< 0.67 mg/l	< 1.0 mg/l
Unsuitable water:	> 1.25 mg/l	> 2.50 mg/l	> 3.75 mg/l
	Sensitive	Semitolerant	Tolerant
	Lemon	Lima bean	Carrot
	Grapefruit	Pepper	Lettuce
	Avocado	Pumpkin	Cabbage
	Orange	Oats	Turnip
	Apricot	Maize	Onion
	Peach	Wheat	Vetch (fodder bean)
	Cherry	Barley	Alfalfa
	Persimmon	Olive	Beetroot
	Grape	Pea	Sugar beet
	Apple	Radish	Date palm
	Pear	Tomato	Athel tamarisk
	Plum	Cotton	Asparagus
	Artichoke	Potato	
	Walnut	Sunflower	

Tison (1957) has given a simple rule of thumb for the assessment of the NaCl content:

< 0.5 g NaCl/l	Water always usable
0.5–1g NaCl/l	Water mostly usable
1–1.5 g NaCl/l	Of only limited use for NaCl-sensitive plants
1.5–2g NaCl/l	Unusable for NaCl-sensitive plants
2.0–2.5 g NaCl/l	Limited use
2.5–3g NaCl/l	Only usable for certain plants
3–4g NaCl/l	Practically unusable
> 4g NaCl/l	Unusable

The ratio of Na^+ ions to the total cation content, stated in meq %, in conjunction with the total dissolved solids expressed either as the sum of the total dissolved solids or as electrical conductance, can be used for the as-

Specific electrical conductance

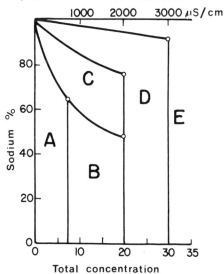

Fig. 89 Diagram for assessment of irrigation water suitability (after Wilcox 1948).
Total concentration is given in meq/l, the sodium fraction in meq % of total cations, electrical conductance in μs/cm. In the classification: A, excellent to good; B, good to
usable; C, usable to doubtful; D, doubtful to
unusable; E, unusable.

sessment of the suitability of a particular water for irrigation (Todd 1960b)
(Fig. 89).

The ability of a water to expel calcium and magnesium by sodium can be
estimated with the aid of the sodium adsorption ratio (SAR) (Richards
1954):

$$SAR = \frac{Na^+}{\sqrt{(Ca^{2+} + Mg^{2+})/2}}$$

The concentration of the ions present is given in meq/l. High SAR values
indicate the risk of a displacement of the alkaline earths. The use of the SAR
value is empirical and has limited value for geochemical applications (Hem
1970).

There is an additional risk from water that has a high hydrogen carbonate
content but relatively low calcium content. Thus water with more than 2.5
meq/l residual sodium carbonate should not be used; 1.25–2.5 meq/l should
not be used if possible; but water with less than 1.25 meq/l may be regarded
for practical purposes as safe.

Residual sodium carbonate is double the amount of hydrogen carbonate or
carbonate after the subtraction of the equivalent content of calcium and
magnesium (Richards 1954). This procedure disregards the separate behavior of Ca^{2+} and Mg^{2+} on precipitation as carbonate and the biological CO_2
production in the soil, which can give rise to considerable amounts of HCO_3^-
ions.

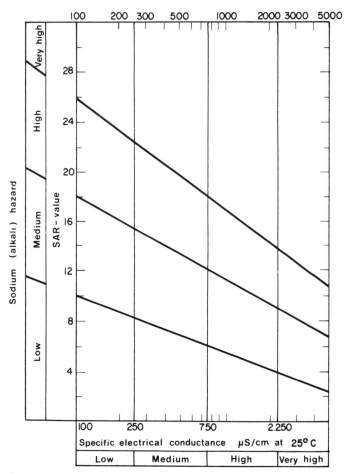

Fig. 90   Diagram for the assessment of suitability of irrigation water (after Richards 1954 and Hem 1970).

Richards (1954) gives a diagram for the assessment of water in which the total dissolved solids (expressed as electrical conductance) are plotted against SAR. The resulting 16 fields give an indication of the probability of possible excessive salinity and undesirable ion exchange (Fig. 90).

4.2.3.3 Industrial Water. The supply of water to homes for drinking, washing, and cooking requires that the standards of potable water for public supply be met; low total hardness is desirable in the interests of reducing the consumption of soap.

Industry and business have varied water quality needs. A thorough treatment of the subject is therefore not possible in this book, and the literature should be consulted (Gmelin 1952, Todd 1960b, Hem 1970). In many cases the limiting standards laid down for potable water do not suffice for industrial purposes. Industrial water for the food and drink, and medicinal drug industries, so far as it comes into direct contact with the produce, must have at least the quality of potable water. For use in the food processing industry for example the limiting values for hydrogen carbonate (300 mg/l), calcium (80 mg/l), and magnesium (40 mg/l) are lower than those for potable water, because many types of vegetables can become hardened when cooked in water with high calcium and magnesium content.

Water used for cooling purposes should not be corrosive, nor should it become corrosive during the cooling process through the temperature-dependent breakdown of the calcium carbonate–carbon dioxide equilibrium. It should not cause furring in the pipes. A severe pollution of water by organic substances leads (especially in factories with recycled cooling water installations) to abundant growth of fungi and algae, which clog pipes and condensers. A pH value of more than 8.3 is detrimental to wooden cooling slats (Press 1965). Temperature is important in the use of cooling water. Occasionally in water-deficient areas cooling water may be injected into the ground after use. Warmed cooling water is however deficient in oxygen and – after corresponding loss by evaporation – also is more contaminated with salts than the original water. With the percolation of such water one has to reckon with the gradual heating of the subsurface (cf. Section 3.2).

The requirements of water for steam generation depend first of all on the type of boiler construction and the working pressure. Generally the water should be as free from suspended matter as possible; it should have low residue on evaporation, low dissolved solids and gases, and no acid reaction. It must also be virtually free from oils, fats, and other organic substances. Nitrate, silica, gypsum, and magnesium chloride may be present to the smallest possible extent; sulfide and aggressive carbon dioxide especially should not be present.

Requirements increase from low to high pressure boilers. Low pressure boilers can use water with total dissolved solids up to 700 mg/l (and 12° hardness – German scale), as in natural water. High pressure boilers demand a maximum total dissolved solids content of 0.5 mg/l (and < 1° German hardness); therefore water treatment is necessary before use (Hem 1970, Press 1965).

Very pure water is needed for atomic energy plants to keep the radioactivity caused by neutron activation of dissolved solids as low as possible (Hem 1970).

Water can be treated for every purpose; however there is a practical limit, especially when large amounts are needed. As a matter of economy it may be necessary to site water-consuming industries in areas where there is an abundant natural supply of suitable water.

References

Adams, C. S. & A. C. Swinnerton (1937): Solubility of limestone. *Trans. Am. Geophys. Union*, **18**, pp. 504–508, Washington, D.C.

Agie, J. (1974): Pollution des nappes aquifers par les sels de deneigement. *Mem. Int. Assoc. Hydrogeol.*, **10**, pp. 7–9.

Albertsen, M. (1977): *Labor- und Felduntersuchungen zum Gasaustausch zwischen Grundwasser und Atmosphäre über natürlichen und verunreinigten Grundwässern.* 145 pp. Ph.D. thesis, Kiel.

Albertsen, M. (1979): Kohlendioxid-Haushalt in der Gasphase der ungesättigten Bodenzone, dargestellt am Beispiel eines Podsols. *Z. Pflanzenernähr. Düng. Bodenkd.*, **142**, pp. 39–56, Weinheim, Bergstrasse.

Albertsen, M. & G. Matthess (1977): Modellversuche zur Bestimmung des diffusionsbedingten Gastransports in rolligen Lockergesteinen und ihre praktische Anwendung bei der Beurteilung belasteter Grundwässer. *Dtsch. Ges. Mineralölwiss. Kohlechem.*, **146**, 58 pp., Hamburg.

Albertsen, M. & G. Matthess (1978): Ground air measurements as a tool for mapping and evaluating organic groundwater pollution zones. *Proceedings of the International Symposium on Groundwater Pollution by Oil Hydrocarbons.* Prague: IAH, pp. 235–251.

Albertson, O. E. (1961): Ammonia nitrogen and the anaerobic environment. *J. Water Pollut. Control Fed.*, **33**, 9, pp. 978–995, Washington, D.C.

Albrecht, B., C. Junge & H. Zakosek (1970): Der N_2O-Gehalt der Bodenluft in drei Bodenprofilen. *Z. Pflanzenernähr. Bodenkd.*, **125**, 3, pp. 205–211, Weinheim, Bergstrasse.

Alekin, O. A. (1962): *Grundlagen der Wasserchemie. Eine Einführung in die Chemie natürlicher Wässer.* 260 pp. Leipzig: VEB Deutsch. Verl.

Alexander, L. T., E. P. Hardy & H. L. Hollister (1960): Radioisotopes in soils: Particularly with reference to Strontium 90. In: R. S. Caldecott & L. A. Snyder, Eds. *A Symposium on Radioisotopes in the Biosphere*, pp. 3–22, Minneapolis.

Alexander, M. (1961): *Introduction to Soil Microbiology.* 472 pp. New York- London: Wiley.

Alexander, M. & B. K. Lustigman (1966): Effect of chemical structure on microbial degradation of substituted benzenes. *J. Agric. Food Chem.* **14**, 4, pp. 410–413, Washington, D.C.

Al'Tovskii, M. E., Z. I. Kuznetsova & V. M. Shvets (1961): *Origin of oil and oil deposits.* 107 pp. New York: Consultants Bureau.

Al'Tovskii, M. E., E. L. Bykova, Z. I. Kuznetsova & V. M. Shvets (1962): Organic substances and microflora of groundwaters and their significance in the processes of oil and gas origin. *Gostoptekhizdat*, 291 pp. Moscow (in Russian).

Ames, L. L. & D. Rai, (1978): *Radionuclide Interactions with Soil and Rock Media.* Richland, Wa.

Amphlett, C. B. (1958): Ion exchange in clay Minerals. *Endeavour*, **17**, 67, pp. 149–155, London.

Amphlett, C. B. (1964): *Inorganic Ion Exchangers*, 141 pp. Amsterdam: Elsevier.

Andersen, J. R. & J. N. Dornbush (1967): A study of the influence of sanitary landfill on groundwater quality. *J. Am. Water Works Assoc.* **59**, pp. 457–470, New York.

Andersen, J. R., and J. N. Dornbusch (1968): Quality changes of shallow groundwater resulting from refuse disposal at a gravel pit. Report on Project SRI 3553, Water Resource Institute, South Dakota State University, Brookings, SD.

Andersen, L. J. (1979.: A semi-automatic level-accurate groundwater sampler. *Geological Survey of Denmark, Yearbook, 1978*, pp. 165–171, Copenhagen.

Anderson, C. C. & H. H. Hinson (1951): Helium-bearing natural gases of the United States – Analyses and analytical methods. U.S. Bureau of Mines Bulletin No. 486, 141 pp., Washington, D.C.

Anrich, H. (1969): Die Darstellung chemischer Analysendaten. In: A. Bentz, Ed., *Lehrbuch der angewandten Geologie*, Vol. 2, Part 2, Stuttgart: Enke, pp. 1426–1440.

D'Ans-Lax (1967): *Taschenbuch für Chemiker und Physiker*, Vol. 1, 1522 pp. Berlin, Heidelberg, New York: Springer.

Apgar, M. A., & D. Langmuir (1971): Ground-water pollution potential of a landfill above the water table. *Ground Water*, **9**, pp. 76–94, National Ground Water Quality Symposium, Denver.

Appleman, C. O. (1927): Percentage of carbon dioxide in soil air. *Soil Sci.*, **24**, pp. 241–245, New Brunswick NJ.

Audoynaud, A. & B. Chauzit (1880): De passage de l'eau et de l'air dans la terre arable. *Forsch. Geb. Agric. Phys.*, **3**, pp. 19–21, Heidelberg (review).

Aurand, K., G. Matthess & R. Wolter (1971): Strontium-90, Ruthenium-106 und Caesium-137 in natürlichen Wässern. *Notizbl, Hess. Landesamt Bodenforsch.*, **99**, pp. 313–333, Wiesbaden.

Aurand, K., W. Kerpen, G. Matthess, R. Wolter & H. Zakosek (1972a): Gefährdung von Trinkwasservorkommen durch radioaktive Kontaminatoren. In: *Radioaktive Stoffe und Trinkwasserversorgung bei nuklearen Katastrophen*, 78 pp., Forsch. Ber. B. Min. Innern, Bonn.

Aurand, K., H. Rühle & H. Schmier (1972b): Grundlagen des Strahlenschutzes. In: *Radioaktive Stoffe und Trinkwasserversorgung bei nuklearen Katastrophen*, 11; 13 pp., Bonn (BM Inn).

Aust, H. & K. Kreysing (1978): Geologische und geotechnische Grundlagen zur Tiefversenkung von flüssigen Abfällen und Abwässern. *Geol. Jahrb*, C **20**, 224 pp., Hannover.

Axt, G. (1965): Das Verhalten von Mischwässern in Bezug auf das Kalk-Kohlensâure-Gleichgewicht. *Chem-Ing.-Tech.*, **37**, 10, p. 1074, Weinheim, Bergstrasse.

Axt, G. (1966): Mischwässer und Kalkaggressivität. *Vom Wasser*, **32**, pp. 423–439, Weinheim, Bergstrasse.

Baars, J. K. (1930): *Over Sulfaatreductie door Bacterien*. 164 pp. Ph.D. thesis, Delft.

Bass Becking, L. G. M. (1959): Geology and microbiology. *Contrib. Mar. Microbiol., New Zealand Oceanogr. Inst.*, **22**, pp. 48–64.

Baas Becking, L. G. M., & D. Moore (1959): The relation between iron and organic matter in sediments. *J. Sediment. Petrol.*, **29**, pp. 454–458, Tulsa. Ok.

Baas Becking, L. G. M. & D. Moore (1960): Iron bacteria in mine-waters. *Nature*, **186**, 4725, p. 660, London.

Baas Becking, L. G. M., J. R. Kaplan & D. Moore (1960): Limits of the natural environment in terms of pH and oxidation-reduction potentials. *J. Geol.*, **68**, pp. 243–284, Chicago.

Back, W. (1963): Preliminary results of a study of calcium carbonate saturation of groundwater in central Florida. *Int. Assoc. Sci. Hydrol. Bull.*, **8**, 3, pp. 43–51, Gentbrugge.

Back, W. (1973): Inorganic geochemical considerations for disposal of municipal sewage by spray irrigation. *Proceedings of the 1973 Workshop*, University of Florida, pp. 187–200.

Back, W. & B. B. Hanshaw (1970): Comparison of chemical hydrogeology of the carbonate peninsulas of Florida and Yucatan. *J. Hydrol.*, **10**, pp. 330–368, Amsterdam.

Back, W. & B. B. Hanshaw (1971): Rates of physical and chemical processes in a carbonate aquifer. In: J. D. Hem, Ed., *Nonequlibrium Systems in Natural Water Chemistry*, American Chemical Society, Advances in Chemistry Series No. 106, pp. 77–93, Washington, D.C.

Back, W. & D. Langmuir (1974): Influence of near-surface reactions and processes on chemical character of groundwater: A review of selected experiences in the United States. *Mem. Int. Assoc. Hydrogeol.* **10**, pp. 31–37.

Baedecker, M. J. & W. Back (1979): Hydrogeological processes and chemical reactions at a landfill. *Groundwater*, **17**, pp. 429–437.

Bahr, H. & W. Zimmermann (1965): Die Wanderung von Detergentien im Boden. *Arch. Hyg. Bakteriol.*, **149**, 7–8, pp. 620–626, Munich.

Balashov, Yu. A., A. B. Ronov, A. A. Migdisov & N. V. Turanskaya (1964): The effect of climate and facies environment on the fractionation of the rare earths during sedimentation. *Geochemistry (USSR), Engl. transl.*, pp. 951–969.

Balke, K.-D., H. Kussmaul & G. Siebert (1973): Chemische und thermische Kontamination des Grundwassers durch Industrieabwässer. *Z. Dtsch. Geol. Ges.* **124**, pp. 447–460, Hannover.

Bär, O. (1924): Versuch einer Lösung des Dolomitproblems auf phasentheoretischer Grundlage. *Senckenbergiana*, **6**, pp. 116–118, Frankfurt am Main.

Bär, O. (1932): Beitrag zum Thema Dolomitentstehung. *Zentralbl. Mineral., Geol., Paleontol.*, A, pp. 46–62, Stuttgart.

Barabanov, L. N. & V. N. Disler (1968): Principal regularities of the formation of nitrogenous thermal waters in the U.S.S.R. and some other countries. *23rd Int. Geol. Congr.*, **17**, pp. 179–184, Prague.

Baranik-Pikowsky, M. A. (1927): Über den Einfluss hoher Salzkonzentrationen auf die Limanbakterien. *Zentrabl. Bakt. Parasitenkd. Infektionskr.*, Part 2, **70**, pp. 373–383, Jena.

Barker, F. B. & R. C. Scott (1958): Uranium and radium in the groundwater of the Llano Estacado, Texas and New Mexico. *Trans. Am. Geophys. Union*, **39**, 3, pp. 459–466, Washington, D.C.

Barnes, I. (1964): Field measurement of alkalinity and pH. U.S. Geological Survey Water Supply Paper No. 1535-H, 17 pp., Washington, D.C.

Barnes, I. (1965): Geochemistry of Birch Creek, Inyo County, California, a travertine depositing creek in an arid climate. *Geochim. Cosmochim. Acta*, **29**, pp. 85–112, Oxford.

Bastisse, E. M. (1951–1953): Dix-huit années d'études lysimetriques appliquées à l'agronomie. *Ann. Agron.*, 1951, pp. 727–781; 1953, pp. 33–76, 331–378, 519–560, 717–780, Paris.

Bates, R. G. (1945): *Electrometric pH Determinations*. 331 pp. New York: Wiley. London: Chapman & Hall.

Bätjer, D. & H. Kuntze (1963): Untersuchungen des Niederschlagswassers im Küstengebiet Ostfrieslands und Oldenburgs. *Küste*, **11**, pp. 34–51, Heide i. Holstein.

Batzel, R. E. (1960): Radioactivity associated with underground nuclear explosions. *J. Geophys. Res.*, **65**, pp. 2897–2902, Richmond, VA.

Bavel, C. H. M. Van (1952): A theory on the soil atmosphere in and around a hemisphere in which soil gases are used or released. *Soil Sci. Soc. Am. Proc.*, **16**, pp. 150–153, Madison, WI.

Baver, L. D. (1966): *Soil Physics*. 3rd ed., 489 pp. New York, London, Sydney: Wiley.

Bayly, I. A. E. & D. Williams (1966): Further chemical observations on some lakes in the south-east of South Australia. *Austr. J. Mar. Freshwater Res.,* **17,** pp. 229–237, Melbourne.

Bear, J. (1972): *Dynamics of Fluids in Porous Media.* American Elsevier Environmental Science Series, 764 pp. New York, London, Amsterdam.

Becksmann, E. (1955): Grundwasserchemismus und Speichergestein. *Z. Dtsch. Geol. Ges.,* **106** (1954), pp. 23–35, Hannover.

Begemann, F. (1961): Der "natürliche" Tritiumhaushalt der Erde und die Frage seiner zeitlichen Variation. Habil. thesis, Mainz, 1961; *Chimia,* **16,** 15 pp.

Beger, H., reedited by J. Gerloff & D. Lüdemann (1966): *Leitfaden der Trink- und Brauchwasserbiologie.* 7 + 360 pp., Stuttgart: Fischer.

Behne, W. (1953): Untersuchungen zur Geochemie des Chlor und Brom. *Geochim. Cosmochim. Acta,* **3,** pp. 186–214, London.

Behre, C. H. & R. K. Summerbell (1934): Oxidation-reduction reactions between natural hydrocarbons and oil-filled waters. *Science,* N.S., **79,** 2037, pp. 39–40, New York.

Belin, R. E. (1959): Radon in the New Zealand geothermal regions. *Geochim. Cosmochim. Acta,* **16,** pp. 181–191, London, New York, Paris, Los Angeles.

Belter, W. G. (1963): Waste management activities of the Atomic Energy Commission. *Groundwater,* **1,** pp. 17–24, Urbana, IL.

Beran, F. & J. A. Guth (1965): Das Verhalten organischer insektizider Stoffe in verschiedenen Böden mit besonderer Berücksichtigung der Möglichkeiten einer Grundwasserkontamination. *Pflanzenschutz berichte,* **33,** 5–8, pp. 65–117, Vienna.

Berner, R. A., J. T. Westrich, R. Graber, J. Smith & C. S. Martens (1978): Inhibition of aragonite precipitation from supersaturated seawater: A laboratory and field study. *Am. J. Sci.* **278,** pp. 816–837, New Haven, CT.

Beyerinck, W. M. (1895): Über *Spirillum desulfuricans* als Ursache von Sulfatreduction. *Zentralbl. Bakteriol. Parasitenkd., Infektionskr.,* Part 2, **1,** pp. 1–9, 49–59, 104–114, Jena.

Bierschenk, W. H. (1961): Observational and field aspects of groundwater flow at Hanford (Washington), in ground disposal of radioactive wastes. *Conference Proceedings,* pp. 147–156, Sanitary Engineering Research Laboratory, University of California, Berkeley.

Biesecker, J. E. & R. George (1966): Stream quality in Appalachia as related to coal-mine drainage, 1965. U.S. Geological Survey Circ. No. 526, Washington, D.C.

Billings, G. K. & H. H. Williams (1967): Distribution of chlorine in terrestrial rocks (a discussion). *Geochim. Cosmochim. Acta,* **31,** p. 2247, London, New York.

Birch, F. (1954): The present state of geothermal investigations. *Geophysics,* **19,** 4, pp. 645–659, Menasha, WI.

Birk, F., R. Geiersbach & W. Müller (1973): Die Auswirkungen der Verkippung und Lagerung von cyanidhaltigen Härtesalzen in Bochum-Gerthe auf das Grund- und Oberflächenwasser. *Z. Dtsch. Geol. Ges.,* **124,** pp. 461–473, Hannover.

Bishop, W. D., R. C. Carter & H. F. Ludwig (1966): Water pollution hazards from refuse-produced carbon dioxide. Third International Conference on Water Pollution Research, Section I, Paper No. 10, 19 pp. Washington, D.C.: Water Pollution Control Federation.

Blake, G. R. & J. B. Page (1948): Direct measurement of gaseous diffusion in soils. *Soil Sci. Soc. Am. Proc.* **13,** pp. 37–42, Madison, WI.

Blanck, E. (1929): Die biologische Verwitterung als Ausfluss der in Zersetzung begriffenen organischen Substanz. *Handb. Bodenlehre,* **2,** pp. 263–297, Berlin.

Blanck, E. & H. Evlia (1932): Ein Beitrag zur Frage nach der Herkunft der im Gestein und Boden zirkulierenden sulfathaltigen Lösungen, sowie zum Kreislauf des Schwefels in der Natur. *Chem. Erde,* **7,** pp. 298–319, Jena.

Blanquet, P. (1946): Détermination de paramètres permettant le calcul de la conductivité de certaines solutions salines complexes. *Ann. Inst. Hydrol. Climatol.,* 17, 62, pp. 131–142, Paris.

Bögli, A. (1964): Die Kalkkorrosion, das zentrale Problem der unterirdischen Verkarstung. *Steir. Beitr. Hydrogeol.,* N.F., 15–16, pp. 75–90, Graz.

Bögli, A. (1969): Neue Anschauungen Über die Rolle von Schichtfugen und Klüften in der karsthydrographischen Entwicklung. *Geol. Rundsch.* 58, 2, pp. 395–408, Stuttgart.

Bogomolov, G. V., G. N. Plotnikova & E. A. Titova (1966): Silica in the subterranean waters of certain regions of the USSR and other countries. *Int. Assoc. Sci. Hydrol. Bull.,* 11, 1, pp. 24–33, Gentbrugge.

Böhnke, B. (1966): Untersuchungen über die Behandlung von cyanhaltigen Abwässern auf Tropfkörpern. *Vom Wasser,* 32, pp. 291–309, Weinheim, Bergstrasse.

Bömer, A. (1905): Beiträge zur chemischen Wasseruntersuchung. *Nahr. Genussm.,* 10, pp. 129–143, Berlin.

Bonde, E. K. & P. Urone (1962): Plant toxicants in underground water in Adams County, Colorado. *Soil Sci.,* 93, pp. 353–356, New Brunswick, NJ.

Borneff, J. (1967): Vorkommen und Bewertung von kanzerogenen Substanzen in Wasser. *Gas-Wasserfach,* 108, 38, pp. 1072–1076, Munich.

Borneff, J. & R. Knerr (1960): Kanzerogene Substanzen in Wasser und Boden. III. Quantitative Ermittlungen zur Löslichkeit, Filtration, Adsorption und Eindringtiefe. *Arch. Hyg. Bakteriol.,* 144, pp. 81–94, Munich, Berlin.

Borneff, J. & H. Kunte (1965): Kanzerogene Substanzen in Wasser und Boden. XVII. Über die Bewertung und Herkunft der polycyclischen, aromatischen Kohlenwasserstoffe im Wasser. *Arch. Hyg. Bakteriol.,* 149, 3–4, pp. 226–243, Munich.

Bösenberg, K. & G. Lüttig (1959): Zur Methodik hydrochemischer Untersuchungen im Wesergebiet. *Geol. Jahrb.,* 76, pp. 579–596, Hannover.

Boulegue, J. (1977): Equilibria in a sulfide rich water from Enghien-les-Bains, France. *Geochim. Cosmochim. Acta,* 41, pp. 1751–1758, Oxford, New York, Paris, Frankfurt.

Boussingault, G. & J. Léwy (1853): Mémoire sur la composition de l'air confiné dans la terre végétale. *Ann. Chim.* (III), 37, pp. 5–50, Paris.

Bouwer, H. (1968):Putting waste water to beneficial use — The Flushing Meadows Project. *Proceedings of the 12th Arizona Watershed Symposium, Phoenix,* pp. 25–30.

Bouyoucos, G. J. & M. M. McCool (1924): The aeration of soils as influenced by air-barometric pressure changes. *Soil Sci.,* 18, pp. 53–63, New Brunswick, NJ.

Bovard, P., A. Grauby & J. Boyer (1963): Rétention et accumulation du [106]Ru + [106]Rh dans les échantillons de divers types de sols. *Colloque International s la Rétention et la Migration d'Ions Radioactifs dans les Sols, Saclay 1962,* pp. 141–144, Paris.

Boynton, D. & W. Reuther (1939): Seasonal variation of oxygen and carbon dioxide in three different orchard soils during 1938 and its possible significance. *Proc. Am. Soc. Hortic. Sci.,* 36, pp. 1–6, Ithaca, NY.

Bradley, W. H. (1929): The occurrence and origin of analcite and meerschaum beds in the Green River formation of Utah, Colorado, and Wyoming. U.S. Geological Survey Professional Paper No. 158-A, pp. 1–7, Washington, D.C.

Brantner, H. (1966): Zur Ökologie manganverwertender Bakterien. *Gas- Wasserfach,* 107, 44, pp. 1244–1246, Munich.

Braun, K., L. Erb, & K. Sauer (1953): Die öffentliche Wasserversorgung in Südbaden. *Gas-Wasserfach,* 94, 10, pp. 281–287, Munich.

Breidenbach, A. W. (1965): Pesticide residues in air and water. *Arch. Environ. Health,* 10, 6, pp. 827–830, Chicago.

Briggs, L. J. (1905): On the adsorption of water vapor and of certain salts in aqueous solution by quartz. *J. Phys. Chem.,* **9**, pp. 617–640, Ithaca. NY.

Bringmann. G. (1969): Physiologische Verfahren der biologischen Wasseranalyse. In S. W. Souci & K.-E. Quentin, Eds., *Handbuch der Lebensmittelchemie,* Vol. 8, Part 2. Berlin, Heidelberg, New York: Springer, pp. 1200–1228.

Bristol, B. M. (1920): On the alga-flora of some desiccated English soils: An important factor in soil biology. *Ann. Bot.,* **34**, pp. 35–80, London.

Brooks, F. A. (1960): Trace and minor elements in Woodbine sub-surface waters of the East Texas basin. *Geochim. Cosmochim. Acta,* **20**, pp. 199–214, Oxford, London, New York, Paris.

Brown, D. J. & J. R. Raymond (1962): Radiologic monitoring of ground water at the Hanford Project. *J. Am. Water Works Assoc.,* **54**, pp. 1201–1212, Richmond, VA.

Brown. E., M. W. Skougstad & Fishman (1970): Methods for collection and analysis of water samples for dissolved minerals and gases. *Techniques of Water Resources Investigation,* No. 5, U.S. Geological Survey, Washington, D.C.

Brown, Ph. M. (1958): The relation of phosphorites to ground water in Beaufort County, North Carolina. *Econ. Geol.,* **53**, 1, pp. 85–101, New Haven, CT.

Brown, R. E., H. M. Parker & J. M. Smith (1956): Disposal of liquid wastes to the ground. *Proceedings of the International Conference on Peaceful Uses of Atomic Energy,* **9**, pp. 669–675, Geneva.

Brown, R. M. (1961): Hydrology of tritium in the Ottawa valley. *Geochim. Cosmochim. Acta,* **21**, pp. 199–216, Oxford, London, New York, Paris.

Brümmer, G., H. S. Grunwaldt & D. Schroeder (1971): Beiträge zur Genese und Klassifizierung der Marschen. III. Gehalte, Oxydationsstufen und Bindungsformen des Schwefels in Koogmarschen. *Z. Pflanzenernähr. Düng. Bodenkd.* **129**, pp. 92–107, Weinheim Bergstrasse.

Buchan, S. (1958): Variations in mineral contents of some ground waters. *Proc. Soc. Water Treat. Exam.,* **7**, pp. 11–29, London.

Buchan, S. & A. Key (1956): Pollution of ground water in Europe. *Bull. WHO,* **14**, pp. 949–1006, Geneva.

Bucherer, H. (1965): Über den mikrobiellen Abbau von Giftstoffen, 4. Mitt.: Über den mikrobiellen Abbau von Phenylacetat, Strichnin, Brucin, Vomicin und Tuborurarin. *Zentralbl. Bakteriol., Parasitenkd., Infektionsr., Hyg.,* Part 2, **119**, 3, pp. 232–238, Stuttgart.

Buchholz, F. (1961): Redoxpotentiale und Sauerstoffgehalt im Grundwasser sandiger Waldböden Nordostdeutschlands. *Z. Pflanzenernähr., Düng., Bodenkd.,* **94**, pp. 154–163, Weinheim, Bergstrasse.

Buchner, A. (1958): Die Schwefelversorgung der westdeutschen Landwirtschaft. *Landwirtsch. Forsch.,* **11**, pp. 79–92, Frankfurt am Main.

Buckingham, E. (1904): Contributions to our knowledge of the aeration of soils. U.S. Department of Agriculture, Bureau of Soils Bulletin No. 25, 52 pp., Washington, D.C.

Bucksteeg, W. (1969a): Charakteristik und Behandlung des Abwassers. In: S. W. Souci & K.-E. Quentin, Eds., *Handbuch der Lebensmittelchemie,* Vol. 8, Part 1. Berlin, Heidelberg, New York: Springer, pp. 486–557.

Bucksteeg, W. (1969b): Abfalldeponien und ihre Auswirkungen auf Grund-und Oberflächenwasser. Gas- Wasserfach, **110**, pp. 529–537, Munich.

Bundesminister des Inneren (1971): Materialien zum Umweltprogramm der Bundesregierung. *Umweltplanung,* 661 + 64 pp. Bonn.

Buschendorf, F., M. Richter & H. W. Walther (1957): Die Blei-Zink-Erzvorkommen des Ruhrgebietes und seiner Umrandung. C. Der Erzgang Christian Levin in den Blei-Erz-

Feldern König Wilhelm III/IV und Rheinstahl (Zechen Christian Levin in Essen-Dellwig und Prosper in Bottrop). *Beih. Geol. Jahrb.,* **28**, 163 pp., Hannover.

Butler, R. G., G. T. Orlob & P. H. McGauhey (1954): Underground movement of bacterial and chemical pollutants. *J. Am. Water Works Assoc.,* **46**, 2, pp. 97–111, Lancaster, PA.

Buttlar, H. von & I. Wendt (1958): Ground-water studies in New Mexico using tritium as a tracer. *Trans. Am. Geophys. Union,* **39**, 4, pp. 660–668, Washington, D.C.

Čadek, J., M. Malkovský & J. Šulcek (1968): Geochemical significance of subsurface waters for the accumulation of ore components. *23. Int. Geol. Congr.,* **6**, pp. 161–168, Prague.

Calvert, D. V. & H. W. Ford (1973): Chemical properties of acid-sulfate soils recently reclaimed from Florida marshland. *Soil Sci. Am., Proc. 37,* pp. 367–370, Madison, WI.

Carlé, W. (1956): Stockwerke und Wanderwege von Mineralwässern in Franken. *Z. Dtsch. Geol. Ges.,* **106**, pp. 118–130, Hannover.

Carlson, S. (1965): Bedeutung, Aufgaben und Ziel der Virusforschung in der Wasserhygiene. *Gas- Wasserfach,* **106**, 12, pp. 325–329, Munich.

Carlston, C. W., L. L. Thatcher & E. C. Rhodehamel (1960): Tritium as a hydrologic tool, the Wharton Tract study. *Int. Assoc. Sci. Hydrol. Publ.,* **52**, pp. 503–512, Gentbrugge.

Carroll, D. (1959): Ion exchange in clays and other minerals. *Bull. Geol. Soc. Am.,* **70**, pp. 749–780, New York.

Carroll, D. (1962): Rainwater as a chemical agent of geologic processes – A Review. U.S. Geological Survey Water Supply Paper No. 1535-G, 18 pp., Washington, D.C.

Chambers, C. W., H. H. Tabak & P. W. Kabler (1963): Degradation of aromatic compounds by phenol-adapted bacteria. *J. Water Pollut. Control. Fed.,* **35**, 12, pp. 1517–1528, Washington, D.C.

Champ, D. R., J. Gulens & R. E. Jackson (1979): Oxidation-reduction sequences in ground water flow systems. *Can. J. Earth Sci.* **16**, pp. 12–23.

Chang, J.-H. (1957): Global distribution of the annual range in soil temperature. *Trans. Am. Geophys. Union,* **38**, 5, pp. 718–723, Washington, D.C.

Chebotarev, I. I. (1955): Metamorphism of natural waters in the crust of weathering. *Geochim. Cosmochim. Acta,* **8**, pp. 23–48, 137–170, and 198–212, London, New York.

Chenais, S. & J. Chenais (1939): *Constantes Physiques de l'Eau,* 167 pp. Paris: Dunod.

Cherry, R. N., D. P. Brown, J. K. Stamer & C. Z. Goetz (1973): Hydrobiochemical effects of spraying waste treatment effluent in St. Petersburg, Florida. *Proceedings of the 1973 Workshop,* University of Florida, pp. 170–186.

Chilingar, G. V. (1956a): Durov's classification of natural waters and chemical composition of atmospheric precipitation in USSR: A review. *Trans. Am. Geophys. Union,* **37**, pp. 193–196, Washington, D.C.

Chalingar, G. V. (1956b): Cl⁻ and SO₄²⁻ content of atmospheric precipitation in USSR: A summary. *Trans. Am. Geophys. Union,* **37**, 4, pp. 410–412, Washington, D.C.

Chilingar, G. V. (1957): Soviet methods of reporting and displaying results of chemical analyses of natural waters and methods of recognizing oil-field waters: A summary. *Trans. Am. Geophys. Union,* **38**, pp. 219–221, Washington, D.C.

Clarke, W. F. (1924): *The Data of Geochemistry,* 5th ed., U.S. Geological Survey Bulletin No. 770, 841 pp., Washington, D.C.

Clements, F. E. (1921): Aeration and air-content. The role of oxygen in root activity. Carnegie Institution, Washington, **315**, 183 pp., Washington, D.C.

Coe, J. J. (1970): Effect of solid waste disposal on groundwater quality. *J. Am. Water Works Assoc.,* **62**, 12, pp. 776–783, New York.

Collins, A. G. (1963): Flame photometric determination of cesium and rubidium in oil field waters. *Anal. Chem.,* **35**, pp. 1258–1261.

Collins, W. D. (1923): Graphic representation of water analysis. *Ind. Eng. Chem.*, **15**, p. 394, New York.

Collins, W. D. & K. T. Williams (1933): Chloride and sulfate in rain water. *Ind. Eng. Chem.*, **25**, pp. 944–945, New York.

Committee on Environmental Improvement (1978): *Cleaning our environment. A Chemical Perspective*, 2nd ed. Washington, D. C.: American Chemical Society.

Committee "Wasser und Mineralöl" (1969): *Beurteilung und Behandlung von Mineralölunfällen auf dem Lande im Hinblick auf den Gewässerschutz*, 138 pp. Bad Godesberg (B.-Min. Gesundheitswesen).

Corey, J. C. & D. Kirkham (1965): Miscible displacement of ^{15}N-tagged nitrate and tritiated water in water-saturated and water-unsaturated soil. *Isotopes and Radiation in Soil-Plant Nutrition Studies*. Vienna: IAEA, pp. 157–170.

Correns, C. W. (1940): Die chemische Verwitterung der Silikate. *Naturwissenschaften*, **28**, pp. 369–376, Berlin.

Correns, C. W. (1956); The geochemistry of the halogens. In: *Physics and Chemistry of the Earth*, Vol. 1, New York: McGraw-Hill, pp. 181–233.

Correns, C. W. (1957): Über die Geochemie des Fluors and Chlors. *Neue Jahrb. Mineral. Abh.*, **91**, pp. 239–256, Stuttgart.

Coste, J. H. & H. L. Wright (1935): The nature of the nucleus in hygroscopic droplets. *Phil. Mag.* (7), **20**, pp. 209–234, London.

Cotton, F. A. & G. Wilkinson (1966): *Advanced Inorganic Chemistry*, 2nd ed., 1136 pp. New York, London, Sydney: Wiley-Interscience.

Craig, H. (1961a): Isotopic variations in meteoric waters. Science, **133**, pp. 1702–1703, Washington, D.C.

Craig, H. (1961b): Standard for reporting concentrations of deuterium and oxygen-18 in natural waters. *Science,* **133**, pp. 1833–1834, Washington, D.C.

Csajághy, G. (1960): A felszin alatti vizek szerves anyagai. *Hidrol. Közl.*, **40**, 4, pp. 324–329, Budapest.

Csanády, M. (1968): Ausbreitung der Schwermetall- und Cyanverunreinigung im Grundwasser. *Hidrol. Közl.*, **48**, 1, pp. 32–38, Budapest.

Culkin, F. (1965): The major constituents of sea water. In: J. P. Riley & G. Skirrow, Eds., *Chemical Oceanography*, Vol. 1. London: Academic Press, pp. 121–161.

Czerwenka, W. & K. Seidel (1965): Neue Wege einer Grundwasseranreicherung in Krefeld. *Gas- Wasserfach*, **106**, pp. 828–833, Munich.

Dalmady, Z. von (1927): Zur graphischen Darstellung der chemischen Zusammensetzung der Mineralwasser. *Z. Ges. Phys. Ther.*, **34**, pp. 144–148.

Dansgaard, W. (1953): The abundance of ^{18}O in atmospheric water and water vapour. *Tellus*, **5**, pp. 461–469, Stockholm.

Dansgaard, W. (1961): The isotopic composition of natural waters. With special reference to the Greenland Ice Cap. Medd. *Grønland*, **165**, 2, 120 pp., Copenhagen.

Davey, C. B. & R. J. Miller (1966): Correlation of temperature-dependent water properties and the growth of bacteria. *Nature*, **209**, 5023, p. 638, London, New York.

Davies, C. W. (1962): Incomplete dissociation in aqueous solutions. In: W. J. Hammer, Ed., *The structure of electrolytic solutions*. New York: Wiley (1959), pp. 19–34; Ion Association, Washington, D.C. (Butterworths).

Davis, G. H. (1961): Geologic control of mineral composition of stream waters of the eastern slope of the southern Coast Ranges, California. U.S. Geological Survey Water Supply Paper No. 1535-B, 30 pp., Washington, D.C.

Davis, G. H., B. R. Payne, T. Dinçer, T. Florkowski & T. Gattinger (1967): Seasonal variations in the tritium content of groundwaters of the Vienna Basin, Austria. *Isotopes in Hydrology,* Vienna: IAEA, pp. 451–473.

Davis, G. H. (1961): Geologic control of mineral composition of stream waters of the eastern slope of the southern Coast Ranges, California. U.S. Geological Survey Water Supply Paper No. 1535-B, 30 pp., Washington, D.C.

Davis, G. H., B. R. Payne, T. Dinçer, T. Florkowski & T. Gattinger (1967): Seasonal variations in the tritium content of groundwaters of the Vienna Basin, Austria. *Isotopes in Hydrology,* Vienna: IAEA, pp. 451–473.

Davis, J. B. (1967): *Petroleum Microbiology,* 604 pp. Amsterdam, London, New York: Elsevier.

Davis, J. J.: Cerium and its relationship to potassium in ecology. In: Schultz & Klement, Eds., *Radioecology.* New York: Reinhold.

Davis, S. N. (1964): Silica in streams and groundwater. *Am. J. Sci.,* **262,** pp. 870–891, New Haven, CT.

Davis, S. N. & R. J. M. DeWiest (1967): *Hydrogeology,* 2nd. ed., 463 pp. New York, London, Sydney: Wiley.

Davis, S. N. & F. R. Hall (1959): Water quality of eastern Stanislaus and northern Merced counties, California. *Stanford Univ. Publ. Geol. Sci.,* **6,** 1, 112 pp., Stanford, CA.

Davison, A. S. (1969): The effect of tipped domestic refuse on ground water quality. *Water Treat. Exam.* **18,** 1, pp. 35–41.

Dean, G. A. (1963): The iodine content of some New Zealand drinking waters with a note on the contribution from sea spray to the iodine in rain. *N. Z. J. Sci.,* **6,** 2, pp. 208–214, Wellington.

Degens, E. T. (1962): Geochemische Untersuchungen von Wässern aus der ägyptischen Sahara. *Geol. Rundsch.,* **52,** pp. 625–639, Stuttgart.

Degens, E. T. & G. V. Chilingar (1967): Diagenesis of subsurface waters. In: G. Larsen & G. V. Chilingar, Eds., *Diagenesis in Sediments. Developments in Sedimentology,* Vol. 8. Amsterdam, London, New York: Elsevier. pp. 477–502.

Degens, E. T. & S. Epstein (1962): Relationship between $^{18}O/^{16}O$ ratios in coexisting carbonates, cherts and diatomites. *Bull. Am. Assoc. Petrol. Geol.,* **46,** pp. 534–542, Tulsa, OK.

Degens, E. T., J. M. Hunt, J. H. Reuter & W. E. Reed (1964): Data on the distribution of amino acids and oxygen isotopes in petroleum brine waters of various geologic ages. *Sedimentology,* **3,** pp. 199–225, Amsterdam, London, New York.

Delecourt, J. (1942): Le titre natronique. First note. *Soc. Belg. Géol.,* **50,** pp. 152–166, Brussels.

Delecourt, J. (1943a): Le titre natronique. Second note. *Soc. Belg. Géol.* **51,** pp. 107–142, Brussels.

Delecourt, J. (1943b): Le titre natronique. Third note. *Soc. Belg. Géol.* **52,** pp. 143–171, Brussels.

Delkeskamp, R. (1908): Die Entstehung der sulfatfreien Mineralquellen. *Kali,* **2,** pp. 349–357, 377–385, Halle (Saale).

Demolon, A. (1948): *Dynamique du Sol.,* 4th ed., 414 pp. Paris.

Denner, J. (1951): Trümmerschuttablagerungen und Schutzgebiete für Wasserversorgungsanlagen. *Gas- Wasserfach,* **92,** pp. 183–184, Munich.

Deutsch, M. (1963): Ground-water contamination and legal controls in Michigan. U.S. Geological Survey Water Supply Paper No. 1961, Washington, D.C.

Deutsch, M. (1965): Natural controls involved in shallow aquifer contamination. *Groundwater,* **3,** 3, pp. 37–40, Urbana, IL.

Deutsche Einheitsverfahren zur Wasser-, Abwasser- und Schlammuntersuchung (1971). *Ed. Fachgr. Wasserchemie G.D. Ch.,* 3rd. ed., Parts 1–6, Weinheim, Bergstrasse: Chemie.

Dhar, N. R. & A. Ram (1933): Formaldehyde in the upper atmosphere. *Nature,* **132,** pp. 819–820, London.

Diederich, G. & G. Matthess (1972): Hydrogeologie. In: Erl.geol.Kt.Hessen 1:25000, Bl. 6217 Zwingenberg a. d. Bergstrasse, Wiesbaden.

DIN 2000: Leitsätze für die zentrale Trinkwasserversorgung (1959): *Fachnormenausschuss Wasserwesen im DNA,* 35 pp. Berlin, Cologne: German Standards Institute.

DIN 2001: Leitsätze für die Einzel-Trinkwasserversorgung (1959): *Fachnormenausschuss Wasserwesen im DNA,* 8 pp. Berlin, Cologne: German Standards Institute.

Dittrich, M. (1903a): Die Quellen des Neckarthales bei Heidelberg in geologisch-chemischer Beziehung. *Mitt. Grossh. Bad. Geol. Landesanst.,* **4,** pp. 65–81, Heidelberg.

Dittrich, M. (1903b): Über die chemischen Beziehungen zwischen Quellwässern und ihren Ursprungsgesteinen. *Mitt. Grossh. Bad. Geol. Landesanst.,* **4,** pp. 199–207, Heidelberg.

Dombrowski, H. J. (1964): Sporenuntersuchungen in den Solen des Steinkohlenbezirks an der Ruhr. *Z. Dtsch. Geol. Ges.,* **116,** 1, pp. 96–101, Hannover.

Dombrowski, H. J. (1969): Mikrobiologische Untersuchung und Beurteilung der Mineral- und Heilwässer. In: S. W. Souci & K.-E. Quentin, Eds., *Handbuch der Lebensmittelchemie,* Vol. 8, Part 2, Berlin, Heidelberg, New York: Springer, pp 1147–1165.

Domenico, P. A. (1972): *Concepts and Models in Groundwater Hydrology,* 405 pp. New York: McGraw-Hill.

Dorsey, N. E. (1940): *Properties of Ordinary Water-Substance.* American Chemical Society Monograph Series, No. 81, 673 pp. New York: Reinhold.

Dowst, R. B. (1967): Highway chloride application and their effects upon water supplies. *J. New Engl. Water Works. Assoc.* **1,** pp. 63–67, Boston.

Drescher, N. (1971): Überlegungen der Pflanzenschutzmittel-Industrie zur Erarbeitung von Unterlagen über die potentielle Grundwasserverunreinigung durch Wirkstoffrückstände. *Schr.-R. Vers. Wasser-, Boden-Lufthyg.,* **34,** pp. 51–56, Stuttgart.

Drewry, W. A. & R. Eliassen (1968): Virus movement in groundwater. *J. Water Pollut. Control Fed.,* **40,** 8, Section 2, pp. 257–271, Washington, D.C.

Drischel, H. (1940): Chlorid-, Sulfat- und Nitratgehalt der atmosphärischen Niederschläge in Bad Reinerz und Oberschreiberhau im Vergleich zu bisher bekannten Werten anderer Orte. *Balneologe,* **7,** pp. 321–334, Berlin.

Dunlap, H. F. & R. R. Hawthorne (1951): The calculation of water resistivities from chemical analyses. *J. Petrol. Technol.* Technical Note No. 67, Section 1, pp. 17; Section 2, pp. 7, Dallas.

Durfor, Ch.N. & E. Becker (1964): Public water supplies of the 100 largest cities in the United States, 1962. U.S. Geological Survey Water Supply Paper No. 1812, 364 pp., Washington, D.C.

Durum, W. H. & J. Haffty (1961): Occurrence of minor elements in water. U.S. Geological Survey Circular No. 445, 11 pp., Washington, D.C.

Durum, W. H., S. G. Heidel & L. J. Tison (1960): World-wide runoff of dissolved solids. *Int. Assoc. Sci. Hydrol.,* **51,** pp. 618–628, Gentbrugge.

Duursma, E. K. (1970): Organic chelation of ^{60}Co and ^{65}Zn by leucine in relation to sorption by sediments. In: D. W. Hood, Ed., *Organic Matter in Natural Waters.* Institute of Marine Science, Occasional Publication No. 1, University of Alaska, College, p. 387.

Ebermayer, E. (1890): Untersuchungen über die Bedeutung des Humus als Bodenbestandteil und über den Einfluss des Waldes, verchiedener Bodenarten und Bodendecken auf die Zusammensetzung der Bodenluft. *Forsch. Geb. Agric.-Phys.,* **13,** pp. 15–49, Heidelberg.

Edmunds, W. M. (1973): Trace element variations across an oxidation-reduction barrier in a limestone aquifer. *Proceedings of the Symposium on Hydrogeochemistry and Biochemistry, Tokyo 1970*, Vol. 1. New York: Clarke, pp. 500–526.

Edmunds, W. M. (1977): Groundwater geochemistry – Controls and process. In: *Papers and Proceedings, Groundwater Quality, Measurement, Prediction and Protection*. Water Research Centre; Medmenham, England, pp. 115–147.

Edmunds, W. M. & A. H. Bath (1976): Centrifuge extraction and chemical analysis of interstitial waters. *Environ. Sci. Technol.*, **10**, pp. 467–472.

Effenberger, E. (1964): Verunreinigungen eines Grundwassers durch Cyanide. *Arch. Hyg., Bakteriol.*, **148**, 4–5, pp. 271–287, Munich.

Egger, F. (1942): Grundwasserbeeinflussung durch industrielle Anlagen. *Gesund.-Ing.*, **65**, pp. 124–126, Munich.

Ehhalt, D. & K. Knott (1965): Kinetische Isotopentrennung bei der Verdampfung von Wasser. *Tellus*, **17**, 3, pp. 389–397, Stockholm.

Eichelsdörfer, D., E. Fischer, U. Hässelbarth, H.-R. Hegi, K. Höll, F. Malz, E. Märki & K. E. Quentin (1969): Physikalische und chemische Untersuchung des Wassers. In: S. W. Souci & K.-E. Quentin, *Handbuch der Lebensmittelchemie*, Vol. 8, Part 1. Berlin, Heidelberg, New York: Springer, pp. 558–770.

Eldridge, E. F. (1963): Irrigation as a source of water pollution. *J. Water Pollut. Control Fed.*, **35**, 5, pp. 614–625, Washington, D.C.

Elion, L. (1927): Formation of hydrogen sulfide by the natural reduction of sulfates. *Ind. Eng. Chem.*, **19**, 12, p. 1368, Washington, D.C.

Ellis, A. J. & W. A. Mahon (1964): Natural hydrothermal systems and experimental hot-water/rock interactions. *Geochim. Cosmochim. Acta*, **28**, pp. 1323–1357, Oxford.

Elmore, D., B. R. Fulton, M. R. Clover, J. R. Marsden, H. E. Gove, H. Naylor, K. H. Purser, L. R. Kilius, R. P. Beukens & A. E. Litherland (1979): Analysis of ^{36}Cl in environmental water samples using an electrostatic accelerator. *Nature*, **227**, 5691, pp. 22–25, London.

Emanuelsson, A., E. Eriksson & H. Egnér (1954): Composition of atmospheric precipitation in Sweden. *Tellus*, **6**, pp. 261–267, Stockholm.

Emmons, W. H. (1917): The enrichment of ore deposits. U.S. Geological Survey Bulletin No. 625, 530 pp., Washington, D.C.

Emrich, G. H. & R. A. Landon (1971): Investigations of the effects of sanitary landfills in coal strip mines on ground water quality. Pennsylvania Bureau of Water Quality Management 30.

Emrich, G. H. & H. F. Lucas, Jr. (1963): Geologic occurrence of natural ^{226}radium in ground water in Illinois. *Int. Assoc. Sci. Hydrol. Bull.*, **8**, 3, pp. 5–19, Gentbrugge.

Engelhardt, W. von (1960): *Der Porenraum der Sedimente*. 207 pp. Berlin, Göttingen, Heidelberg: Springer.

Engler, G. & H. Höfer (1909): *Das Erdöl*, **2**, 967 pp., Leipzig.

Enthalpies Libres de Formation Standards, à 25°C. (1955): Technical Report No., 28, Centre d'Etude de la Corrosion, pp. 1–9, Brussels.

Epstein, S. & T. Mayeda (1953): Variation of ^{18}O content of waters from natural sources. *Geochim. Cosmochim. Acta*, **4**, pp. 213–224, London.

Eriksson, E. (1952): Composition of atmospheric precipitation. *Tellus*, **4**, pp. 215–232, 280–303, Stockholm.

Eriksson, E. (1958): The chemical climate and saline soils in the arid zone. *UNESCO Climatol., Rev. Res., Arid Zone Res.*, **10**, pp. 147–180, Paris.

Eriksson, E. (1963): Atmospheric tritium as a tool for the study of certain hydrologic aspects of river basins. *Tellus*, **15**, pp. 303–308, Stockholm.

Eriksson, E. (1965): An account of the major pulses of tritium and their effects in the atmosphere. *Tellus,* **17**, 1, pp. 118–130, Stockholm.

Eriksson, E. (1970): The importance of investigating global background pollution. Meteorological aspects of air pollution. Technical Note No. 106, WHO No. 251, TP 139, pp. 31–54, Geneva.

Étard, A. & L. Olivier (1882): De la réduction des sulfates par les êtres vivants. *C. R. Acad. Sci.,* **95**, pp. 846–849, Paris.

Evans, D. D. & D. Kirkham (1950): Measurement of the air permeability of soil in situ. *Soil Sci. Soc. Am. Proc.,* **14**, pp. 65–73, Madison, WI.

Evans, E. J. (1958): Chemical investigations of the movement of fission products in soil. AECL No. 667, CRER-792, 23 pp. Chalk River, Ontario: Atomic Energy Canada Ltd.

Everdingen, R. O. Van (1970): The Paint Pots, Kootenay National Park, British Columbia — Acid spring water with extreme heavy-metal content. *Can. J. Earth Sci.,* **7**, 3, pp. 831–852, Ottawa.

Everdingen, R. O. Van & J. A. Banner (1971): Precipitation of heavy metals from natural and synthetic acidic aqueous solution during neutralisation with limestone. Inland Water Branch, Department of Energy, Mines and Resources, Technical Bulletin No. 35, 21 pp., Ottawa, Canada.

Everdingen, R. O. Van & R. A. Freeze (1971): Subsurface disposal of waste in Canada. Inland Waters Branch, Department of Environment, Technical Bulletin No. 49, 64 pp., Ottawa, Canada.

Exler, H. J. (1972): Ausbreitung und Reichweite von Grundwasserverunreinigungen im Unterstrom einer Mülldeponie. *Gas- Wasserfach,* **113**, pp. 101–112, Munich.

Exler, H. J. (1979): Hydrogeologische Untersuchungsergebnisse im Unterstrom der Mülldeponie Grosslappen. *Gas- Wasserfach.,* **120**, pp. 13–21, Munich.

Fair, G. M. et al. (1971): *Elements of Water Supply and Wastewater Disposal,* 2nd ed., 752 pp. New York: Wiley.

Fairbridge, R. H. W. (1967): Phases of diagenesis and authigenesis. In: G. Larsen & G. V. Chilingar, Eds., *Diagenesis in Sediments. Developments in Sedimentology,* Vol. 8. Amsterdam, London, New York: Elsevier, pp. 19–89.

Fairs, R. A. & B. H. Parks (1973): *Radioisotope Laboratory Techniques,* 3rd ed., 312 pp. London: Butterworths.

Farkasdi, G., A. Golwer, K. H. Knoll, G. Matthess & W. Schneider (1969): Mikrobiologische und hygienische Untersuchungen von Grundwasserverunreinigungen im Unterstrom von Abfallplätzen. *Städtehygiene,* **20**, pp. 25–31, Uelzen, Hamburg.

Fast, H. & K. Sauer (1958): Die chemische Zusammensetzung südbadischer Grundwasser, Herkunftsfragen und Versuch einer Typologie. *Vom Wasser,* **25**, pp. 48–81, Weinheim, Bergstrasse.

Faust, S. D. & O. M. Aly (1964): Water pollution by organic pesticides. *J. Am. Water Works Assoc.,* **56**, 3, pp. 267–279, New York.

Fauth, H. (1969): Feldmethoden der Untersuchung. In: A. Bentz, Ed., *Lehrbuch der angewandten Geologie,* Vol. 2, Part 2. Stuttgart: Enke, pp. 1421–1426.

Feely, H. W. & J. L. Kulp (1957): The origin of Gulf Coast salt dome sulfur deposits. *Bull. Am. Assoc. Petrol. Geol.,* **41**, pp. 1802–1853, Tulsa, OK.

Feitknecht, W. & F. Held (1944): Über die Hydroxychloride des Magnesiums. *Helv. Chim. Acta,* **27**, pp. 1480–1495, Basle.

Fergusson, G. J. (1965): Radiocarbon and tritium in the upper troposphere. *Proceeding of the Sixth International Conference on Radiocarbon and Tritium Dating, Pullman, Washing-*

ton, 11.6. 1965, Vol. 7, AEC Rep. Conf. 650652, Division of Technical Information, National Bureau of Standards, pp. 525–540, Springfield, VA.

Feth, J. H. (1966): Nitrogen compounds in natural water – A review. *Water Resourc. Res.*, **2**, 1, pp. 41–58, Washington, D.C.

Feth, J. H., S. M. Rogers & C. E. Roberson (1961): Aqua de Ney California, a spring of unique chemical character. *Geochim. Cosmochim. Acta*, **22**, pp. 75–86. Oxford, London, New York, Paris.

Feulner, A. J. & J. H. Hubble (1960): Occurrence of strontium in the surface and groundwaters of Champaign County, Ohio. *Econ. Geol.*, **55**, pp. 176–186, New Haven, CT.

Feulner, A. J. & R. G. Schupp (1963): Seasonal changes in the chemical quality of shallow groundwater in northwestern Alaska. U.S. Geological Survey Professional Paper No. 475-B, pp. 189–191, Washington, D.C.

Finkelnburg, W. (1967): *Einführung in die Atomphysik.* 11th–12th ed., 12 + 525 pp. Berlin: Springer.

Finkenwirth, A. (1968): Injection des eaux résiduaires industrielles en profondeur dans le sous-sol. *Trib. Cebedeau*, **21**, pp. 452–460, Liège.

Flathe, H., K. Deppermann & J. Homilius (1961): Die Widerstandsmethode. In: A. Bentz, Ed., *Lehrbuch der angewandten Geologie*, Vol. 1, Stuttgart: Enke, pp. 725–770.

Fleck, H. (1876): Kohlensäuregehalt der Bodenluft in Dresden. *Jahresber. Forschr. Agric.-Chem.*, **16–17**, pp. 159–160, Berlin.

Fodor, J. von (1875): Experimentelle Untersuchungen über Boden und Bodengase. *Dtsch. Vierteljahresschr. Öff. Gesundheitspflege*, **7**, pp. 205–237, Braunschweig.

Förstner, U. & G. Müller (1974): *Schwermetalle in Flüssen und Seen als Ausdruck der Umweltverschmutzung*, 225 pp. Berlin, Heidelberg, New York: Springer.

Foster, J. W. (1962): Hydrocarbons as substrates for microorganisms. *Anton van Leeuwenhoek*, **28**, pp. 241–274, Amsterdam.

Foster, M. D. (1950): The origin of high sodium bicarbonate waters in the Atlantic and Gulf coastal plains. *Geochim. Cosmochim. Acta*, **1**, pp. 33–48, London.

Frank, W. H. (1965): Die künstliche Grundwasseranreicherung an der Ruhr in ihrer geschichtlichen Entwicklung und zukünftigen Bedeutung. *Gas- Wasserfach*, **106** pp. 1095–1101, Munich.

Frank, W. H. (1969): Beurteilungsgrundsätze und Anforderungen an Trink- und Betriebswasser. In: S. W. Souci & K.-E. Quentin, Eds., *Handbuch der Lebensmittelchemie*, Vol. 8. Berlin, Heidelberg, New York: Springer, pp. 794–861.

Frear, G. L. & J. Johnston (1929): Solubility of calcium carbonate (calcite) in certain aqueous solutions at 25°C. *J. Am. Chem. Soc.*, **51**, pp. 2082–2093, Easton, PA.

Freeze, R. A. & J. A. Cherry (1979): *Groundwater*, 604 pp. Englewood Cliffs, NJ: Prentice-Hall.

Fresenius, L. & O. Fuchs (1930): Zur Berechnung der Mineralwasseranalysen.- *Z. Analy. Chem.*, **82**, pp. 226–234.

Fresenius, W. & K.-E. Quentin with collaboration of W. Schneider, R. E. Fresenius & G. Schretzenmayr (1969): Untersuchung der Mineral- und Heilwässer. In: S. W. Souci & K.-E. Quentin, Eds., *Handbuch der Lebensmittelchemie*, Vol. 8. Berlin, Heidelberg, New York: Springer, pp. 862–1042.

Fresenius, W. & W. Schneider (1972): Analyse der Rückstände von Müllverbrennungsanlagen. *Müll- Abfall-Abwässer*, **19**, pp. 5–10, Düren.

Frey, R. (1933): Les analyses des eaux et leur interprétation géologique. Serv. Min. et Carte Géol. Maroc, Notes et Mémoires, No. 26, 68 pp., Rabat.

Fricke, K. (1953): Der Schwermetallgehalt der Mineralquellen. *Erzmetall, Z. Erzbergbau Metallhüttenwes.,* **6,** pp. 257–266, Stuttgart.

Fricke, K. & H. Werner (1957): Geochemische Untersuchungen von Mineralwässern auf Kupfer, Blei und Zink in Nordrhein-Westfalen und angrenzenden Gebieten (Vorläufige Mitt.). *Heilbad Kurort* (1957), pp. 45–46, Gütersloh.

Fried, J. J. (1975): *Groundwater Pollution,* 330 pp. Amsterdam: Elsevier.

Friedman, I. (1953): Deuterium content of natural waters and other substances. *Geochim. Cosmochim. Acta,* **4,** pp. 89–103, London: Pergamon.

Fritz, P., G. Matthess & R. M. Brown (1976): Deuterium and oxygen-18 as indicators of leachwater movement from a sanitary landfill. Panel Proceedings Series, STI/PUB No. 429. Vienna: IAEA, pp. 131–142.

Fuhs, G. W. (1961a): Der mikrobielle Abbau von Kohlenwasserstoffen. *Arch. Mikrobiol.,* **39,** pp. 374–422, Berlin, Göttingen, Heidelberg.

Fuhs, G. W. (1961b): Grundzüge des mikrobiellen Abbaus von Kohlenwasserstoffen, ein Beitrag zum bakteriellen Abbau von Ölen. *Wasserwirtschaft,* **51,** pp. 277–280, Stuttgart.

Furtak, H. & H. R. Langguth (1967): Zur hydrochemischen Kennzeichnung von Grundwässern und Grundwassertypen mittels Kennzahlen. *Int. Assoc. Hydrogeol. Mem.,* **7,** pp. 89–96, Hannover.

Gad, G. & E. Fürstenau (1954); Eine Betriebsmethode zur Bestimmung des Fluors im Wasser und Ermittlung des Fluorspiegels im westdeutschen Raum. *Gesund.-Ing.,* **75,** pp. 352–356, Munich.

Gahl, R. & B. Anderson (1928): Sulphate reducing bacteria in California oil waters. *Zentralbl Bakt. Parasitenkd. Infektionskr.,* **73,** Part 2, pp. 331–338, Jena.

Gambell, A. W. & D. W. Fisher (1964): Occurrence of sulfate and nitrate in rainfall. *J. Geophys. Res.,* **69,** 20, pp. 4203–4210, Washington, D.C.

Garcia, F. G., S. G. Garcia & M. C. Sanchez (1956): The alkali soils of the lower valley of the Guadalquivir: Physico-chemical properties and nature of their clay fraction. *6 Congr. Sci. Sol., Rapp.,* **B,** 1, pp. 185–191, Paris.

Garrels, R. M. (1960): *Mineral equilibria at low temperature and pressure,* 254 pp. New York: Harper & Row.

Garrels, R. M. (1967): Genesis of ground waters from igneous rocks. In: P. H. Abelson; Ed., *Researches in Geochemistry,* Vol. 2, pp. 405–420.

Garrels, R. M. & C. L. Christ (1965): *Solutions, Minerals and Equilibria,* 450 pp. New York, Evanston, London: Harper & Row (Tokyo: Weatherhill).

Garrels, R. M., M. E. Thompson & R. Siever (1960): Stability of some carbonates at 25°C and one atmosphere total pressure. *Am. J. Sci.,* **258,** pp. 412–418, New Haven, CT.

Gat, J. R. & Y. Tzur (1967): Modification of the isotopic composition of rainwater by processes which occur before groundwater recharge. *Isotopes in Hydrology,* IAEA Proceedings Series, Vienna: IAEA, pp. 49–60.

Geiger, R. (1950): *Das Klima der bodennahen Luftschicht.* 3rd ed., 460 pp. Brunswick: Vieweg.

George, W. O. & W. W. Hastings (1951): Nitrate in the groundwater of Texas. *Trans. Am. Geophys. Union,* **32,** pp. 450–456, Washington, D.C.

Georgii, H.-W. (1963): Oxides of nitrogen and ammonia in the atmosphere. *J. Geophys. Res.,* **68,** 13, pp. 3963–3970, Washington, D.C.

Georgii, H.-W. (1965): Untersuchungen über Ausregnen und Auswaschen atmosphärischer Spurenstoffe durch Wolken und Niederschlag. *Ber. Dtsch. Wetterd.,* **14,** 100, 23 pp., Offenbach/M. (German Weather Service.)

Gerb, L. (1953): Reduzierte Wässer. *Gas- Wasserfach,* **94,** pp. 87–92, 157–161, Munich.

Gerb, L. (1958): Grundwassertypen. *Vom Wasser*, **25**, pp. 16–47, Weinheim, Bergstrasse.

Gericke, S. & B. Kurmies (1957): Pflanzennährstoffe in den atmosphärischen Niederschlägen. *Phosphorsäure*, **17**, pp. 279–300, Essen.

Germanier, R. & K. Wuhrmann (1963): Über den aeroben mikrobiellen Abbau aromatischer Nitroverbindungen. *Pathol. Microbiol.*, **26**, 5, pp. 569–578, Basle, New York.

Geyh, M. A. (1970): Carbon-14 concentration of lime in soils and aspects of the carbon-14 dating of groundwater. *Isotope Hydrology*. Vienna: IAEA, pp. 215–223.

Geyh, M. A. & O. F. Kuckelkorn (1969): Zur Gliederung eines Grundwasserkörpers mit Hilfe von ^{14}C- und ^{3}H-Konzentrations–Bestimmungen an Wasserproben. *Gas- Wasserfach*, **101**, pp. 1394–1397, Munich.

Giebler, G. (1960): Die dritte Chemische Wasserstatistik der Wasserwerke in der Bundesrepublik Deutschland und West-Berlin und ihre Ergebnisse. *Gas- Wasserfach*, **101**, 34, pp. 860–864, Munich.

Giesecke, F. (1930): Das Verhalten des Bodens gegen Luft. In: E. Blanck, Ed., *Handbuch den Bodenlehre*, Vol. 6, Berlin: Springer, pp. 253–342.

Gillham, R. W., J. A. Cherry & J. F. Pickens (1975): Mass transport in shallow groundwater flow systems. *Canadian Hydrology Symposium – 1975*, p. 367, *Conference Proceedings*, Winnipeg, 1975, National Research Council of Canada, Associate Committee on Hydrology, Ottawa.

Ginsburg, I. I. (1963): *Grundlagen und Verfahren geochemischer Sucharbeiten auf Lagerstätten der Buntmetalle und seltenen Metalle*, 339 pp. Berlin: Akademie-Verl.

Ginsburg-Karagitscheva, T. L. (1933): Microflora of oil waters and oil-bearing formations and biochemical processes caused by it. *Bull. Am. Assoc. Petrol. Geol.*, **17**, 1, pp. 52–65, Tulsa, OK.

Ginter, R. L. (1930): Causative agents of sulphate reduction in oil well waters. *Bull. Am. Assoc. Petrol. Geol.*, **14**, pp. 139–152, Tulsa, OK.

Girard, R. (1935): *Essai de Représentation Graphique des Analyses d'Eau*. Association Générale des Hygienistes et Techniciens Municipaux, 19 pp.

Glasstone, S. (1947): *Thermodynamics for Chemists*, 522 pp., New York: Van Nostrand.

Glasstone, S. (1949): *An Introduction to Electrochemistry*, 4th ed., 557 pp. Toronto, New York, London: Van Nostrand.

Glathe, H., K. H. Knoll, & A. A. M. Makawi (1963a): Die Lebensfähigkeit von *Escherichia coli* in verschiedenen Bodenarten. *Z. Pflanzenernärh. Düng. Bodenkd.*, **100**, 2, pp. 142–150, Weinheim, Bergstrasse.

Glathe, H., K. H. Knoll, & A. A. M. Makawi (1963b): Das Verhalten von *Salmonellen* in verschiedenen Bodenarten. *Z. Pflanzenernähr. Düng. Bodenkd.*, **100**, 3, pp. 224–233, Weinheim Bergstrasse.

Glueckauf, E. (1961): *Atomic Energy Waste: Its Nature, Use and Disposal*, 420 pp. London: Butterworths.

Gmelins Handbuch der anorganischen Chemie (1952): Syst.-Nr. 3 *Sauerstoff*, Lfg.2, Vorkommen, Technologie, 8th ed., 300 pp. Weinheim, Bergstrasse: Chemie.

Gmelins Handbuch der anorganischen Chemie (1953): 8th ed., System-Nr. 9 *Schwefel*, Part A, 2, 510 pp. Weinheim, Bergstrasse: Chemie.

Goguel, R. (1965): Ein Beitrag zum Chemismus von Gas- und Flüssigkeitseinschlüssen in gesteinsbildenden Mineralen. *Nachr. Akad. Wiss. Göttingen*, II. *Math.-Phys. Kl.* (1964), **21**, pp. 267–278, Göttingen.

Gold, V. (1956): *pH-Measurements*, 125 pp. London: Methuen, New York: Wiley.

Golding, R. M. & M. G. Speer (1961): Alkali ion analysis of thermal water in New Zealand. *N. Z. Sci.*, **4**, pp. 203–213. Wellington.

Goldschmidt, V. M. (1937): The principles of distribution of chemical elements in minerals and rocks. *J. Chem. Soc.* pp. 655–673, London.

Golwer, A. (1973): Beeinflussung des Grundwassers durch Strassen. *Z. Dtsch. Geol. Ges.,* **124,** pp. 435–446, Hannover.

Golwer, A. & G. Matthess (1968): Research on groundwater contaminated by deposits of solid waste. *Int. Assoc. Sci. Hydrol. Publ.,* **78,** pp. 129–133.

Golwer, A. & G. Matthess (1972): Die Bedeutung des Gasaustausches in der Grundluft für die Selbstreinigungsvorgänge in verunreinigten Grundwässern. *Z. Dtsch. Geol. Ges.,* **123,** pp. 29–38, Hannover.

Golwer, A. & W. Schneider (1973): Belastung des Bodens und des unterirdischen Wassers durch Strassenverkehr. *Gas- Wasserfach,* **114,** pp. 154–165, Munich.

Golwer, A., K. H. Knoll, G. Matthess, W. Schneider & K. H. Wallhäusser (1972): Mikroorganismen im Unterstrom eines Abfallplatzes. *Gesund.-Ing.,* **93,** pp. 142–151, Munich.

Golwer, A., K. H. Knoll, G. Matthess, W. Schneider & K. H. Wallhäusser (1973): Biochemical processes under anaerobic and aerobic conditions in groundwater contaminated by solid waste. *Proceedings of the Symposium on Hydrogeochemistry and Biochemistry, Tokyo, 1970,* Vol. 2, Washington, D.C.: Clarke, pp. 344–357.

Golwer, A., K. H. Knoll, G. Matthess, W. Schneider & K. H. Wallhäusser (1976): Belastung und Verunreinigung des Grundwassers durch feste Abfallstoffe. *Abh. Hess. Landesamt. Bodenforsch.,* **73,** 131 pp., Wiesbaden.

Golwer, A., G. Matthess & W. Schneider (1970): Selbstreinigungsvorgänge im aeroben und anaeroben Grundwasserbereich. *Vom Wasser,* **36,** (1969), pp. 64–92, Weinheim, Bergstrasse.

Golwer, A., G. Matthess & W. Schneider (1971): Einflüsse von Abfalldeponien auf das Grundwasser. *Städtetag,* **24,** 2, pp. 119–124, Stuttgart.

Gorham, E. (1955): On the acidity and salinity of rain. *Geochim. Cosmochim. Acta,* **7,** pp. 231–239, London, New York.

Gorham, E. (1961): Factors influencing supply of major ions to inland waters, with special reference to the atmosphere. *Geol. Soc. Am. Bull.,* **72,** pp. 795–840, New York.

Gorup-Besanetz, E. von (1872): Über dolomitische Quellen des Frankenjura. *Justus Liebigs Ann.,* **8,** pp. 230–240, Leipzig.

Gottschaldt, N. & W. Winter (1967): Stand der Reinigungsmöglichkeiten tensidhaltiger Abwässer. *Fortschr. Wasserchem. Grenzgeb.,* **5,** pp. 284–301, Berlin.

Grečin, I. P. & Čen Jun'-schen (1960): The influence of various CO_2-concentrations in the soil air on the oxidation-reduction conditions. *Počovedenie,* **7,** (1960), pp. 106–110, Moscow (in Russian).

Grim, R. E. (1968): *Clay Mineralogy,* 596 pp. New York: McGraw-Hill.

Grisak, G. E. & R. E. Jackson (1978): An appraisal of the hydrogeological processes involved in shallow subsurface radioactive waste management in Canadian terrain. Science Series No. 84, Inland Waters Directorate, 194 pp., Ottawa.

Grisek, G. E., R. E. Jackson & J. F. Pickens (1978): Monitoring ground water quality: The technical difficulties *Proceedings of the American Water Resources Association Conference,* San Francisco.

Gucker, F. T. & R. L. Seifert (1967): *Physical Chemistry,* 824 pp. London: English Universities Press.

Guillerd, A. (1941): Contrôle de l'analyse d'une eau minérale par résistivité électrique. *Ann. Inst. Hydrol. Climatol.,* **13,** 48, pp. 131–141, Paris.

Gundlach, H. (1965): Untersuchungen an einigen Schwefelquellen in Griechenland. *Geol. Jahrb.,* **83,** pp. 411–430, Hannover.

Gunter, B. D. & B. C. Musgrave (1966): Gas chromatographic measurements of hydrothermal emanations at Yellowstone National Park. *Geochim. Cosmochim. Acta,* **30** pp. 1175–1189, Oxford.

Gurevich, M. S. (1962): The role of microorganisms in producing the chemical composition of ground water. In: S. I. Kuznetsov, Ed., *Geologic Activity of Microorganisms,* Vol. 9, *Transactions of the Institute of Microbiology.* New York: Consultants Bureau, pp. 65–75.

Gurewitsch, W. J. (1962): Über die Verbreitung von Brom in chlorhaltigen Wässern. *Z. Angew. Geol.,* **8,** pp. 36–37, Berlin.

Gustafson, P. F., L. D. Marinelli & S. S. Brar (1958): Natural and fission-produced gamma-ray emitting radioactivity in soil. *Science,* **127,** pp. 1240–1242, Washington, D.C.

Haaren, F. W. J. Van (1966): Chemische Aspekte der künstlichen Grundwasseranreicherung. *Veröff. Hydrol. Forschungsabt. Dortmunder Stadtwerke AG,* **9,** pp. 27–37, Dortmund.

Haberer, K. (1969a): Physikalische und chemische Eigenschaften des Wassers. In: S. W. Souci & K.-E. Quentin, Eds., *Handbuch der Lebensmittelchemie,* Vol. 8, Part 1. Berlin, Heidelberg, New York: Springer, pp. 1–50.

Haberer, K. (1969b): Nachweis radioaktiver Stoffe in Trink- und Betriebswasser. In: S. W. Souci & K.-E. Quentin, Eds., *Handbuch der Lebensmittelchemie,* Vol. 8, Part 1. Berlin, Heidelberg, New York: Springer, pp. 771–793.

Haccius, B. & O. Helfrich (1958): Untersuchungen zur mikrobiellen Benzoloxydation. II. Beschreibung und systematische Stellung benzolabbauender Mikroorganismen. *Arch. Mikrobiol.,* **28,** pp. 394–403, Berlin.

Hahn, J. (1972): Diagenetisch bedingte Veränderungen im Chemismus intrudierter Meerwässer und ihre Beziehungen zum Chemismus von Tiefengrundwässern in Nordwestdeutschland. *Geol. Jahrb.,* **90,** pp. 245–264, Hannover.

Hannén, F. (1892): Untersuchungen über den Einflüss der physikalischen Beschaffenheit des Bodens auf die Diffusion der Kohlensäure. *Forsch.-Geb. Agric. Phys.,* **15,** pp. 6–25. Heidelberg.

Hanshaw, B.B. & E-An Zen (1965): Osmotic equilibrium and overthrust faulting. *Geol. Soc. Am. Bull.,* **76,** pp. 1379–1386, New York.

Harned, H. S. & B. B. Owen (1958): *The Physical Chemistry of Electrolytic Solutions,* 3rd ed., 31 + 803 pp. New York: Reinhold.

Harned, H. S. & S. R. Scholes, Jr. (1941): The ionization constant of HCO_3^- from 0–50°. *J. Am. Chem. Soc.,* **63,** pp. 1706–1709, Washington, D.C.

Harrassowitz, H. (1933): Die alkalischen Quellen in ihrer geochemischen Bedeutung. *Z. Kurortwiss.,* **2,** pp. 211–216, Berlin.

Harrassowitz, H. (1935): Die deutschen Chlor-Calcium-Quellen. *Kali,* **29,** pp. 75–80, Halle (Saale).

Harth, H. (1965): Zum Problem der Anreicherung der Gewässer mit Kaliumsalzen und [40]K. *Dtsch. Gewässerk. Mitt., Sonderh.,* pp. 4–7, Coblenz, Rhein.

Harth, H. (1969): Der Einflüsse von Land- und Forstwirtschaft auf den Grundwasserchemismus. *Dtsch. Gewässerk. Mitt., Sonderh.,* pp. 58–62, Coblenz Rhein.

Hartmann, L. (1966): Effect of surfactants on soil bacteria. *Bull. Environ. Contamin. Toxicol.,* **1,** 6, pp. 219–224, New York.

Hartmann, L. & H. Mosebach (1966): Adsorptionsversuche mit neuen Detergentien. *Tenside,* **3,** 10, pp. 349–354, Munich.

Hässelbarth, U. (1963): Das Kalk-Kohlensäure-Gleichgewicht in natürlichen Wässern unter Berücksichtigung des Eigen- und Fremdelektrolyt-Einflusses. *Gas- Wasserfach,* **104,** 4, pp. 89–93, 6, pp. 157–160, Munich.

Hässelbarth, U. (1969): Bestimmung der Aggressivität (Marmorlösevermögen). In: S. W. Souci

& K.-E. Quentin, Eds., *Handbuch der Lebensmittelchemie,* Vol. 8, Part 1. Berlin, Heidelberg, New York: Springer, pp. 721–723.

Hässelbarth, V. & D. Lüdemann (1967): Die biologische Verockerung von Brunnen durch Massenentwicklung von Eisen- und Manganbakterien. *Bohrtech. Brunnenbau Rohrleitungsbau,* **18,** pp. 363–368, 401–406, Berlin.

Haupt, H. (1935): Schädlicher Einflüsse von Ascheablagerungen auf Grundwasser. *Gas- Wasserfach,* **78,** pp. 526–528, Munich.

Häusler, J. (1969): The succession of microbial processes in the anaerobic decomposition of organic compounds. In: S. H. Jenkins, Ed., *Advances in Water Pollution Research.* Oxford: Pergamon, pp. 407–415.

Hawkes, H. E. (1957): Principles of geochemical prospecting. Contribution to geochemical prospecting for minerals. U.S. Geological Survey Bulletin No. 1000-F, pp. 225–355, Washington, D.C.

Haxel, O. & G. Schumann (1962): Erzeugung radioaktiver Kernarten durch die kosmische Strahlung. In: H. Israel & A. Krebs, Eds., *Nuclear Radiation in Geophysics,* Berlin, Göttingen, Heidelberg: Springer, pp. 97–135,

Hecht, G. (1964): Über das Vorkommen natriumhydrogenkarbonathaltiger Wässer in Thüringen. *Z. Angew. Geol.,* **10,** pp. 250–255, Berlin.

Heicklen, J. (1976): *Atmospheric Chemistry,* 406 pp. New York, San Francisco, London: Academic Press.

Heimath, B. (1933): Untersuchungen über Schwefelsäurevorkommen in saurem Waldboden. *Z. Pflanzenernähr. Düng. Bodenkd.,* **A31,** pp. 229–251, Berlin.

Heitele, H. (1968): Versickerungsvorgänge in der Buntsandsteinzone des südlichen Saarlandes. Dissertation, 100 pp. Tübingen.

Hem, J. D. (1959): Study and interpretation of the chemical characteristics of natural water. U.S. Geological Survey Water Supply Paper No. 1473, 269 pp., Washington, D.C.

Hem, J. D. (1960a): Restraints on dissolved ferrous iron imposed by bicarbonate, redox potential, and pH. U.S. Geological Survey Water Supply Paper No. 1459-B, pp. 33–55, Washington, D.C.

Hem, J. D. (1960b): Some chemical relationships among sulfur species and dissolved ferrous iron. U.S. Geological Survey Water Supply Paper No. 1459-C, pp. 57–73, Washington, D.C.

Hem, J. D. (1961a): Stability field diagrams as aids in iron chemistry studies. *J. Am. Water Works Assoc.,* **53,** 2, pp. 221–228, New York.

Hem, J. D. (1961b): Calculation and use of ion activity. Geochemistry of water. U.S. Geological Survey Water Supply Paper No. 1535-C, 17 pp., Washington, D.C.

Hem, J. D. (1970): Study and interpretation of the chemical characteristics of natural water, 2nd ed., U.S. Geological Survey Water Supply Paper No. 1473, 363 pp., Washington, D.C.

Hem, J. D. (1972): Chemistry and occurrence of cadmium and zinc in surface and groundwater. *Water Resour. Res.,* **8,** pp. 661–679.

Hem, J. D. (1977): Reactions of metal ions at surfaces of hydrous iron oxide. *Geochim. Cosmochim. Acta,* **41,** pp. 527–538, Oxford, New York, Paris, Frankfurt: Pergamon.

Hem, J. D. & W. H. Cropper (1959): Survey of ferrous-ferric chemical equilibria and redox potentials. U.S. Geological Survey Water Supply Paper No. 1459-A, pp. 1–30, Washington, D.C.

Hendee, W. R. (1970): *Medical Radiation Physics,* 599 pp. Chicago: Year Book Medical Publ.

Hendricks, S. B. (1941): Base exchange of the clay mineral montmorillonite for organic actions and its dependence upon adsorption due to Van der Waals forces. *J. Phys. Chem.,* **45,** pp. 65–81, Ithaca, NY.

Hendricks, S. B. (1955): Necessary, convenient, commonplace. In: *Water, Yearbook of Agriculture*, pp. 9–14, U.S. Department of Agriculture, Washington, D.C.

Hendrickson, G. E. & R. S. Jones (1952): Geology and ground-water resources of Eddy County, New Mexico. New Mexico Bureau of Mines and Mineral Resources, Ground-water Report No. 3, 169 pp.

Henzel, N. & O. Strebel (1967): Modelluntersuchungen über die Tiefenverlagerung von Fallout in verschiedenen Böden. *Z. Geophys.*, **33**, pp. 33–47, Würzburg.

Herbst, W. (1959): Studien zur radioaktiven Kontamination der menschlichen Umwelt. 1. Boden und Pflanzenwelt. *Atompraxis*, **5**, pp. 280–284, Karlsruhe.

Herbst, W. (1961): Befunde und Bemerkungen zum Eingang von Radiostrontium in Boden. Vegetation und Biocyklen. *Strahlenschutz*, **18**, pp. 122–128, Munich (B.-Min. Atomkernenergie u. Wasserwirtsch.).

Herrmann, A. G. (1961): Über die Einwirkung Cu-, Sn-, Pb- und Mn-haltiger Erdölwässer auf die Stassfurt-Serie des Südharzbezirkes. *Neue Jahrb. Mineral. Mh.*, pp. 60–67, Stuttgart.

Heyl, K. E. (1954): *Hydrochemische Untersuchungen im Gebiet des Siegerländer Erzbergbaus*, 72 pp. Dissertation, Department of Natural Science and Mathematics, Ruprecht-Karl University of Heidelberg, Heidelberg.

Heyman, J. J. & A. H. Molof (1967): Initiation of biodegradation of surfactants. *J. Water Pollut. Control Fed.*, **39**, 1, pp. 50–62, Washington, D.C.

Higgins, G. H. (1959): Evaluation of the ground-water contamination hazard from underground nuclear explosions. *J. Geophys. Res.*, **64**, pp. 1509–1519, Baltimore.

Hill, D. W. & P. L. McCarty (1967): Anaerobic degradation of selected chlorinated hydrocarbon pesticides. *J. Water Pollut. Control Fed.*, **39**, 8, pp. 1259–1277, Washington, D.C.

Hill, R. A. (1940): Geochemical patterns in Coachella Valley. *Trans. Am. Geophys. Union*, **21**, Part 1, pp. 46–53, Washington, D.C.

Hintz, E. & L. Grünhut (1907): Besondere Grundsätze für die Darstellung der chemischen Analysenergebnisse. In: *Deutsches Bäderbuch*, pp. L–LXIV, Leipzig: Weber.

Hise, C. R. Van (1904): *A Treatise on Metamorphism*. U.S. Geological Survey Monograph No. 47, 1286 pp., Washington, D.C.

Hodgman, C. D., R. C. Weast & S. M. Selby (1958): *Handbook of Chemistry and Physics*, 39th ed., 3213 pp. Cleveland: Chemical Rubber Publ. Co.

Hoering, T. (1957): The isotopic composition of the ammonia and the nitrate ion in rain. *Geochim. Cosmochim. Acta*, **12**, pp. 97–102, London.

Hoffmann, G. (1963): Die höchsten und die tiefsten Temperaturen auf der Erde. *Umschau*, **63**, 1, pp. 16–18, Frankfurt am Main.

Hoffmann, R. W. & G. W. Brindley (1960): Adsorption of non-ionic aliphatic molecules from aqueous solution on montmorillonite clay. Organic studies. II. *Geochim. Cosmochim. Acta*, **20**, pp. 15–29, Oxford, London, New York, Paris.

Holden, W. S. (1970): *Water Treatment and Examination*, 513 pp. London: Churchill.

Höll, K. (1963): Chemische Untersuchungen von Lysimeter-Abläufen aus gewachsenen Böden bestimmter Pflanzengesellschaften. *Vom Wasser*, **30**, pp. 65–80, Weinheim, Bergstrasse.

Höll, K. (1965): Das schwere Wasser in den Heilquellen. *Heilbad Kurort*, **17**, 7, pp. 136–139, Gütersloh.

Höll, K. (1970): Sickerwasser-Beschaffenheit in verschiedenen naturbelassenen Böden. *Z. Dtsch. Geol. Ges., Sonderh. Hydrogeol. Hydrogeochem.*, pp. 129–137, Hannover.

Höll, K. with contributions of H. Peter and D. Lüdemann (1972): *Water: Examination, Assessment, Conditioning, Chemistry, Bacteriology, Biology*, 389 pp. Berlin, New York: de Gruyter.

Holland, H. D., T. V. Kirsipu, J. S. Huebner & U. M. Oxburgh (1964): On some aspects of the chemical evolution of cave waters. *J. Geol.,* **72**, pp. 36–67, Chicago.

Holluta, J. (1960): Geruchs- und Geschmacksbeeinträchtigung des Trinkwassers – Ursachen und Bekämpfung. *Gas- Wasserfach,* **101**, 40, pp. 1018–1023, 1070–1076, Munich.

Holluta, J., L. Bauer & W. Kölle (1968): Über die Einwirkung steigender Flusswasserverschmutzung auf die Wasserqualität und die Kapazität der Uferfiltrate. *Gas- Wasserfach,* **109**, 50, pp. 1406–1409, Munich.

Hölting, B. (1966): Die Mineralquellen in Bad Wildungen und Kleinern (Landkreis Waldeck, Hessen). *Abh. Hess. Landesamt, Bodenforsch.,* **53**, 59 pp., Wiesbaden.

Holtzem, H. & J. Schwibach (1967): Probleme der Beseitigung radioaktiver Abfälle in Deutschland. *Atomwirtschaft,* **12**, pp. 413–417, Düsseldorf.

Horn, M. K. & J. A. Adams (1966): Computer-derived geochemical balances and element abundances. *Geochim. Cosmochim. Acta,* **30**, pp. 279–297, Oxford, London, New York, Paris.

Hostetler, P. B. (1964): The degree of saturation of magnesium and calcium carbonate minerals in natural waters. *Int. Assoc. Hydrol. Sci.,* **64**, pp. 34–49.

Hughes, G. M., R. A. Landon & R. N. Farvolden (1969): Hydrogeologic data from four landfills in northeastern Illinois. Environmental Geology Notes No. 26, Illinois State Geological Survey, Urbana.

Hughes, G., J. J. Tremblay, H. Anger & J. D'Cruz (1971a): Pollution of groundwater due to municipal dumps. Inland Water Branch Technical Bulletin No. 42, 98 pp., Ottawa.

Hughes, G. M., R. A. Landon & R. N. Farvolden (1971b): Hydrogeology of solid waste disposal sites in northeastern Illinois. Illinois State Geological Survey and U.S. Environmental Protection Agency.

Huisman, L. (1972): *Groundwater Recovery,* 336 pp. London: Macmillan.

Huisman, L. & F. W. J. Van Haaren (1966): Treatment of water before infiltration and modification of its quality during its passage underground. *Int. Water Supply Assoc., 7th Congr.,* **1**, Special Subject No. 3, 43 pp., Barcelona.

Husmann, K. (1963): Beitrag zur Frage des Abbaues von Detergentien bei der Langsamfiltration. *Veröff. Inst. Siedlungswasserwirtsch. T. H. Hannover,* **12**, 103 pp., Hannover.

Husmann, S. (1966a): Versuch einer ökologischen Gliederung des interstitiellen Grundwassers in Lebensbereiche eigener Prägung. *Arch. Hydrobiol.,* **62**, 2, pp. 231–268, Stuttgart.

Husmann, S. (1966b): Die Organismengemeinschaften der Sandlückensysteme in natürlichen Biotopen und Langsamsandfiltern. *Veröff. Hydrol. Forschungsabt. Dortmunder Wasserwerke AG,* **9**, pp. 93–113, Dortmund.

Husmann, S. (1968): Langsamfilter als Biotopmodelle der experimentalökologischen Grundwasserforschung. *Gewässer Abwässer,* **46**, pp. 20–49, Düsseldorf.

Hutchinson, G. E. (1957): *A Treatise on Limnology,* Vol. 1, *Geography, Physics and Chemistry,* 1015 pp. New York: Wiley.

Institut für Radiohydrometrie (1967): *Jahresbericht 1966,* 70 pp., Munich.

Institut für Radiohydrometrie (1969): *Jahresbericht 1968,* 84 pp., Munich.

Internationale Kommission zum Schutze des Rheins gegen Verunreinigung (1967): *Bericht über die physikalisch- chemische Untersuchung des Rheinwassers, V, 1961–1965,* 148 pp., Coblenz, Rhein.

Ivanov, M. V. (1966): Die Anwendung von ^{35}S für die Untersuchung der Intensität der mikrobiologischen Sulfatreduktion im Grundwasser. *Abh. Dtsch. Akad. Wiss. Berlin, Kl. Chem. Geol. Biol.,* 1965, **2**, pp. 205–210, Berlin.

Jäckli, H. & K. Kleiber (1943): Temperaturstudien an Gebirgsquellen. *Eclog. Geol. Helv.,* **36**, pp. 7–15, Basle.

Jackson, R. E. & K. J. Inch (1976): Development of methods for sampling, preserving and analyzing dissolved chemical species and other aqueous parameters in ground waters. Annual Progress Report, Short Research Notes, 1975–1976, Hydrological Research Division, Water Resources Branch, Inland Waters Directorate, Report Series No. 45, pp. 66–75, Ottawa.

Jackson, R. E. & K. J. Inch (1980): Hydrogeochemical processes affecting the migration of radionuclides in a fluvial sand aquifer at the Chalk River Nuclear Laboratories. NHRI, Paper No. 7, Science Series No. 104, Ottawa.

Jackson, R. E., W. F. Merritt, D. R. Champ, J. Gulens & K. J. Inch (1980): The distribution coefficient as a geochemical measure of the mobility of contaminants in a groundwater flow system. Panel Proceedings Series, STI/PUB/518, pp. 209–225, Vienna (IAEA).

Jacobs, M. S. Schmorak (1960): Salt water encroachment in the coastal plain of Israel. *Int. Assoc. Sci. Hydrol. Publ.* **52**, pp. 408–423, Gentbrugge.

Jacobshagen, V. (1961): Die Isotopenzusammensetzung natürlicher Wässer und ihre Änderungen im Wasserkreislauf. *Geol. Rundsch.*, **51**, pp. 281–290, Stuttgart.

Jacobshagen, V. & K. O. Münnich (1964): ^{14}C-Altersbestimmungen und andere Isotopen – Untersuchungen an Thermalsolen des Ruhrkarbons. *Z. Dtsch. Geol. Ges.*, **116**, 1, 160 pp., Hannover.

Jenne, E. A. (1968): Control of Mn, Fe, Ni, Cu, and Zn concentration in soils and waters: Significant role of hydrous Mn and Fe oxides. In: *Trace Inorganics in Water*, American Chemical Society Publication No. 73, pp. 337–387, Washington, D.C.

Jenne, E. A. (1977): Trace element sorption by sediments and soils – Sites and processes. In: W. Chappel & K. Petersen, Eds., *Symposium on Molybdenum in the Environment*, Vol. 2. New York: Dekker, Chapter 5, pp. 425–523.

Jenne, E. A. & J. S. Wahlberg (1968): Role of certain stream-sediment components in radioion sorption. U.S. Geological Survey Professional Paper No. 433-F, Washington, D.C.

Jeris, J. S. & P. L. McCarty (1965): The biochemistry of methane fermentation using ^{14}C tracers. *J. Water Pollut. Control. Fed.* **37**, 2, pp. 178–192, Washington, D.C.

Jones, P. H. (1968): Geochemical hydrodynamics – A possible key to the hydrology of certain aquifer systems in the northern part of the Gulf of Mexico Basin. *23rd Int. Geol. Congr.*, **17**, pp. 113–125, Prague.

Jong, E. de & H. J. V. Schappert (1972): Calculation of soil respiration and activity fron CO_2-profiles in the soil. *Soil Sci.*, **113**, pp. 328–333, New Brunswick, NJ.

Jordan, D. G. (1962): Ground water contamination in Indiana. *J. Am. Water Works Assoc.*, **54**, pp. 1213–1220, New York.

Jourain, G. (1960): Moyens et résultats d'étude de la radioactivité due au radon dans les eaux naturelles. *Geochim. Cosmochim. Acta*, **20**, pp. 51–82, Oxford, London, New York, Paris.

Junge, C. (1954): Die Rolle der Aerosole und der gasförmigen Beimengungen der Luft im Spurenstoffhaushalt der Troposphäre. *Tellus*, **5**, pp. 1–26, Stockholm.

Junge, C. E. (1958): The distribution of ammonia and nitrate in rainwater over the United States. *Trans. Am. Geophys. Union.*, **39**, 2, pp. 241–248, Washington, D.C.

Junge, E. (1963): *Air Chemistry and Radioactivity*, 382 pp. New York, London: Academic Press.

Kalle, K. (1943): *Der Stoffhaushalt des Meeres*, 263 pp. Leipzig: Akademie Verl.

Kanz, E. (1960): Über das Verschwinden von Keimen im Grundwasser des diluvialen Schotterbodens. *Arch. Hyg. Bakteriol.*, **144**, pp. 375–401, Munich, Berlin.

Kappelmeyer, O. (1961): Geothermik. In: A. Bentz, Ed., *Lehrbuch der angewandten Geologie*, Vol. 1, Stuttgart: Enke, pp. 863–889.

Kappelmeyer, O. (1968): Beiträge zur Erschliessung von Thermalwässern und natürlichen Dampfvorkommen. *Geol. Jahrb.*, **85**, pp. 783–808, Hannover.

Kappelmeyer, O. & R. Haenel (1974): Geothermics with special reference to application. *Geoexploration Monographs*, Vol. 1, 4, 238 pp. Berlin, Stuttgart: Borntraeger.

Kappen, H. & M. Zapfe (1917): Über Wasserstoffionenkonzentrationen in Auszügen von Moorböden und von moor- und rohhumusbildenden Pflanzen. *Landw. Versuchsstn.*, **90**, pp. 321–374, Berlin.

Karbach, L. (1961): Untersuchungen über den Einflüsse der Bodenmikroorganismen auf die Redoxverhältnisse im Boden. *Landw. Forsch.*, **14**, pp. 64–69, Frankfurt am Main.

Karlson, P. (1970): *Kurzes Lehrbuch der Biochemie für Mediziner und Naturwissenschaftler*, 381 pp. Stuttgart: Thieme.

Karrenberg, H., W. Niehoff, F. Preul & W. Richter (1958): Groundwater maps developed in the geological surveys of Niedersachsen and Nordrhein-Westfalen of the Federal Republic of Germany. Internat. Assoc. Sci. Hydrol. General Ass., II, pp. 54–61, Gentbrugge.

Käss, W. (1964): Ergebnisse der chemischen Untersuchungen von Solen des Steinkohlenbezirks an der Ruhr. *Z. Dtsch. Geol. Ges.*, **116**, 1, pp. 244–253, Hannover.

Käss, W. (1965): Sind Härtegrade entbehrlich? Eine Anregung. *Dtsch. Gewässerk. Mit.*, **9**, 3, pp. 63–64, Coblenz, Rhein.

Käss, W. (1967): Zur Geochemie einiger neuerschlossener Buntsandstein- Mineralwässer am Schwarzwald-Ostrand. *Jahrb. Geol. Landesamt, Baden-Württemberg*, **9**, pp. 81–104, Freiburg i. Br.

Käss, W. (1969): Gefährdung des Grundwassers durch Mineralöle, Problematik und Experimente. *Heilbad Kurort*, **21**, 2, pp. 12–17, Gütersloh.

Keilhack, K. (1935): *Lehrbuch der Grundwasser- und Quellenkunde*, 3rd ed., 575 pp. Berlin: Borntraeger.

Keller, G. (1942): Geologischer Untergrund, chemische Verunreinigungsanzeiger und bakteriologischer Befund bei erbohrten Grundwässern. *Gesund.-Ing.*, **65**, pp. 180–191, 206–211, Munich.

Keller, G. (1958): Abwasserbeseitigung als hydrogeologisches Problem. *Z. Dtsch. Geol. Ges.*, **109**, 2, pp. 504–518, Hannover.

Kempf, T. (1970): Die Bedeutung der Bestimmung organischer Stoffe im Wasser. *Schr.-R. Ver. Wasser-, Boden- Lufthyg.*, **33**, pp. 105–115, Berlin, Dahlem.

Kennedy, V. C. (1956): Geochemical studies in the southwestern Wisconsin zinc-lead area. U.S. Geological Survey Bulletin No. 1000-E, pp. 187–223, Washington, D.C.

Kenny, A. W., R. N. Crooks & J. R. W. Kerr (1966): Radium, radon and daughter products in certain drinking waters in Great Britain. *J. Inst. Water Eng.* **20**, 2, pp. 123–134, London.

Keswick, B. H. & C. P. Gerba (1980): Viruses in groundwater. *Environ. Sci. Technol.*, **14**, 11, pp. 1290–1297, Washington, D.C.

Klechkovsky, V. M., L. N. Sokolova & G. N. Tselishcheva (1958): The sorption of microquantities of strontium and cesium in soils. *Proceedings of the 2nd UN Conference on Peaceful Uses of Atomic Energy*, Vol. 18, pp. 486–493, Geneva.

Kloke, A. (1961): Die Wanderung von ^{90}Sr im Boden. *Naturwissenschaften*, **48**, 21, 674 pp., Berlin, Göttingen, Heidelberg.

Klotter, H.-E. & E. Hantge (1969): Abfallbeseitigung und Grundwasserschutz. *Müll Abfall.* **1**, pp. 1–8, Berlin, Munich, Bielefeld.

Klotter, H. E. & W. Langer (1964): Bisheriger Zustand der Abfallablagerung und seine Folgen. *Müll Abfall.*, 2nd. part, code number 4510, pp. 1–3, Berlin: E. Schmidt.

Klotz, I. M. (1950): *Chemical Thermodynamics*, 369 pp. Englewood Cliffs, NJ: Prentice-Hall.

Klut-Olszewski (1945): *Untersuchung des Wassers an Ort und Stelle, seine Beurteilung und Aufbereitung*, 9th ed., 281 pp. Berlin: Springer.

Knetsch, G., A. Shata, E. Degens, K. O. Münnich, J. C. Vogel & M. M. Shazly (1962): Untersuchungen an Grundwässern der Ost-Sahara. *Geol. Rundsch.,* **52,** 2, pp. 587–610, Stuttgart.

Knie, K. (1966): Cyan, Vorkommen und Untersuchung. *Österr. Abwasser-Rundsch.,* **11,** pp. 58–61, Vienna.

Knie, K. & H. Gams (1960): Zum Chemismus der Brunnenwässer im Seewinkel. *Wasser Abwasser,* **1960,** pp. 56–81, Vienna.

Knoll, K. H. (1969): Hygienische Bedeutung natürlicher Selbstreinigungsvorgänge für die Grundwasserbeschaffenheit im Bereich von Abfalldeponien. *Müll Abfall,* **1,** 2, pp. 35–41, Berlin, Bielefeld, Munich.

Knoop, E. & D. Schroeder (1958): Der ^{90}Sr-Gehalt einiger Böden Schleswig-Holsteins. *Naturwissenschaften,* **45,** 18, pp. 436–437, Berlin, Göttingen, Heidelberg.

Knorr, M. (1960): Die hygienische Beurteilung resistenter Schadstoffe im Boden und Grundwasser. *Gesund.-Ing.,* **87,** 11, pp. 326–336, Munich.

Knutsson, G. & H. G. Forsberg (1967): Laboratory evaluation of ^{51}Cr-EDTA as a tracer for groundwater flow. *Isotopes in Hydrology.* Vienna: IAEA, pp. 629–652.

Köhler, H. (1937): Studien über Nebelfrost und Schneebildung und über den Chlorgehalt des Nebelfrostes, des Schnees und des Seewassers im Halddegebiet. *Bull. Geol. Inst. Univ. Upsala,* **26,** pp. 279–308, Upsala.

Kohout, F. A. (1960): Flow pattern of fresh and saltwater in the Biscayne aquifer of the Miami area, Florida. *Int. Assoc. Sci. Hydrol. Publ.* **52,** pp. 440–448, Gentbrugge.

Kollatsch, D. (1968): Das "Anaerob-Aerob-Verfahren" in der biologischen Abwasserreinigung. *Wasser, Luft Betrieb,* **12,** 4, pp. 236–242, Mainz.

Kölle, W. & H. Sontheimer (1968): Wassergefährdende Substanzen. *Gas-Wasser-Wärme,* **22,** 11, 226–236, Vienna.

Kölle, W. & H. Sontheimer (1969): Die Problematik der Grundwasserverschmutzung durch Mineralölsubstanzen. *Brennstoff-Chemie,* **50,** 4, pp. 123–129, Essen.

König, J. (1893): *Chemie der menschlichen Nahrungs- und Genussmittel,* Vol. 2, 3rd ed., 1385 pp. Berlin: Springer.

Kooijmans, L. H. L. (1966): Occurrence, significance and control of organisms in distribution systems. International Water Supply Association, Seventh Congress, General Report, Vol. 3, 36 pp., Barcelona.

Koppe, P. (1966): Identifizierung der Hauptgeruchsstoffe im Uferfiltrat des Mittel- und Niederrheins. *Vom Wasser,* **32,** pp. 33–68, Weinheim, Bergstrasse.

Koppe, P. & G. Giebler (1965): Über die Gefährdung einer öffentlichen Wasserversorgung durch die Verunreinigung des Grundwassers mit Arsen. *Städtehygiene,* **16,** 11, pp. 241–245, Hamburg.

Kortüm, G. (1966a): *Einführung in die chemische Thermodynamik,* 5th ed., 484 pp. Göttingen: Vandenhoek & Rupprecht; Weinheim, Bergstrasse: Chemie.

Kortüm, G. (1966b) with collaboration of W. Vogel: *Lehrbuch der Elektrochemie,* 4th ed., 592 pp. Weinheim, Bergstrasse: Chemie.

Kraft, D. & K. H. Knoll (1970): Die Belastung von Abwasseranlagen durch galvanische Abwässer mit besonderer Berücksichtigung der Gebührenberechnung. *Müll-Abfall-Abwasser,* **14,** 4 pp., Düren, Rheinland.

Kramer, D. (1958): Die Bodenreinigung, dargestellt am Beispiel der Zuckerfabrikabwässer. *Wasserwirtschaft-Wassertechn.,* **8,** pp. 549–558, Berlin.

Kramer, J. R. (1968): Mineral-water equilibria in silicate weathering. *23rd Int. Geol. Congr.,* **6,** pp. 149–160, Prague.

Krauskopf, K. B. (1956a): Dissolution and precipitation of silica at low temperatures. *Geochim. Cosmochim. Acta,* **10,** pp. 1–26, London, New York, Paris.

Krauskopf, K. B. (1956b): Factors controlling the concentrations of thirteen rare metals in sea-water. *Geochim. Cosmochim. Acta,* **9,** pp. 1–32, London, New York.

Krauskopf, K. B. (1967): *Introduction to Geochemistry,* 145 pp. New York: McGraw-Hill.

Krauskopf, K. B. (1969): Thermodynamics used in geochemistry. In: K. H. Wedepohl. Ed., *Handbook of Geochemistry,* Vol. 1, Berlin, Heidelberg, New York: Springer, pp. 37–77.

Krauss, H.-L. (1961): Zur Struktur des Wassers. *Naturwiss. Rundsch.,* **14,** pp. 176–181, Stuttgart.

Kreitler, C. W. & D. C. Jones (1975): Natural soil nitrate: The cause of the nitrate contamination of ground water in Runnels County, Texas. *Groundwater,* **13,** pp. 53–61.

Krejci-Graf, K. (1934): Geochemie der Naturgase. *Kali,* **28,** pp. 249–252, 261–265, 275–278, 287–290, Halle (Saale).

Krejci-Graf, K. (1963a): Diagnostik der Salinitätsfazies der Ölwässer. *Fortschr. Geol. Rheinl. Westf.,* **10,** pp. 367–448, Krefeld.

Krejci-Graf, K. (1963b): Über rumänische Ölfeldwässer. *Geol. Mitt.* **2,** 4, pp. 351–391, Aachen.

Kretzschmar, R. (1964): *Untersuchungen über die Verlagerung von Ammonium-, Nitrat-, Chlorid- und Sulfationen im Boden der Niederterrasse des Rheins bei Bonn und ihr Abwandern in tiefere Bodenschichten.* 227 pp. Ph.D. thesis, Bonn.

Kristensen, K. J. & H. Enoch (1964): Soil air composition and oxygen diffusion rate in soil columns at different heights above a water table. *Ber. 8th Int. Congr. Soil Sci.,* **2,** pp. 159–170, Bucharest.

Krumbein, W. C. & R. M. Garrels (1962): Origin and classification of chemical sediments in terms of pH and oxidation-reduction potentials. *J. Geol.,* **60,** 1, pp. 1–33, Chicago.

Kruse, H. (1965): Kanzerogene Stoffe und die Trinkwasserversorgung. *Gas- Wasserfach,* **106,** 12, pp. 318–325, Munich.

Ku, H. F. H., B. G. Katz, D. J. Sulam and R. K. Krulikas (1978): Scavenging of chromium and cadmium by aquifer material – South Farmingdale – Massapequa area, Long Island, New York. *Groundwater,* **17,** pp. 112–119.

Kühn, R. (1964): Chemische Gesichtspunkte zur Frage der Herkunft der Solen im Ruhrgebiet. *Z. Dtsch. Geol. Ges.,* **116,** 1, pp. 254–255, Hannover.

Kupke, H. (1963): Hydrochemische Probleme bei der Grundwassererschliessung, dargestellt am Beispiel des Wasserwerkes Erkner. *Ber. Geol. Ges. DDR,* **8,** 4, pp. 404–408, Berlin.

Kurbatov, J. D., J. L. Kulp & E. Mack, Jr. (1945): Adsorption of strontium and barium ions and their exchange on hydrous ferric oxide. *J. Am. Chem. Soc.,* **67,** pp. 1923–1929, Easton, PA.

Kurmies, B. (1957): Über den Schwefelhaushalt des Bodens. *Phosphorsäure,* **17,** 5–6, pp. 258–278, Essen.

Küster, F. W., A. Thiel & K. Fischbeck (1969): *Logarithmische Rechentafeln,* 100th ed., 310 pp. Berlin: de Gruyter.

Kuznetsov, S. J. (1962): Geologic activity of microorganisms. *Trans. Inst. Microbiol.,* **9,** 12 pp., New York.

Kuznetsov, S. J., M. V. Ivanov & N. N. Lyalikova (1963): *Introduction to Geological Microbiology,* 252 pp. New York, San Francisco, Toronto, London: McGraw-Hill.

Laguna, W. De (1962): Engineering geology of radioactive waste disposal. In: *Reviews in Engineering Geology,* Vol. 1. New York: Geological Society of America, pp. 129–160.

Lahmann, E. (1970): Luftverunreinigung durch Schwefeldioxid in Städten der Bundesrepublik Deutschland. Ergebnisse von 1968–1969. *Bundesgesundheitsblatt,* **26,** pp. 375–378, Berlin.

Lai, S. M. & J. J. Jurinak (1972): Cation adsorption in one-dimensional flow through soils: A numerical solution. *Water Resour. Res.* **8**, 1, p. 99.

Lal, D., V. N. Nijampurkar & S. Rama (1970): Silicon-32 hydrology. *Isotope Hydrology, 1970,* Vienna: IAEA, pp. 847–868.

Lamar, J. E. & R. S. Shrode (1953): Water soluble salts in limestones and dolomites. *Econ. Geol.,* **48**, pp. 97–112, New Haven, CT.

Landolt-Börnstein (1923): *Physikalisch-chemische Tabellen,* Vol. 1, 5th ed., 784 pp. Berlin: Springer.

Landolt-Börnstein (1931): *Physikalisch-chemische Tabellen,* 5th ed., 2.Erg.-B., 1, 506 pp. Berlin: Springer.

Lane, A. C. (1908): Mine waters and their field assay. *Bull. Geol. Soc. Am.,* **19**, pp. 502–512, New York.

Lang, A. & H. Bruns (1940): Über die Verunreinigung des Grundwassers durch chemische Stoffe. *Gas- Wasserfach,* **83**, pp. 6–9, Munich, Berlin.

Lang, W. B. (1941): New source for sodium sulphate in New Mexico. *Bull. Am. Assoc. Petrol. Geol.,* **25**, pp. 152–160, Tulsa, OK.

Langbein, W. B. (1961): Salinity and hydrology of closed lakes. U.S. Geological Survey Professional Paper No. 412, 20 pp., Washington, D.C.

Langelier, W. F. (1936): The analytical control of anti-corrosion water treatment. *J. Am. Water Works Assoc.,* **28**, pp. 1500–1521, Baltimore.

Langelier, W. F. & H. F. Ludwig (1942): Graphical methods for indicating the mineral character of natural waters. *J. Am. Water Works Assoc.,* **34**, 3, pp. 335–352, Baltimore.

Langer, W. (1963): Feste Abfallstoffe und Wasserhaushalt. *Städtetag,* **7**, pp. 1–14, Stuttgart.

Langmuir, D. (1969): Iron in ground-water of the Magothy and Raritan Formations in Camden and Burlington Counties, New Jersey. New Jersey Research Circular No. 19, New Jersey Department of Conservation and Economic Development.

Langmuir, D. (1971a): Particle size effect on the reaction Goethite = Hematite + Water. *Am. J. Sci.,* **271**, pp. 147–156.

Langmuir, D. (1971b): The geochemistry of some carbonate groundwaters in central Pennsylvania. *Geochim. Cosmochim. Acta.* **35**, pp. 1023–1045, Oxford, London, New York Paris.

Langmuir, D. & R. L. Jacobson (1973): Undersized lots invite polluted well water. College of Earth and Mineral Science, Pennsylvania State University, University Park.

La Rivière, J. W. M. (1955): The production of surface-active compounds by microorganisms and its possible significance in oil recovery. I and II. *Amton v. Leeuwenhoek,* **21**, pp. 1–8, 9–26, Amsterdam.

Larson, T. E. & A. M. Buswell (1942): Calcium carbonate saturation index and alkalinity interpretations. *J. Am. Water Works Assoc.,* **34**, pp. 1667–1684, Baltimore.

Larson, T. E. & I. Hettick (1956): Mineral composition of rainwater. *Tellus,* **8**, pp. 191–201, Stockholm.

Latimer, W. M. (1953): *The Oxidation States of the Elements and Their Potentials in Aqueous Solutions,* 2nd ed., 392 pp. Englewood Cliffs, NJ: Prentice-Hall.

Leenheer, J. A., R. L. Malcolm, P. W. McKinley & L. A. Eccles (1974): Occurrence of dissolved organic carbon in selected ground water samples in the United States. *J. Res. U.S. Geol. Surv.,* **2**, 3, 361 pp., Washington, D.C.

Leggat, E. R., J. F. Blakey & B. C. Massey (1972): Liquid waste disposal at the Linfield Disposal Site, Dallas, Texas. U.S. Geological Survey Open-File Report, Washington, D.C.

LeGrand, H. E. (1958): Chemical character of water in the igneous and metamorphic rocks of North Carolina. *Econ. Geol.,* **53**, 2, pp. 178–189, New Haven, CT.

LeGrand, H. E. (1965): Patterns of contaminated zones of water in the ground. *Water Resourc. Res.*, 1, 1, pp. 83–95, Washington, D.C.

Lemon, E. R. & A. E. Erickson (1952): The measurement of oxygen diffusion in the soil with a platinum microelectrode. *Soil Sci. Soc. Am. Proc.*, 16, pp. 160–163, Madison, WI.

Leutwein, F. (1960): Geochemische Prospektion. *Erzmetall, Z. Erzbergbau Metallhüttenwes.*, 13, pp. 493–498, Stuttgart.

Lewis, G. N. & M. Randall (1921): The activity coefficient of strong electrolytes. *J. Am. Chem. Soc.*, 43, pp. 1112–1154, Easton, PA.

Lichtenstein, E. P., K. R. Schulz, R. F. Skrentny & Y. Tsukano (1966): Toxicity and fate of insecticide residues in water. *Arch. Environ. Health*, 12, 2, pp. 199–212, Chicago.

Liesegang, W. (1927): Die Untersuchung atmosphärischer Niederschläge. *Kl. Mitt. Ver. Wasser-, Boden-, Lufthyg.*, 3, pp. 317–327, Berlin, Dahlem.

Liesegang, W. (1933): Über einige gesundheitlich wichtige gas- und dampfförmige Verunreinigungen der atmosphärischen Luft. *Kl. Mitt. Wasser-, Boden-, Luftyg.*, 9, pp. 286–296, Berlin, Dahlem.

Linderoth, C. E. & D. W. Pearce (1961): Operating practices and experiences at Hanford (Washington) in ground disposal of radioactive wastes. *Conference Proceedings*, pp. 7–15, Sanitary Engineering Research Laboratory, University of California, Berkeley.

Lindgren, W. (1933): *Mineral Deposits*, 4th ed., 930 pp. New York, London: McGraw-Hill.

Lingelbach, H. & H. Kühn (1965): Auswirkungen von Herbiziden auf Grund- und Oberflässer. *Fortschr. Wasserchem. Grenzgeb.*, 3, pp. 153–157, Berlin: Akademie-Verl.

Lingelbach, H., R. Saalbreiter & H. Kühn (1962): Verschmutzung von Trinkwasserversorgungsanlagen durch Gaswasser. *Z. Gesamt. Hyg. Grenzgeb.*, 8, 10, pp. 761–767, Berlin.

Lininger, R. L., R. A. Duce, J. W. Winchester & W. R. Matson (1966): Chlorine, bromine, iodine and lead in aerosols from Cambridge, Massachusetts. *J. Geophys. Res.*, 71, 10, pp. 2457–2463, Richmond, VA.

Livingstone, D. A. (1963): Chemical composition of rivers and lakes. U.S. Geological Survey Professional Paper No. 440-G, 64 pp., Washington, D.C.

Loewengart, S. (1958): Geochemistry of waters in northern and central Israel and the origin of their salts. *Bull. Res. Counc. Isr.*, 7G, pp. 176–205, Jerusalem.

Loewengart, S. (1961): Airborne salts – The major source of the salinity of waters in Israel. *Bull. Res. Counc. Isr.*, 10G, 1–4, pp. 183–206, Jerusalem.

Logan, J. (1961): Estimation of electrical conductivity from chemical analysis of natural waters. *J. Geophys. Res.*, 66, pp. 2479–2483, Richmond, VA.

Löhnert, E. (1969): Grundwasserverunreinigungen im Elbe-Tal (Freie und Hansestadt Hamburg). *Gas- Wasserfach*, 110, pp. 1171–1177, Munich.

Löhnert, E. (1970): Grundwasserchemismus und Kationentausch im norddeutschen Flachland. *Z. Dtsch. Geol. Ges.*, Sonderh. Hydrogeol. Hydrogeochem., pp. 139–159, Hannover.

Löhnert, E. (1972): Contribution to the geochemistry of groundwater in northern Germany. *Geol. Mijnbouw*, 51, pp. 63–70, s'Gravenhage.

Lohr, E. W. & S. K. Love (1954): The industrial utility of public water supplies of the United States: Part 1. States east of the Mississippi River. U.S. Geological Survey Water Supply Paper No. 1299, 639 pp., Washington, D.C.

Luck, W. (1965): Über die Assoziation des flüssigen Wassers. *Fortschr. Chem. Forsch.*, 4 pp. 653–781, Berlin, Göttingen, Heidelberg: Springer.

Luck, W. A. P., Ed. (1974): *Structure of Water and Aqueous Solutions. Proceedings of the International Symposium, Marburg 1973*, 590 pp. Weinheim, Bergstrasse: Verl. Chemie, Physik Verl.

Lüdemann, D. (1969): Ökologische Verfahren der biologischen Wasseranalyse. In: S. W. Souci & K.-E. Quentin, Eds., *Handbuch der Lebensmittelchemie,* Vol. 8, Part 2. Berlin, Heidelberg, New York: Springer, pp. 1193–1199.

Lüdemann, D. (1970): Biologie des Wassers. In: K. Höll, Ed., *Wasser: Untersuchung, Beurteilung, Aufbereitung, Chemie, Bakteriologie, Biologie.* Berlin: de Gruyter, pp. 379–416.

Luedecke, C. (1899): Die Boden- und Wasserverhältnisse der Provinz Rheinhessen, des Rheingaus und Taunus. *Abh. Grossh. Hess. Geol. Landesamt,* 3, 4, pp. 149–298, Darmstadt.

Luedecke, C. (1901): Die Boden- und Wasserverhältnisse des Odenwaldes und seiner Umgebung. *Abh. Grossh. Hess. Geol. Landesamt,* 4, 1, 183 pp., Darmstadt.

Lundegårdh, H. (1924): *Der Kreislauf der Kohlensäure in der Natur. Ein Beitrag zur Pflanzenokologie und zur landwirtschaftlichen Düngungslehre,* 308 pp. Jena: Fischer.

Lundegårdh, H. (1927): Carbon dioxide evolution of soil and crop growth. *Soil Sci.,* 23, pp. 417–453, New Brunswick, NJ.

Lüning, O. & E. Heinsen (1934): Hohe Carbonathärte als Anzeiger von Grundwasserverschmutzung. *Z. Unters. Lebensmit.,* 67, pp. 627–638, Berlin.

Lupton, J. E., R. F. Weiss & H. Craig (1977): Mantle helium in the Red Sea brines. *Nature,* 266, pp. 244, London.

MacLean, R. D. (1969): The effect of tipped domestic refuse on groundwater quality: a survey in North Kent. *Water Treat Exam.,* 18, pp. 18–34.

Maehler, C. Z. & A. E. Greenberg (1962): Identification of petroleum industry wastes in groundwaters. *J. Water Pollut. Control Fed.,* 34, 12, pp. 1262–1267, Washington, D.C.

Manov, G. G., R. G. Bates, W. J. Hamer & S. F. Acree (1943): Values of the constants in the Debye-Hückel equation for activity coefficients. *J. Am. Chem. Soc.,* 65, pp. 1765–1767, Washington, D.C.

Man's Impact on the Global Environment (1971): *Report of the Study of Critical Environmental Problems,* 3rd ed., 219 pp. Cambridge, MA: MIT Press.

Marshall, C. E. & W. E. Bergman (1942): The electrochemical properties of mineral membranes. Measurement of potassium-ion activities in colloidal clays. *J. Phys. Chem.,* 46, pp. 52–61, Ithaca, NY.

Marter, W. L. (1967): Ground waste disposal practices at the Savannah River Plant. *Disposal of Radioactive Wastes into the Ground.* Vienna: IAEA, pp. 95–107.

Martin, M. (1958): Relation entre la resistivité des eaux et leur composition chimique. *Rev. Inst. Fr. Petrol. Ann. Combust. Liq.,* 13, 6, pp. 987–996, Paris.

Martini, H. J. (1952): Über Auftreten, Herkunft und wasserwirtschaftliche Bedeutung von versalzenem Grundwasser in Nordwestdeutschland. *Wasserwirtsch., Sonderh.,* 1951, pp. 31–34, Stuttgart.

Matthess, G. (1958): Geologische und hydrochemische Untersuchungen in der östlichen Vorderpfalz zwischen Worms und Speyer. *Notizbl, Hess. Landesamt Bodenforsch.,* 86, pp. 335–378, Wiesbaden.

Matthess, G. (1961): Die Herkunft der Sulfat-Ionen im Grundwasser. *Abh. Hess. Landesamt Bodenforsch.,* 35, 85 pp., Wiesbaden.

Matthess, G. (1971): Hydrogeologie. In: Erl. geol.Kte.Hessen 1:25000 Bl. 5623 Schlüchtern, pp. 158–184, Wiesbaden.

Matthess, G. (1972a): Hydrogeologic criteria for the self-purification of polluted groundwaters. *Int. Geol. Congr.,* 11, pp. 296–302, Montreal.

Matthess, G. (1972b): Bleigehalte in Gestein, Boden und Grundwasser. In: *Blei und Umwelt,* Berlin, Dahlem: Ver.Wa-Bo.Lu, pp. 21–27.

Matthess, G. (1974): Heavy metals as trace constituents in natural groundwaters and polluted. *Geol. Mijngouw,* 53, 4, pp. 149–155, s'Gravenhage.

Matthess, G. (1976): Effects of man's activities on groundwater quality. *Hydrol. Sci. Bull.,* **21,** pp. 617–628, Reading, PA.

Matthess G., & K. Hamann (1966): Biogene Schwankungen des Sulfatgehaltes von Grundwässern. *Gas- Wasserfach,* **107,** 18, pp. 480–484, Munich.

Matthess, G. & A. Pekdeger (1980): Chemisch-biochemische Umsetzungen bei der Grundwasserneubildung. *Gas- Wasserfach,* **121,** pp. 214–219, Munich.

Matthess, G. & A. Pekdeger (1981): Concepts of a survival and transport model of pathogenic parasitic bacteria and viruses in groundwater. *Proceedings of the International Symposium on Quality of Groundwater, Nordwijkerhout,* Amsterdam: Elsevier.

Matthess, G. & J.-D. Thews (1963): Hydrogeologie. Erl. geol. Kte. Hessen 1:25000, Bl. 5223 Queck, pp. 245–281, Wiesbaden.

Matthess, G., K. O. Münnich & C. Sonntag (1976): Practical problems of groundwater model ages for groundwater protection studies. Panel Proceeding Series, STI/PUB/429, Vienna: IAEA, pp. 185–194.

Matthess, G., L. Thilo, W. Roether, & K. O. Münnich (1968): Tritiumgehalte im Wasser tieferer Grundwasserstockwerke. *Gas- Wasserfach,* **109,** 14, pp. 353–355, Munich.

Matthess, G., A. Pekdeger & H. D. Schulz (1977): Geochemical-biogeochemical processes in seepage water during groundwater recharge. *Proceedings of the Second International Symposium on Water-Rock Interaction,* Vol. 1, pp. 146–155, Strasbourg.

Matthess, G., A. Pekdeger, H. D. Schulz, H. Rast & W. Rauert (1979): Tritium tracing in hydrogeological studies using model lysimeters. *Isotope Hydrology,* Vol. 2. Vienna: IAEA, pp. 769–785.

Mattson, S. (1927): Anionic and cationic adsorption by soil colloidal materials of varying SiO_2/ $Al_2O_3 + Fe_2O_3$ ratio. *Trans. 1st Int. Congr. Soil Sci.,* **2,** pp. 199–211, Washington, D.C.

Maucha, R. (1932): Hydrochemische Methoden in der Limnologie. *Binnengewässer,* **XXII,** 173 pp., Stuttgart (Schweizerbart).

Mayer, S. W. & E. R. Tompkins (1947): Ion exchange as a separation method: A theoretical analysis of the column separation process. *J. Am. Chem. Soc.,* **69,** pp. 2866–2874, Washington, D.C.

Mazor, E. (1962): Radon and radium content of some Israeli water sources and a hypothesis on underground reservoirs on brines, oils and gases in the Rift Valley. *Geochim. Cosmochim. Acta,* **26,** pp. 765–786, Oxford, London, New York, Paris: Pergamon.

Mazor, E. (1968): Genesis of mineral waters in the Tiberias–Dead Sea–Arava Rift Valley, Israel. *23rd Int. Geol. Congr.,* **17,** pp. 65–80, Prague.

Mazor, E. (1972): Paleotemperatures and other hydrological parameters deduced from noble gases dissolved in groundwaters; Jordan Rift Valley, Israel. *Geochim. Cosmochim. Acta,* **36,** pp. 1321–1336, Oxford.

Mazor, E. & R. O. Fournier (1973): More on noble gases in Yellowstone National Park hot waters. *Geochim. Cosmochim. Acta,* **37,** pp. 515–525, Oxford.

Mazor, E. & G. J. Wasserburg (1965): Helium, neon, argon, krypton and xenon in gas emanations from Yellowstone and Lassen Volcanic National Parks. *Geochim. Cosmochim. Acta,* **29,** pp. 443–454, Oxford.

McAuliffe, C. (1966): Solubility in water of paraffin, cycloparaffin, olefin, acetylene, cycloolefin, and aromatic hydrocarbons. *J. Phys. Chem.,* **70,** pp. 1267–1275, Easton, PA.

McGauhey, P. H., R. B. Krone & J. H. Winneberger (1966): Soil mantle as a wastewater treatment system. Sanitary Engineering Research Laboratory Report, pp. 66–67, University of California, Berkeley.

McKelvey, V. E., D. I. Everhart & R. M. Garrels (1955): Origin of uranium deposits. *Econ. Geol.,* 50th anniversary volume, Part 1, pp. 464–533, Lancaster, PA.

McNellis, M. & C. O. Morgan (1969): Modified Piper diagrams by the digital computer. State Geological Survey, University of Kansas, Lawrence, Special Distribution Publication No. 43, 36 pp.

Medsker, L. L., D. Jenkins & J. F. Thomas (1968): Odorous compounds in natural waters. *Environ. Sci. Technol.*, **2**, 6, pp. 461–464, Washington, D.C.

Meetham, A. R. (1950): Natural removal of pollution from the atmosphere. *Q. J. R. Meteorol. Soc.*, **76**, pp. 359–371, London.

Meisl, S. (1970): Petrologische Studien im Grenzbereich Diagenese-Metamorphose. *Abh. Hess. Landesamt Bodenforsch.*, **57**, 93 pp., Wiesbaden.

Meissner, B. (1953): Über den biologischen Abbau der Phenole. *Wasserwirtschaft-Wassertechn.*, **3**, pp. 470–473, Berlin.

Mekhtieva, V. L. (1962): Distribution of microorganisms in sheet-waters of the Volga region near Kuibyshev and the regions contiguous to it. *Geokhimiya*, **8**, pp. 707–719, Moscow.

Melnick, J. L., C. P. Gerba & C. Wallis (1978): Viruses in water. *Bull. WHO*, **56**, 4, pp. 499–508, Geneva.

Merrit, W. F. & C. A. Mawson (1967): Experiences with ground disposal at Chalk River. *Disposal of Radioactive Wastes into the Ground*. Vienna: IAEA, pp. 79–93.

Meyer, L. (1864): Chemische Untersuchung der Thermen zu Landeck in der Grafschaft Glatz. *J. Prakt. Chem.*, **91**, pp. 1–15, Leipzig.

Meyer, R. (1960): Auswertung zweigipfliger Überlebenskurven der *Salmonella paratyphi* B im Grundwasser. *Arch. Hyg. Bakteriol.*, **144**, pp. 564–568, Munich, Berlin.

Michel, G. (1965): Zur Mineralisation des tiefen Grundwassers in Nordrhein-Westfalen, Deutschland. *J. Hydrol.*, **3**, 2, pp. 73–87, Amsterdam.

Michel, G. (1968): Grundwasser vom Natrium-Hydrogencarbonat-Chlorid-Typ im Nordosten des Münsterschen Beckens (Nordrhein-Westfalen). *Bohrtech. Brunnenbau Rohrleitungsbau*, **19**, 1, pp. 5–14, Berlin.

Michel, G. & K. H. Rüller (1964): Hydrochemische Untersuchungen des Grubenwassers der Zechen der Hüttenwerke Oberhausen AG. *Bergbau-Archiv*, **25**, 4, pp. 21–27, Essen.

Milde, G. & H.-U. Mollweide (1969): Die wichtigsten Verschmutzungsgefahren für Grundwasserlagerstätten. *Z. Angew. Geol.*, **15**, pp. 17–25, Berlin.

Milde, G. & H.-U. Mollweide (1970): Hydrogeologische Faktoren bei der Grundwasserverunreinigung. *Wasserwirtschaft- Wassertechn.*, **20**, pp. 234–237, Berlin.

Miller, J. C. (1973): Nitrate contamination of the water-table aquifer by septic tank systems in the Coastal Plain of Delaware. Paper presented at the Rural Environment Conference, Warren, VT.

Miller, J. P. (1961): Solutes in small streams draining single rock types. Sangre de Cristo Range, New Mexico. U.S. Geological Survey Water Supply Paper No. 1535-F, 23 pp., Washington, D.C.

Mills, R. A. Van & R. C. Wells (1919): The evaporation and concentration of waters associated with petroleum and natural gas. U.S. Geological Survey Bulletin No. 693, 104 pp., Washington, D.C.

Mink, J. F. (1964): Groundwater temperatures in a tropical island environment. *J. Geophys. Res.*, **69**, 24, pp. 5225–5230, Richmond, VA.

Mintz, D. M. (1966): Modern theory of filtration. *Int. Water Supply Assoc.*, *7th Congr.*, **1**, Special Subject No. 10, 32 pp., Barcelona.

Moeller, J. (1879): Über die freie Kohlensäure im Boden. *Forsch. Geb. Agric. Phys.*, **2**, pp. 329–338, Heidelberg.

Mollweide, H.-U. (1971): Zur Frage der Beeinflussung des Grundwassers durch die Ablagerung fester Rückstandsstoffe. *Z. Ges. Hyg. Grenzgeb.*, **17**, 4, pp. 261–264, Berlin.

Morgan, C. O. & M. McNellis (1969): Stiff diagrams of water-quality data programmed for the digital-computer. State Geological Survey, University of Kansas, Lawrence, Special Distribution Publication No. 43, 27 pp.

Motts, W. S. & M. Saines (1969): The occurrence and characteristics of ground-water contamination in Massachusetts. University of Massachusetts, Amherst, Water Research Center, Publication No. 7.

Muffler, L. J. P. & D. E. White (1968): Origin of CO_2 in the Salton Sea geothermal system, southeastern California, U.S.A. *23rd Int. Geol. Congr.*, **17**, pp. 185–194, Prague.

Muhlert, F. (1930): *Der Kohlenschwefel*, 139 pp. Halle (Saale): Knapp.

Müller, G. (1971): Einheitliche Anforderungen an die Beschaffenheit, Untersuchung und Beurteilung von Trinkwasser in Europa. *Schr.-R. Ver. Wasser-, Boden- u. Lufthy.*, **14b**, 2nd ed., 50 pp., Stuttgart.

Müller, J. (1952): Bedeutsame Feststellungen bei Grundwasserverunreinigungen durch Benzin. *Gas- Wasserfach*, **93**, pp. 205–209, Munich.

Münnich, K. O. (1963a): Atombomben-Tritium als Indikator in der Hydrologie. *Phys. Bl.* **19**, pp. 418–421, Mosbach/Bd., (Physik).

Münnich, K. O. (1963b): Der Kreislauf des Radiokohlenstoffs in der Natur. *Naturwissenschaften*, **50**, pp. 211–218, Berlin, Heidelberg, New York: Springer.

Münnich, K. O. (1968): Isotopen-Datierung von Grundwasser. *Naturwissenschaften*, **55**, 4, pp. 158–163, Berlin, Heidelberg, New York: Springer.

Münnich, K. O. & W. Roether (1963): A comparison of carbon-14 and tritium ages of groundwater. *Radioisotopes in Hydrology*. Vienna: IAEA, pp. 397–406.

Münnich, K. O., W. Roether, & L. Thilo (1967): Dating of groundwater with tritium and ^{14}C. *Isotopes in Hydrology*. Vienna: IAEA, pp. 305–320.

Murray, J. & R. Irvine (1895): On the chemical changes which take place in the composition of the sea-water associated with blue muds on the floor of the ocean. *Trans. R. Soc. Edinburgh*, **37**, pp. 481–508, Edinburgh.

Nace, R. L. (1960): Water management, agriculture and ground water supplies. U.S. Geological Survey Circular No. 415, 12 pp., Washington, D.C.

Neumayr, V. (1979): *Schwermetallspuren und Ursachen ihrer Verbreitung in Grundwässern an der Westküste von Schleswig-Holstein*. Ph.D. thesis, Kiel.

Nielsen, H. & D. Rambow (1969): S-Isotopenuntersuchungen an Sulfaten hessischer Mineralwässer. *Notizbl. Hess. Landesamt Bodenforsch.*, **97**, pp. 352–366, Wiesbaden.

Nöring, F. (1951a): Ausgewählte Fragen der Grundwasserchemie in Beziehung zu Oberfläche und Untergrund. *Z. Dtsch. Geol. Ges.*, **102**, pp. 123–128, Hannover.

Nöring, F. (1951b): Einflüsse der Kunstdüngung auf den Chemismus des Grundwassers. *Gesund.-Ing.*, **72**, pp. 190–191, Munich.

Nöring, F. (1954): Chemische und physikalische Erscheinungen bei infiltriertem Grundwasser. *Int. Assoc. Sci. Hydrol.*, **37**, pp. 113–117, Gentbrugge.

Nöring, F. (1958): Contamination of ground water by oil wells. *Int. Assoc. Hydrol. Sci. C.R. Rapp. Assoc. Gen. Toronto 1957*, **2**, pp. 277–278, Gentbrugge.

Nöring, F. (1966): Unsere Reserven an Wasser, besonders Grundwasser und ihre Nutzung für die Trinkwasserversorgung. DVGW- Broschüre "Neue Aspekte zur Wassergewinnung," pp. 10–15, Munich.

Nöring, F., G. Farkasdi, A. Golwer, K. H. Knoll, G. Matthess & W. Schneider (1968): Über die Abbauvorgänge von Grundwasserverunreinigungen im Unterstrom von Abfalldeponien. *Gas- Wasserfach*, **109**, pp. 137–142, Munich.

Nowak, H. & F. Preul (1971): Untersuchungen über Blei- und Zinkgehalte in Gewässern des Westharzes. *Beih. Geol. Jahrb.*, **105**, 68 pp., Hannover.

Oborn, E. T. (1960): A survey of pertinent biochemical literature. Chemistry of iron in natural water. U.S. Geological Survey Water Supply Paper No. 1459-F, pp. 111–190, Washington, D.C.

Orborn, E. T. & J. D. Hem (1960): Microbiologic factors in the solution and transport of iron. U.S. Geological Survey Water Supply Paper No. 1459-H, pp. 213–235, Washington, D.C.

Ødum, H. & W. Christensen (1936): Danske Grundvandstyper og deres geologiske optraeden. *Dan. Geol. Unders.* (III), **26**, 184 pp., Copenhagen.

Oehler, K. E. (1969): with contrib. of K. Haberer & K.-E. Quentin: Technologie des Trink- und Betriebswassers. In: S. W. Souci & K.-E. Quentin, Eds., *Handbuch der Lebensmittelchemie*, Vol. 8, Part 1. Berlin, Heidelberg, New York: Springer, pp. 248–433.

Okamoto, G., T. Okura & K. Goto (1957): Properties of silica in water. *Geochim. Cosmochim. Acta*, **12**, pp. 123–132, London.

Orlow, J. E. (1930): Über ein wahres Mass der Aggressivität natürlicher Wässer. *Z. Anorg. u Allg. Chem.*, **191**, pp. 87–103, Leipzig.

Orlow, J. E. (1931): Über den Einflüsse neutraler Elektrolyte auf die Aggressivität des Wassers gegenüber Calciumcarbonat. *Z. Anorg. Allg. Chem.*, **200**, pp. 87–104, Leipzig.

Osnizkaja, L. K. (1958): Microbiological study of oil field waters of the East Karnat District. *Mikrobiologija*, **27**, pp. 478–483, Moscow (in Russian with English abstract).

Ottermann, A. & G. Krzysch (1965): Der Gehalt der Niederschläge an Stickstoff, Phosphor, Kalium, Calcium und Sulfatschwefel (II. Beitrag). *Z. Pflanzenernähr. Düng. Bodenkd.*, **111**, pp. 122–131, Weinheim Bergstrasse.

Overman, A. R. (1973): Landspreading municipal effluent and sludge in Florida. *Proceedings of the 1973 Workshop*, University Florida.

Paces, T. (1972): Chemical characteristics and equilibration in natural water–felsic rock–CO_2 system. *Geochim. Cosmochim. Acta*, **36**, pp. 217–240.

Paces, T. (1973): Steady-state kinetics and equilibrium between ground water and granitic rocks. *Geochim. Cosmochim. Acta*, **37**, pp. 2641–2663.

Packer, P. E. (1967): Forest treatment effects on water quality. In: W. E. Sopper & H. W. Lull, Eds., *Forest Hydrology*, Vol. 2. Oxford: Pergamon, pp. 687–699.

Palmer, C. (1911): The geochemical interpretation of water analyses. U.S. Geological Survey Bulletin No. 479, 31 pp., Washington, D.C.

Palmer, C. (1924): California oil field waters. *Econ. Geol.*, **19**, pp. 623–635, New Haven, CT.

Parizek, R. R. (1973): Site selection criteria for wastewater disposal soils and hydrogeologic considerations. In: W. E. Sopper & L. T. Kardos, Eds., *Recycling Treated Municipal Waste-Water and Sludge through Forest and Cropland*. University Park: Pennsylvania State University, pp. 95–147.

Parizek, R. R. & E. A. Myers (1968): Recharge of groundwater from renovated sewage effluent by spray irrigation. *Proceedings of the Fourth American Water Resources Conference*, New York, pp. 426–443.

Parizek, R. R., L. T. Kardos, W. E. Sopper, E. A. Myers, E. D. Davis, M. A. Farrell & J. B. Nesbitt (1967): Waste water renovation and conservation. Pennsylvania State University Studies, No. 23, University Park.

Parizek, R. R., W. B. White & D. Langmuir (1971): *Hydrogeology and Geochemistry of Folded and Faulted Rocks of the Central Appalachian Type and Related Land-Use Problems*, Guidebook of the Geological Society of America.

Parks, G. (1975): Adsoprtion in the marine environment. In: J. P. Riley & G. Skirrow, Eds., *Chemical Oceanography*, Vol. 1, 2nd ed. New York: Academic Press, pp. 241–308.

Patteisky, K. (1952): Die thermalen Salzsolen des Ruhrgebietes und ihre Quellgase. *Z. Dtsch. Geol. Ges.*, **104**, pp. 532–533, Hannover.

Patterson, R. J., S. K. Frape, L. S. Dykes & R. A. McLeod (1978): A coring and squeezing technique for the detailed study of subsurface water chemistry. *Can. J. Earth Sci.*, **15**, pp. 162–169.

Paul, M. (1939): Erfahrungen mit einem neuen geothermischen Aufschlussverfahren. *Z. Geophys.*, **15**, pp. 88–93, Brunswick.

Pauling, L. (1968): *Die Natur der chemischen Bindung,* 3rd ed., 620 pp. Weinheim, Bergstrasse: Chemie.

Pearson, F. J. & B. B. Hanshaw (1970): Sources of dissolved carbonate species in groundwater and their effects on carbon-14 dating. *Isotope Hydrology 1970.* Vienna: IAEA, pp. 271–286.

Pekdeger, A. (1977): *Labor- und Felduntersuchungen zur Genese der Sicker- und Grundwasserbeschaffenheit,* 229 pp. Ph.D. thesis, Kiel.

Pekdeger, A. (1979): Labor- und Felduntersuchungen zur Genese der Sicker- und Grundwasserbeschaffenheit. *Meyniana,* **31**, pp. 25–57, Kiel.

Penman, H. L. (1940a): Gas and vapour movements in the soil. I. The diffusion of vapours through porous solids. *J. Agric. Sci.,* **30**, pp. 437–462, London.

Penman, H. L. (1940b): Gas and vapour movements in the soil. II. The diffusion of carbon dioxide through porous solids. *J. Agric. Sci.,* **30**, pp. 570–581, London.

Pentcheva, E. N. (1967a): Sur certaines particularités hydrogéochimiques des éléments alcalins rares. *Bull. Soc. Fr. Mineral. Cristalogs.,* **40**, pp. 402–406, Paris.

Pentcheva, E. N. (1967b): Sur les particularités hydrochimiques de certains oligoéléments des eaux naturelles. *Ann. Idrol.,* **5**, pp. 90–113, Parma.

Perel'Man, A. I. (1963): The aquatic transport and the biologic absorption of elements in the soil. In: M. Schoeller, *Recherches sur l'Acquisition de la Composition Chimique des Eaux Souterraines.* Bordeaux: pp. 179–187.

Perlmutter, N. M., M. Lieber & H. L. Frauenthal (1963): Movement of waterborne cadmium and hexavalent chromium wastes in South Farmingdale, Nassau County, Long Island, New York, U.S. Geological Survey Professional Paper No. 475-C, pp. 179–184, Washington, D.C.

Perry, J. H. (1950): *Chemical Engineers' Handbook,* 3rd ed., 1942 pp. New York, Toronto, London: McGraw-Hill.

Peter, H. (1970): Bakteriologie des Trinkwassers. In: K. Höll, Ed., *Wasser: Untersuchung, Beurteilung, Aufbereitung, Chemie, Bakteriologie, Biologie.* Berlin: de Gruyter, pp. 301–375.

Pettenkofer, M. von (1871): Über den Kohlensäuregehalt der Grundluft im Geröllboden von München in verschiedenen Tiefen und zu verschiedenen Zeiten. *Z. Biol.,* **7**, pp. 395–417, Munich.

Pettenkofer, M. von (1973): *Z. Biol.,* **9**, pp. 250–257, Munich.

Pettyjohn, W. A. (1972): *Water Quality in a Stressed Environment.* Minneapolis: Burgess Publ. Co.

Pfaff, C. (1937): Über Lysimeter-Versuche. *Forschungsdienst, N. F., Dtsch. Landwirtsch. Randsch., Sonderh.* **6**, pp. 102–114, Neudamm.

Pfeiffer, D. (1962): Hydrologische Messungen in der Praxis des Geologen. Einfache Verfahren. *Bohrtech. Brunnenbau Rohrleitungsbau,* **13**, pp. 53–60, 96–104, 147–162, Berlin.

Pfeilsticker, K. (1937): Die Spektralanalyse in der Wasserchemie. *Vom Wasser,* **11**, pp. 238–250, Berlin.

Philipp, G. & K.-E. Quentin (1969): Zur toxischen Wirkung von Pestiziden auf Abwassermikroorganismen. *Z. Wasser Abwasser Forsch.,* **2**, 1, pp. 23–24, Munich.

Pierau, H. (1967): Kritische Bemerkungen zum "Schichttorten-Modell" der geordneten Ablagerung. *Städtetag,* **10,** pp. 528–584, Stuttgart.

Pilpel, N. (1968): Das natürliche Schicksal von Öl auf dem Meere. *Endeavour,* **27,** 100, p. 11–13, London.

Pinder, G. F. (1973): A Galerkin–finite element simulation of groundwater contamination on Long Island. *Water Resour. Res.,* **9,** pp. 1657–1669, Washington, D.C.

Pinneker, E. V. (1967): Die Bestimmung der Fliessrichtung und Fliessgeschwindigkeit unterirdischer Wässer in tiefen Schnichten der Tafelländer. *Steir. Beitr. Hydrogeol.,* **18–19,** pp. 289–310, Graz.

Pinneker, E. V. (1968): The problem of the formation of underground concentrated brines. *23rd Int. Geol. Congr.,* **17,** pp. 95–99, Prague.

Piper, A. M. (1944): A graphic procedure in the geochemical interpretation of water analyses. *Trans. Am. Geophys. Union,* **25,** 6, pp. 914–928, Washington, D.C.

Piper, A. M., A. A. Garret et al. (1953): Native and contaminated ground waters in the Long Beach–Santa Ana area, California. U.S. Geological Survey Water Supply Paper No. 1136, 320 pp., Washington, D.C.

Pittner, K. (1952): Der Schwefeldioxidgehalt der Luft und die Schäden in der Umgebung von Industrieschornsteinen. *Arb Sozialfürsorge,* **7,** pp. 337–341, Villingen.

Plauchud, E. (1877): Recherches sur la formation des eaux sulfureuses naturelles. *C.R. Acad. Sci.,* **84,** pp. 235–237, Paris.

Plauchud, E. (1882): Sur la réduction des sulfates par les sulfuraires et sur la formation des sulfures métalliques naturels. *C.R. Acad. Sci.,* **95,** pp. 1363–1367, Paris.

Pluhowski, E. J. & I. H. Kantrowitz (1963): Influence of land-surface conditions on groundwater temperatures in southwestern Suffolk County, Long Island, New York. U.S. Geological Survey Professional Paper No. 475-B, pp. 186–188, Washington, D.C.

Plummer, L. N. & T. M. L. Wigley (1976): The dissolution of calcite in CO_2-saturated solutions at 25°C and 1 atmosphere total pressure. *Geochim. Cosmochim. Acta,* **40,** pp. 191–202.

Plummer, L. N., B. F. Jones & A. H. Truesdell (1976): WATEQF – A FORTRAN IV version of WATEQ, a computer program for calculating chemical equilibrium of natural waters. *U.S. Geol. Surv. Water Res. Invest.,* **76–13,** 61 pp., Washington, D.C.

Poland, J. F., A. A. Garrett & A. Sinnott (1959): Geology, hydrology, and chemical character of ground waters in the Torrance–Santa Monica area, California. U.S. Geological Survey Water Supply Paper No. 1461, 425 pp., Washington, D.C.

Porter, J. R. (1950): *Bacterial Chemistry and Physiology,* 5th ed., 1073 pp. New York: Wiley, London: Chapman & Hall.

Potilitzin, A. (1882): Zusammensetzung des die Naphtha begleitenden und aus Schlamm Vulkanen ausströmenden Wassers. *Ber. Dtsch. Chem. Ges.,* **15,** p. 3099, Berlin. (Original in *J. Russ. Phys.-Chem. Soc.,* **1,** p. 300, 1882.)

Pourbaix, M. (1945): Etude graphique du traitement des eaux par la chaux. *Bull. Soc. Chim. Belg.,* **54,** pp. 10–42, Ghent.

Pourbaix, M. J. N. (1949): *Thermodynamics of Dilute Aqueous Solutions,* 136, pp. London.

Preul, H. C. & G. J. Schroepfer (1968): Travel of nitrogen in soils. *J. Water Pollut. Control Fed.,* **40,** 1, pp. 30–48, Washington, D.C.

Prinz, E. (1923): *Handbuch der Hydrologie,* 2nd ed., 422 pp. Berlin: Springer.

Proctor, J. F. & I. W. Marine (1965): Geologic, hydrologic and safety considerations in the storage of radioactive wastes in a vault excavated in crystalline rock. *Nuclear Sci. Eng.,* **22,** 3, pp. 350–365, New York.

Puchelt, H. (1964): Zur Geochemie des Grubenwassers im Ruhrgebiet. *Z. Dtsch. Geol. Ges.,* **116,** pp.' 167–203, Hannover.

Quentin, K.-E. (1957): Der Fluorgehalt bayerischer Wässer. 1. Mitteilung. *Gesund.-Ing.,* **78,** 21–22, pp. 329–333, Munich.

Quentin, K.-E. (1969): Beurteilungsgrundsätze und Anforderungen an Mineral- und Heilwässer. In: S. W. Souci & K.-E. Quentin, Eds., *Handbuch der Lebensmittelchemie,* Vol. 8, Part 2. Berlin, Heidelberg, New York: Springer, pp. 1043–1056.

Quentin, K.-E. & L. Feiler (1968): Selen in Grund- und Oberflächenwässern. *Vom Wasser,* **34,** pp. 19–30, Weinheim, Bergstrasse.

Quentin, K.-E. & F. Pachmayr (1962): Colorimetrische Sulfidbestimmung im Wasser über Methylenblau. *Vom Wasser,* **28,** pp. 79–93, Weinheim, Bergstrasse.

Quentin, K.-E., L. Weil & P. Udluft (1973): Grundwasserverunreinigungen durch organische Umweltchemikalien. *Z. Dtsch. Geol. Ges.,* **124,** pp. 417–424, Hannover.

Rand, M. C., A. E. Greenberg & M. A. Franson (1975): *Standard Methods for the Examination of Water and Wastewater,* 14th ed. Washington, D.C.

Raney, W. A. (1950): Field measurement of oxygen diffusion through soil. *Soil Sci. Soc. Am. Proc.,* **14** (1949), pp. 61–63, Madison, WI.

Rankama, K. (1954): *Isotope Geology,* 535 pp. New York: McGraw-Hill; London: Pergamon.

Rankama, K. & T. G. Sahama (1960): *Geochemistry,* 2nd ed., 912 pp. Chicago.

Reardon, E. J. & P. Fritz (1978): Computer modelling of groundwater ^{13}C and ^{14}C isotope compositions. *J. Hydrol.,* **36,** pp. 201–224, Amsterdam.

Reichardt, E. (1880): *Grundlagen zur Beurteilung des Trinkwassers,* 3rd ed., 170 pp. Halle (Saale).

Reichert, S. O. (1962): Radionuclides in groundwater at the Savannah River Plant waste disposal facilities. *J. Geophys. Res.,* **67,** 11, pp. 4363–4374, Richmond, VA.

Reimann, K. (1969): Untersuchungen über den Mechanismus der toxischen Hemmung des Belebtschlammes durch ein Schwermetallion, dargestellt am Beispiel des Kobalts. *Z. Wasser Abwasserforsch.,* **2,** 1, pp. 25–35, Munich.

Reimer, H. & T. Rossi (1970): Zur Emission von Chlorwasserstoff bei der Verbrennung von Hausmüll. *Müll Abfall,* **2,** 3, pp. 71–74, Berlin, Bielefeld, Munich.

Reissig, H. (1966): Untersuchungen über die ^{90}Sr-Auswaschung in Böden. *Chem. Erde,* **25,** 3, pp. 204–229, Jena.

Reiter, R. (1964): with contribution of B. Rajewski: *Felder, Ströme und Aerosole in der unteren Troposphäre,* 24 + 603 pp. Darmstadt: Steinkopf.

Renick, B. C. (1924a): Base exchange in ground water by silicates as illustrated in Montana. U.S. Geological Survey Water Supply Paper No. 520-D, pp. 53–72, Washington, D.C.

Renick, B. C. (1924b): Some geochemical relations of ground water and associated natural gas in the Lance Formation, Montana. *J. Geol.,* **32,** pp. 668–684, Chicago.

Reploh, H. (1969): Mikrobiologische Untersuchung und Beurteilung des Trink- und Betriebswassers. In: S. W.Souci & K.-E. Quentin, Eds., *Handbuch der Lebensmittelchemie,* Vol. 8, Part 2. Berlin, Heidelberg, New York: Springer, pp. 1079–1146.

Richards, E. H. (1917): Dissolved oxygen in rain water. *J. Agric. Sci.,* **8,** 3, pp. 331–337, London.

Richards, L. A. (1954): *Diagnosis and Improvement of Saline and Alkali Soils. U.S. Department of Agriculture Handbook,* Vol. 60, 160 pp. Washington, D.C.

Richter, D. (1965): Über den Verbleib des Fallout-^{90}Sr im Boden. *Chem. Erde,* **24,** 1, pp. 67–76, Jena.

Richter, W. & H. Flathe (1954): Die Versalzung von küstennahen Grundwässern, dargestellt an einem Teil der deutschen Nordseeküste. *Int. Assoc. Sci. Hydrol. Publ.* **37**, pp. 118–130, Rome.

Richter, W. & R. Wager (1969) with contributions of H. Anrich, K. Deppermann, H. J. Dürbaum, H. Fauth, M. Geyh, & E. Groba: *Hydrogeologie.* In: A. Bentz: Ed., *Lehrbuch der angewandten Geologie,* Vol. 2, Part 2. Stuttgart: Enke, pp. 1357–1546.

Ricke, W. (1961): Ein Beitrag zur Geochemie des Schwefels. *Geochim. Cosmochim. Acta,* **21**, pp. 35–80, Oxford, London, New York.

Riehm, H. (1961): Die Bestimmung der Pflanzennährstoffe im Regenwasser und in der Luft unter besonderer Berücksichtigung der Stickstoffverbindungen. *Agrochimica,* **5**, pp. 174–188, Pisa.

Riffenburg, H. B. (1925): Chemical character of ground waters of the Northern Great Plains. U.S. Geological Survey Water Supply Paper No. 560-B, pp. 31–52, Washington, D.C.

Robie, R. A., B. S. Hemingway & J. R. Fisher (1978): Thermodynamic properties of minerals and related substances at 298.15 K and 1 bar (10^5 pascals) pressure and at higher temperature. U.S. Geological Survey Bulletin No. 1452, 456 pp., Washington, D.C.

Robinson, R. A. & R. H. Stokes (1949): The role of hydration in the Debye-Hückel theory. *Ann. N.Y. Acad. Sci.,* **51**, 4, pp. 593–604, New York.

Robinson, R. A. & R. H. Stokes (1970): *Electrolyte Solutions,* 2nd ed. (rev.), 571 pp. London: Butterworths.

Robinson, W. O., H. Bastron & K. J. Murata (1958): Biogeochemistry of the rare-earth elements with particular reference to hickory trees. *Geochim. Cosmochim. Acta,* **14**, pp. 55–67, Oxford.

Rodis, H. G. & R. Schneider (1960): Occurrence of ground waters of low hardness and of high chloride content in Lyon County, Minnesota. U.S. Geological Survey Circular No. 423, 2 pp., Washington, D.C.

Roether, W. (1967a): *Tritium im Wasserkreislauf,* 67 pp. Ph.D. thesis, Heidelberg.

Roether, W. (1967b): Estimating the tritium input to groundwater from wine samples: Groundwater and direct run-off contribution to Central European surface waters. *Isotopes in Hydrology.* Vienna: IAEA, pp. 73–91.

Roether, W. (1970): Tritium und Kohlenstoff-14 im Wasserkreislauf. *Z. Dtsch. Geol. Ges., Sonderh. Hydrogeol. Hydrogeochem.,* pp. 183–192, Hannover.

Rogers, A. S. (1958): Physical behavior and geologic control of radon in mountain streams. U.S. Geological Survey Bulletin No. 1052-E, pp. 187–211, Washington, D.C.

Rogers, G. S. (1917): Chemical relations of the oil-field waters in San Joaquin Valley, California. U.S. Geological Survey Bulletin No. 653, 110 pp., Washington, D.C.

Röhrer, F. (1933): Über den Nitratgehalt der Tiefenwässer. *Geol. Rundsch.,* **23a**, Salomon-Calvi-Festschrift, pp. 315–331, Stuttgart.

Romell, L.-G. (1922): Luftväxlingen i marken som ekologisk faktor. *Medd. Statens Skogsförsöksanstalt,* **19**, 2, pp. 125–359, Stockholm.

Rossini, F. D., D. D. Wagman, W. H. Evans, S. Levine & I. Jaffe (1952): Selected values of chemical thermodynamic properties. U.S. National Bureau of Standards, Circular No. 500, 1268 pp., Washington, D.C.

Rossini, F. D., K. S. Pitzer, R. L. Arnett, R. M. Braun & G. C. Pimentel (1953): Selected values of physical and thermodynamic properties of hydrocarbons and related compounds. *Tables of the American Petroleum Institute Research Project,* Vol. 44, 1050 pp. Pittsburgh: Carnegie Press.

Rössler, B. (1951): Beeinflussung des Grundwassers durch Müll und Schuttablagerungen. *Vom Wasser,* **18**, pp. 43–60, Weinheim, Bergstrasse.

Roy, C. J. (1945): Silica in natural waters. *Am. J. Sci.,* **243**, pp. 393–403, New Haven, CT.

Rubin, A. J. (1974): *Chemistry of Water Supply, Treatment, and Distribution.* 446 pp. Ann Arbor, MI.

Runnels, D. D. (1969): Diagenesis, chemical sediments, and the mixing of natural waters. *J. Sediment. Petrol.,* **39,** pp. 1188–1201, Tulsa, OK.

Russel, E. J. & A. Appleyard (1915): The atmosphere of the soil: Its composition and the causes of variation. *J. Agric. Sci.,* **7,** 1, pp. 1–48, London.

Rutten, M. G. (1949): Exchange of cations in some Dutch subterranean waters. I. On water rich in sodium from the subsoil of the Dutch dunes. II. Underground waters from the South Limburg coalmine district. *Geol. Mijnbouw,* **11,** pp. 139–145, 165–171, 'sGravenhage.

Sachs, J. (1860): Physiologische Mitteilungen verschiedenen Inhalts. II. Auflösung des Marmors durch Maiswurzeln. *Bot. Zeit.,* **18,** pp. 117–119, Leipzig.

Sachs, J. (1865): *Handbuch der Experimentalphysiologie der Pflanzen,* 514 pp. Leipzig: Engelmann.

Sarles, W. B., W. C. Frazier, J. B. Wilson & S. G. Knight (1956): *Microbiology,* 2nd ed., 491 pp. New York: Harper & Row.

Scharpff, H.-J. (1972): *Die Mineralwässer der Wetterau (Hessen). Hydrogeologische und hydrochemische Untersuchungen im Niederschlagsgebiet der Nidda,* 256 pp. Ph.D. thesis, Darmstadt.

Schatenstein, A. I. (1960) with cooperation of E. A. Jakowlewa, E. N. Swjaginzewa, J. M. Warschawski, E. A. Israilewitsch & N. M. Dychno: *Isotopenanalyse des Wassers,* 270 pp. Berlin: VEB Verl. Wiss.

Scheffer, F. & P. Schachtschabel (1976): *Lehrbuch der Bodenkunde,* 9th ed., 394 pp. Stuttgart: Enke.

Schinzel, A. (1968): Das Verhalten schwer abbaubarer Substanzen im Boden und Grundwasser. *Gas-Wasser-Wärme,* **22,** 2, pp. 23–32, Vienna.

Schloesing, T. (1870): Analyse des eaux contenues dans les terres arables. *C.R. Acad. Sci.,* **70,** pp. 98–102, Paris.

Schloesing, T. (1873): Etude de la nitrification dans les sols. *C.R. Acad. Sci.,* **77,** pp. 203–207, Paris.

Schloesing, T. (1889): Sur l'atmosphère contenue dans le sol. *C.R. Acad. Sci.,* **109,** pp. 673–676, Paris.

Schmalz, B. L. (1961): Operating practices, experiences, and problems at the National Reactor Testing Station (Idaho), in ground disposal of radioactive wastes. *Conference Proceedings,* pp. 17–33, Sanitary Engineering Research Laboratory, University of California, Berkeley.

Schmidt, B. & W. D. Kampf (1961): Quantitative Bestimmung der Nitrat- und Nitritreduktion durch Bakterien aus Oberflächenwasser. *Gesund.-Ing.,* **82,** pp. 339–343, Munich.

Schmidt, K. (1963): Die Abbauleistungen der Bakterienflora bei der Langsamsandfiltration und ihre Beeinflussung durch die Rohwasserqualität und andere Umwelteinflüsse. Ph.D. thesis, reprinted in *Veröff. Hydrolog. Forsch. Abt. Dortmunder Stadtwerke* AG, **5,** 170 pp., Dortmund.

Schmidt, K. D. (1974): Nitrates and groundwater management in the Fresno urban area. *J. Am. Water Works Assoc.,* **66,** pp. 146–148.

Schmidt, W. & P. Lehmann (1929): Versuche zue Bodenatmung. *Sitzb. Akad. Wiss.,* Part IIa, **138,** pp. 823–852, Vienna.

Schmitz, W. (1956): Der Mineralgehalt der Oberflächengewässer des Fulda-Eder-Flussgebietes. *Ber. Limnol. Flusstat. Freudenthal, Aussenst. Hydrobiol. Anst. Max-Planck-Ges.,* **7,** pp. 43–60, (Max-Planck-Ges. Dok.-Stelle).

Schneider, H. (1964): *Geohydrologie Nordwestfalens,* 264 pp. Berlin, Konradshöhe: Schmidt.

Schneider, R. (1962): An application of thermometry to the study of ground water. U.S. Geological Survey Paper No. 1544-B, 16 pp., Washington, D.C.

Schoeller, H. (1934): Les échanges de bases dans les eaux souterraines vadoses: Trois exemples en Tunisie. *Bull. Soc. Geol. Fr.*, 4, 5, pp. 389–420, Paris.

Schoeller, H. (1935): Utilité de la notion des échanges de bases pour la comparaison des eaux souterraines. *Bull. Soc. Géol. Fr.*, 5, 5, pp. 651–657, Paris.

Schoeller, H. (1938): Notions sur la corrosion interne des canalisations d'eau. *Ann. Ponts Chaussées Mem.* 8, pp. 199–282, Paris.

Schoeller, H. (1941): L'influence du climat sur la composition chimique des eaux souterraines vadoses. *Bull. Soc. Géol. Fr.*, 11, pp. 267–289, Paris.

Schoeller, H. (1945): Etude sur la température des eaux souterraines (cas des nappes). *P. V. Seances Soc. Sci. Phys. Nat.*, Bordeaux, pp. 45–50.

Schoeller, H. (1949): *Cours d'Hydrogéologie*, 364 pp. Rueil: Ecole National Supérieure des Pétroles.

Schoeller, H. (1950): Les variations de la teneur en gaz carbonique des eaux souterraines en fonction de l'altitude. *C.R. Acad. Sci.*, 230, pp. 560–561, Paris.

Schoeller, H. (1951): Relation entre la concentration en chlore des eaux souterraines et les échanges de bases avec les terrains qui les renferment. *C.R. Acad. Sci.*, 232, pp. 1432–1434, Paris.

Schoeller, H. (1956): Géochimie des eaux souterraines. Application aux eaux de gisements de petrole. *Rev. Inst. Pétrol. Ann. Combust. Liq.*, 10, (1955), pp. 181–213, 219–246, 507–552, 671–719, 823–874, Paris.

Schoeller, H. (1962): *Les Eaux Souterraines*, 642 pp. Paris: Masson.

Schoeller, M. (1963): *Recherches sur l'Acquisition de la Composition Chimique des Eaux Souterraines*, 231 pp. Bordeaux: Drouillard.

Schofield, R. K. (1940): Clay mineral structures and their physical significance. *Trans. Br. Ceram. Soc.*, 39, pp. 147–158, Stoke-on-Trent.

Schofield, R. K. (1949): Calculation of surface area of clays from measurements of negative adsorption. *Trans. Br. Ceram. Soc.*, 48, pp. 207–213, Stoke-on-Trent.

Schubert, J. (1930): Das Verhalten des Bodens gegen Wärme. In: E. Blanck, Ed., *Handbuch der Bodenlehre*, Vol. 6. Berlin: Springer, pp. 342–375.

Schulz, H. D. (1970): *Chemische Vorgänge beim Übergang vom Sickerwasser zum Grundwasser*, 114, pp. Ph.D. thesis, Aachen.

Schulz, H. D. (1977): Die Grundwasserbeschaffenheit der Geest Schleswig-Holsteins Eine statistische Auswertung. *Bes. Mitt. Dtsch. Gewässerk. Jahrb.* 40, 141 pp., Kiel.

Schulze, W. W. & K. Haberer (1966): Das Verhalten von Radionukliden im Boden, eine Literaturstudie. *Vom Wasser*, 32, pp. 69–127, Weinheim, Bergstrasse.

Schuphan, W. (1971): Potentielle und reale Gefahren von Düngungsmassnahmen und Pestizideinsatz für die Umwelt. *Schr.-R. Ver. Wasser-, Boden- Lufthyg.*, 34, pp. 35–50, Stuttgart.

Schwartz, H. G. (1967): Microbial degradation of pesticides in aqueous solution. *J. Water Pollut. Control Fed.*, 19, pp. 1701–1714, Washington, D.C.

Schwartz, W. (1958): Die Bakterien des Schwefelkreislaufes und ihre Lebensbedingungen. *Freiberger Forsch.-H.*, C44, pp. 1–13, Berlin.

Schweisfurth, R. (1966): Eisen- und Mangan-Mikroben im Wasserwerksbetrieb. Veröff. Abt. u. Lehrstuhl Wasserchemie, 1. Vortr.R., H. 1, pp. 199–217, Karlsruhe.

Schweisfurth, (1968): Untersuchungen über manganoxidierende und -reduzierende Mikroorganismen. *Mitt. Int. Ver. Limnol.*, 14, pp. 179–186, Stuttgart.

Schwille, F. (1953a): Natriumhydrogenkarbonat- und Natriumchlorid- Wässer im tieferen Untergrund des Mainzer Beckens. *Notizbl. Hess. Landesamt Bodenforsch.*, **81**, pp. 314–335, Wiesbaden.

Schwille, F. (1953b): Chloride und Nitrate in den Grundwässern Rheinhessens und des Rheingaues. *Gas- Wasserfach*, **94**, 14, pp. 184–188, Munich.

Schwille, F. (1955): Ionenumtausch und der Chemismus von Grund- und Mineralwässern. *Z. Dtsch. Geol. Ges.*, **106**, 1, (1954), pp. 16–22, Hannover.

Schwille, F. (1957); Das Härtedreieck. *Gas- Wasserfach*, **98**, 12, pp. 280–282, Munich.

Schwille, F. (1962): Nitrate im Grundwasser. *Dtsch Gewässerkd. Mitt.*, **6**, 2, pp. 25–32, Coblenz, Rhein.

Schwille, F. (1964): Die "hydrologischen" Grundlagen für die Untersuchung, Beurteilung Sanierung von Mineralölkontaminationen des Untergrundes. *Dtsch. Gewässerkd. Mitt.*, **8**, 1, pp. 1–16, Coblenz, Rhein.

Schwille, F. (1966): Die Kontamination des Untergrundes durch Mineralöl – Ein hydrologisches Problem. *Dtsch. Gewässerkd. Mitt.*, **10**, 6, pp. 194–207, Coblenz, Rhein.

Schwille, F. (1969): Hohe Nitratgehalte in den Brunnenwässern der Moseltalaue zwischen Trier und Koblenz. *Gas- Wasserfach*, **110**, 2, pp. 35–44, Munich.

Schwille, F. (1976): Anthropogenically reduced groundwaters. *Hydrol. Sci. Bull.*, **21**, 4, pp. 629–645, Reading, PA.

Schwille, F. & C. Vorreyer (1969): Durch Mineralöl "reduzierte" Grundwässer. *Gas-Wasserfach*, **110**, 44, pp. 1225–1232, Munich.

Schwille, F. & D. Weisflog (1968): *Die Wasserstoffionenkonzentration der Grundwässer*, 7 pp. Coblenz: Bundesanstalt für Gewässerkunde.

Schwoerbel, J. (1961): Entstehung von Grundwasser – Arten bei Süsswassermilben (Hydrachnellae) und die Bedeutung der parasitischen Larvenphase. *Naturwissenschaften*, **48**, pp. 309–310, Berlin.

Scott, R. C. & F. B. Barker (1961): Ground-water sources containing high concentrations of radium. U.S. Geological Survey Professional Paper No. 424-D, pp. 357–359, Washington, D.C.

Scott, R. C. & F. B. Barker (1962): Data on uranium and radium in ground water, in the United States 1954 to 1959. U.S. Geological Survey Professional Paper No. 426, 115 pp., Washington, D.C.

Seelmann-Eggebert, W., G. Pfennig & H. Münzel (1968): *Nuklidkarte*, 3rd ed., 28 pp. Bonn: B.-Min. Wiss. Forsch.

Semmler, W. (1958): Die Halden – ein hydrologisches Problem. *Schlägel Eisen*, **9**, pp. 694–698, Düsseldorf.

Semmler, W. (1960): Verdorbenes Grundwasser. *Schlägel Eisen*, **8**, pp. 534–541, Düsseldorf.

Semmler, W. (1961): Unreines Grundwasser und Grundwasserentzug als vorübergehender Bergschaden. *Bergbauwissenschaften*, **8**, pp. 565–568, Goslar.

Shedlowsky, T. & D. A. MacInnes (1935): The first ionization constant of carbonic acid, 0 to 38°, from conductance measurements. *J. Am. Chem. Soc.*, **57**, pp. 1705–1710, Easton, PA.

Shyets, V. M. (1964): Organic matter in ground waters. *Int. Assoc. Hydrogeol. Mem. 5*, pp. 414–423, Athens.

Siebert, G. & H. Werner (1969): Bergeverkippung und Grundwasserbeeinflussung am Niederrhein. *Fortschr. Geol. Rheinld. Westf.* 17, pp. 263–278, Krefeld.

Sieper, H. (1971): Bestimmung und Abbau von Dithianon, Pyridinitril und Flurenol in Böden und Gewässern. *Schr.-R. Ver. Wasser- Boden- Lufthyg.*, **34**, pp. 109–115, Stuttgart.

Sierp, F. (1939a): Häusliches und städtisches Abwasser. In: S. W. Souci & K.-E. Quentin, Eds., *Handbuch der Lebensmittelchemie*, Vol. 8, Part 1. Berlin: Springer, pp. 209–470.

Sierp, F. (1939b): Gewerbliche und industrielle Abwässer. In: S. W. Souci & K.-E. Quentin, Eds., *Handbuch der Lebensmittelchemie,* Vol. 8, Part 1. Berlin: Springer, pp. 471–670.

Silvey, W. D. (1967): Occurrence of selected minor elements in the waters of California. U.S. Geological Survey Water Supply Paper No. 1535-L, 25 pp., Washington, D.C.

Skougstad, M. W. & F. B. Barker (1960): Occurrence and behaviour of natural and radioactive strontium in water. *Public Works,* **91,** pp. 88–90, New York.

Skougstad, M. W. & C. A. Horr (1960): Occurrence of strontium in natural water. U.S. Geological Survey Circular No. 420, 6 pp., Washington, D.C.

Skougstad, M. W., M. J. Fishman, L. C. Friedman, D. E. Erdmann & S. S. Duncan (1970): Methods for determination of inorganic substances in water and fluvial sediments. *Techniques of Water Resources Investigations.* U.S. Geological Survey, Washington, D.C.

Slichter, C. S. (1902): The motions of underground waters. U.S. Geological Survey Water Supply Paper, 67, 106 pp., Washington, D.C.

Smirnova, Z. S. (1961): Composition of species and some physiological properties of bacteria, which may act as indicators for oil and gas exploration. *Mikrobioligija,* **03,** pp. 684–687, Moscow (in Russian with English abstract).

Smirnow, S. S. (1954): *Die Oxydationszone sulfidischer Lagerstätten,* 312 pp. Berlin: Akademie Verl.

Smith, B. M., W. N. Grune, F. B. Higgins, Jr. & J. G. Terrill, Jr. (1961): Natural radioactivity in ground water supplies in Maine and New Hampshire. *J. Am. Water Works Assoc.,* **53,** pp. 75–88, New York.

Smith, D. B., P. L. Wearn, H. J. Richards & P. C. Rowe (1970): Water movement in the unsaturated zone of high and low permeability strata using natural tritium. *Isotope Hydrology 1970.* Vienna: IAEA, pp. 73–87.

Smith, F. G. (1953): Review of physico-chemical data on the state of supercritical fluids. *Econ. Geol.,* **48,** 1, pp. 14–38, New Haven, CT.

Smith, M. J. E. & M. T. Sutton Bowman (1930): Note on the destruction of alkali carbonates by organic acids in the ground waters of oilfields. *Congrès International des Mines, Metallurgie et Géologie Appliquée,* Section Geologia, VI. Section, pp. 259–261, Lüttich.

Smolensky, P. (1877): Über den Kohlensäuregehalt der Grundluft. *Z. Biol.,* **13,** pp. 383–394, Munich.

Sobolev, J. A., L. M. Chomcik, J. M. Bazenov, N. V. Polochina & A. G. Nazarjuk (1967): The experimental disposal of low-level radioactive wastes in clay soil at the Moscow disposal centre. *Disposal of Radioactive Wastes into the Ground.* Vienna: IAEA, pp. 37–47, (in Russian).

Sofer, Z., & J. R. Gat (1975): The isotope composition of evaporating brines: Isotope activity ratio in saline solutions. *Earth Planet. Sci. Lett.,* **26,** pp. 179–186, Amsterdam.

Söhngen, N. L. (1913): Benzin, Petroleum, Paraffinöl und Paraffin als Kohlenstoff- und Energiequelle für Mikroben. *Zentralbl. Bakt. Parasitenkd. Infektionsk.,* Part 2, **37,** pp. 595–608, Jena.

Sopper, W. E. & L. T. Kardos, Eds. (1973): *Recycling Treated Municipal Waste-water and Sludge Through Forest and Cropland.* University Park: Pennsylvania State University Press.

Sosman, R. B. (1924): Notes on the discussion of the papers presented in the symposium on hot springs. *J. Geol.,* **32,** pp. 464–468, Chicago.

Spengler, G. (1964): with collaboration of G. Michalczyk: Die Schwefeloxyde in Rauchgasen und in der Atmosphäre. *Ein Problem der Luftreinhaltung,* 152 pp. Düsseldorf: VDI-Verlag.

Spitsyn, V. I., M. K. Pimenov, F. P. Yudin & V. D. Balukova (1967): Scientific prerequisites for utilizing deep-lying formations for burying liquid radioactive wastes. *Disposal of Radioactive Wastes into the Ground.* Vienna: IAEA, pp. 563–576.

Stabler, H. (1911): The industrial application of water analyses. U.S. Geological Survey Water Supply Paper No. 274, pp. 165–181, Washington, D.C.

Stainer, R. Y., E. A. Adelberg & J. L. Ingraham (1976): *Microbial World,* 871 pp. Englewood Cliffs, NJ: Prentice-Hall.

Starkey, R. L. & K. M. Wight (1945): *Anaerobic Corrosion of Iron in Soil.* American Gas Association; Report of Technical Section, 108 pp., New York.

Stead, F. W. (1963): Tritium distribution in groundwater around large underground fusion explosions. Science, **142,** pp. 1163–1165, New York.

Steck, W. (1971): Sättigungskonzentrationen von Mineralölprodukten in Wasser. *Sem. Fortbild. Sachverst änd. Mineralölunf.,* 6 pp., Coblenz, Rhein.

Sticher, H. & R. Bach (1966): Fundamentals in the chemical weathering of silicates. *Soils Fert.,* **29,** 4, pp. 321–325, Harpenden.

Stiff, H. A., Jr. (1951): The interpretation of chemical water analysis by means of patterns *J. Petrol. Technol.,* **3,** 10, Technical Note 84, Section 1, pp. 15–16, Section 2, p. 3, Dallas.

Stoklasa, J. & A. Ernest (1905): Über den Ursprung, die Menge und die Bedeutung des Kohlendioxyds im Boden. *Zentralbl. Bakt. Parasitenkd. Infektionsk.,* Part 2, **14,** pp. 723–736, Jena.

Strahlenschutzverordnung (First) (1965): Bundesgesetzblatt I, Anlage II, pp. 1673–1678.

Strahlenschutzverordnung (2) (1976): Bundesgesetzblatt I, 125, pp. 2905–2995.

Stratmann, H. (1955): Untersuchungen über den Schwefeldioxydgehalt bodennaher Luftschichten in der Umgebung von Steinkohlen- Kraftwerken. *Mitt. Ver. Grosskesselbesitz.,* **37,** pp. 705–714.

Strauch, D. (1964): Veterinärhygienische Untersuchungen bei der Verwertung fester und flüssiger Siedlungsabfälle. *Schr.-R. Gebiet Öffl. Gesundheitswes.,* **18,** 114 pp., Stuttgart.

Strell, M. (1955): *Wasser und Abwasser, Reinhaltung der Gewässer,* 352 pp. Munich: Oldenbourg.

Stremme, H. E. (1950): Die Schwefelsäure im Säurehaushalt der Waldböden. *Z. Pflanzenernähr. Düng., Bodenkd.* **50,** 95, pp. 89–99, Weinheim, Bergstrasse, Berlin.

Strohecker, R. (1936): Ein neuer Weg zur Ermittlung der Angriffslust (Aggressivität) von Wässern. *Z. Anal. Chem.,* pp. 321–328, Munich: Bergmann.

Ströhl, G. W. (1966): Über einen Fall weitreichender Grundwasserverunreinigung durch Pestizide und Detergentien. *Gesund.-Ing.,* **87,** 4, pp. 108–114, Munich.

Stumm, W. (1961): Discussion of Hem, J. D., Stability field diagrams as aids in iron chemistry studies. *J. Am. Water Works Assoc.,* **53,** 2, pp. 228–232, Richmond, VA.

Stumm, W. & J. J. Morgan (1970): *Aquatic Chemistry,* 583 pp. New York, London, Sydney, Toronto: Wiley.

Stumper, R. (1934): La kinétique et la catalyse de la décomposition du bicarbonate de calcium en solution aqueuse. *Chim. Ind.,* **32,** pp. 1023–1037, Paris.

Stundl, K. (1956): Behinderung der bakteriellen Abbauvorgänge im Boden. *Gas-Wasser-Wärme,* **10,** 12, pp. 317–321, Vienna.

Stundl, K. (1964): Abbau organischer Stoffe durch Bodenorganismen. *Gas-Wasser-Wärme,* **18,** 9, pp. 196–201, Vienna.

Stundl, K. (1965): Zur Biologie der Filterwirkung des Bodens. *Mitt. Österr. Sanitätsverwaltg.,* **66,** 5, pp. 167–171, Vienna.

Stundl, K. (1967): Versuche über Bodenfiltration zur Bemessung der Sicherungsmassnahmen für ein Grundwasserwerk. *Österr. Wasserwirtsch.,* **19,** pp. 20–26, Vienna.

Stundl, K. (1968): Abbauvorgänge im Boden und ihr Einflüsse auf die Grundwasserqualität. *Gas-Wasser-Wärme*, **22**, 7, pp. 142–147, Vienna.

Sturm, G. & F. J. Bibo (1965): Nitratgehalte des Trinkwassers, unter besonderer Berücksichtigung der Verhältnisse im Rheingaukreis. *Gas- Wasserfach*, **106**, 12, p. 332–334, Munich.

Suarez, D. L. & D. Langmuir (1976): Heavy metal relationships in a Pennsylvania soil. *Geochim. Cosmochim. Acta*, **40**, pp. 589–598, Oxford, New York, Paris, Frankfurt.

Suess, E. (1909): *Das Antlitz der Erde*, Vol. 3, Part 2. Vienna, Leipzig: Tempsky, Freytag, pp. 789–1158.

Sugawara, K., T. Koyama & K. Terada (1958): Coprecipitation of iodide ions by some metallic hydrated oxides with special reference to iodide accumulation in bottom water layers and in interstitial water of muds in some Japanese lakes. *J. Earth Sci., Nagoya Univ.*, **6**, pp. 52–61, Nagoya.

Sugawara, K., S. Okabe & M. Tanaka (1961): Geochemistry of molybdenum in natural waters. II. *J. Earth Sci., Nagoya Univ.*, **9**, 1, pp. 114–128, Nagoya.

Sugisake, R. (1961): Measurement of effective flow velocity of ground water by means of dissolved gases. *Am. J. Sci.*, **259**, pp. 144–153, New Haven, CT.

Swain, F. M. (1970): *Nonmarine Organic Geochemistry*, 445 pp. Cambridge: Cambridge University Press.

Tamura, T. (1964): Selective ion exchange reactions from cesium and strontium by soil minerals. *Comptes Rendus du Colloque International sur la Rétention et la Migration d'Ions Radioactifs dans les Sols, Saclay, 1962*, pp. 95–104, Paris.

Tamura, T. (1972): Sorption phenomena significant in radioactive-waste disposal. In: T. D. Cook, Ed., *Underground Waste Management and Environment Implications*. Ann. Association of Petroleum Geologists, Memoirs, Vol. 18, pp. 318–330, Tulsa, OK.

Tarrant, K. R. & J. O'G. Tatton (1968): Organochlorine pesticides in rainwater in the British Isles. *Nature*, **219**, pp. 725–727, London.

Tausz, J. & M. Peter (1919): Neue Methode der Kohlenwasserstoff- Analyse mit Hilfe von Bakterien. Zentralbl. *Bakt. Parasitenkd., Infektionsk.*, Part 2, **49**, pp. 497–554, Jena.

Tepper, E. Z. (1963): On humic substance degrading bacteria of the autochthonic microflora of the soil. *Mikrobiologija*, **32**, 4, pp. 655–662, Moscow (in Russian with English abstract).

Thews, J.-D. (1971): Summenwirkung von Verunreinigungen in Trinkwasserschutzgebieten. *Gas- Wasserfach*, **112**, 2, pp. 101–103, Munich: Oldenbourg.

Thews, J.-D. (1972): Zur Typologie der Grundwasserbeschaffenheit im Taunus und Taunusvorland. *Abh. Hess. Landesamt Bodenforsch.*, **63**, 42 pp., Wiesbaden.

Thilo, L. (1969): *Der Einflüsse des Karbonataustausches auf die ^{14}C-Datierung von Grundwasser*, 63 pp. Ph.D. thesis, Heidelberg.

Thilo, L. & K. O. Münnich (1970): Reliability of carbon-14 dating of groundwater: Effect of carbonate exchange. *Isotope Hydrology 1970*. Vienna: IAEA, pp. 259–270.

Thimann, K. V. (1964): *Das Leben der Bakterien*, 2nd ed., German transl., 875 pp. Jena: Fischer.

Thompson, H. S. (1850): On the absorbent power of soils. *J. R. Agric. Soc. Engl.*, **11**, pp. 68–74, London.

Tillmans, J. (1932): *Die chemische Untersuchung von Wasser und Abwasser*, 2nd ed., 252 pp. Halle (Saale): Knapp.

Tillmans, J. & O. Heublein (1912): Über die kohlensauren Kalk angreifende Kohlensäure der natürlichen Wässer. *Gesund.-Ing.*, **35**, pp. 669–677, Munich, Berlin.

Tison, J. (1957): Hydrology (including irrigation). *Arid Zone Res.*, **9**, pp. 66–106, Paris (UNESCO).

Todd, D. K. (1960a): Salt water intrusion of coastal aquifers in the United States. *Int. Assoc. Sci. Hydrol. Publ.*, **52**, pp. 452–461, Gentbrugge.

Todd, D. K. (1960b): *Ground Water Hydrology*, 2nd ed., 336 pp. New York, London: Wiley.

Tokarev, A. N. & A. V. Shcherbakov (1956): *Radiohydrogeology*. Moscow: State Publ. Sci. (Technical Literature by U.S. Atomic Energy Commission, AEC-tr-4100), 346 pp.

Toler, L. G. & S. J. Pollock (1974): Retention of chloride in the unsaturated zone. *U.S. Geol. Surv. J. Res.*, **2**, pp. 119–123, Washington, D.C.

Torrey, A. E. & F. A. Kohout (1956): Geology and groundwater resources of the lower Yellowstone River Valley, between Glendive and Sidney, Montana. U.S. Geological Survey Water Supply Paper No. 1355, 92 pp., Washington, D.C.

Toth, J. (1966): Mapping and interpretation of field phenomena for groundwater reconnaissance in a prairie environment, Alberta, Canada. *Int. Assoc. Sci. Hydrol. Bull.*, **11**, 2, pp. 20–68, Gentbrugge.

Trueb, E. (1962): Hydrologische Interpretation von Temperaturbeobachtungen in Grundwasserströmen. *Schweiz. Z. Vermess., Kulturtech. Photogramm.*, **7**, pp. 191–200, Winterthur.

Truelsen, C. (1965): Brunnendurchmesser(Bohrdurchmesser). In: E. Bieske, Ed., *Handbuch des Brunnenbaus*, Vol. 2. Berlin, Konradshöhe; R. Schmidt, pp. 302–316.

Truesdell, A. H. (1972): Ion exchange. In: R. W. Fairbridge, Ed., *The Encyclopaedia of Geochemistry and Environmental Sciences*, 591 pp. New York: Van Nostrand Reinhold Co.

Tukey, H. B., Jr. & H. B. Tukey (1959): Nutrient loss from above-ground plant parts by leaching. *Atompraxis*, **5**, pp. 213–218, Karlsruhe.

Turekian, K. K. (1969): The oceans, streams, and atmosphere. In: K. H. Wedepohl, Ed., *Handbook of Geochemistry*, Vol. 1. Berlin, Heidelberg, New York: Springer, pp. 297–323.

Turekian, K. K. & J. L. Kulp (1956): The geochemistry of strontium. *Geochim. Cosmochim. Acta*, **10**, pp. 245–296, London, New York, Paris.

Udluft, H. (1953): Über eine neue Darstellungsweise von Mineralwasseranalysen. II. *Notizbl. Hess. Landesamt Bodenforsch.*, **81**, pp. 308–313, Wiesbaden.

Udluft, H. (1957): Zur graphischen Darstellung von Mineralwasseranalysen und von Wasseranalysen. *Heilbad Kurort*, **9**, 10, pp. 173–176, Gütersloh.

Ulbrich, R. (1957): Die Herkunft der Nitrate und Chloride in Grundwässern der Umgebung von Würzburg und Grundwässern der Rhön. *Gesund.-Ing.*, **78**, 5–6, pp. 80–82, Munich.

Universität Heidelberg, II. Physikalisches Institut (1971): *Jahresbericht 1970*, Heidelberg.

U.S. Environmental Protection Agency (1974): *Methods for Chemical Analysis of Water and Wastes*. Washington, D.C.

U.S. Public Health Service (1962): *Drinking Water Standards*. US Public Health Service Publication, No. 956, 61 pp., Washington, D.C.

Valyashko, M. G. & N. K. Vlasova (1965): On the processes of formation of calcium-chloride brines. *Geochem. Int.*, **2**, 1, pp. 25–36, Washington, D.C.

Versluys, J. (1915): Chemische werkingen in den ondergrond der duinen. *Verl. Afd. Natuurkd. K. Akad. Wetensch.*, **18**, pp. 1671–1676, Amsterdam.

Vickery, R. C. (1953): *Chemistry of the Lanthanons*, 296 pp. London: Butterworths.

Vinogradov, V. I. (1957): Migration of molybdenum in the zone of weathering. *Geokhimiya*, **1957**, 2, pp. 120–126, Moscow.

Vogel, H. U. von (1956): with collaboration of W. Bubam & H. Nahme: *Chemiker-Kalender.*, 560 pp. Berlin, Göttingen, Heidelberg: Springer.

Vogel, J. C. (1970): Carbon-14 dating. *Isotope Hydrology 1970*. Vienna: IAEA, pp. 225–239.

Vogel, J. H. (1913): *Die Abwasser aus der Kaliindustrie, ihre Beseitigung sowie ihre Einwirkung in und an den Wasserläufen*, 591 pp. Berlin: Borntraeger.

Völz, H. & W. Schwartz (1962): Über die regionale Verteilung kohlenwasserstoffoxydierender Bakterien in verschiedenen Böden. *Erdöl Kohle*, 15, pp. 426–429, Hamburg.

Vorob'Eva, G. I. (1957): The role of *Pseudomonas* in petroleum microbiology. *Chem. Abstr.*, 51, 12475 g.

Wahlberg, J. S. & M. J. Fishman (1962): Adsorption of cesium on clay minerals. U.S. Geological Survey Bulletin No. 1140-A, 30 pp., Washington, D.C.

Waksman, S. A. & F. G. Tenney (1927): The composition of natural organic materials and their decomposition in the soil. II. Influence of age of plant upon the rapidity and nature of its decomposition – Rye plants. *Soil Sci.*, 24, pp. 317–333, New Brunswick, NJ.

Walger, E. & H. D. Schulz (1976): Fortran IV Programme zur Daten- Aufbereitung bei chemischen Wasseranalysen. *Abh. Hess. Landesamt Bodenforsch.* 73, pp. 105–130, Wiesbaden.

Walker, A. C., U. B. Bray & J. Johnston (1927): Equilibrium in solutions of alkali carbonates. *J. Am. Chem. Soc.*, 49, pp. 1235–1256, Easton, PA.

Wallhäusser, K. H. (1951a): Die antibiotischen Beziehungen einer natürlichen Mikroflora. *Arch. Mikrobiol.*, 16, pp. 201–236, Berlin, Heidelberg: Springer.

Wallhäusser, K. H. (1951b): Untersuchungen über das antagonistische Verhalten von Mikroorganismen am natürlichen Standort. *Arch. Mikrobiol.*, 16, pp. 237–251, Berlin, Heidelberg: Springer.

Wallhäusser, K. H. (1965): Mikrobiologische Untersuchungen über Flutwässer bei der Erdölgewinnung. *Erdöl Kohle*, 18, pp. 357–360, Hamburg.

Wallhäusser, K. H. (1967): Ölabbauende Mikroorganismen in Natur und Technik. *Helgoländer Wiss. Meeresunters.*, 16, pp. 328–335, Hamburg.

Walton, A. (1963): The distribution in soils of radioactivity from weapons tests. *J. Geophys. Res.*, 68, 5, pp. 1485–1496, Richmond, VA.

Wandt, K. (1960): Hydrochemische Untersuchung von Grundwasser diluvialer und tertiärer Schichten in Schleswig-Holstein. *Meyniana*, 9, pp. 98–129, Kiel.

Wasmer, H. R. (1969): Ablagerung fester Abfallstoffe und Gewässerschutz. *Gas, Wasser, Abwasser*, 49, pp. 136–142, Zurich.

Waterton, T. (1969): The effect of tipped domestic refuse on ground water quality. *Water Treat. Exam.*, 18, 1, pp. 15–17.

Weber, G. (1961): Chemisch-bakteriologische Untersuchungen an einem Trinkwasserstau und staunahem Grundwasser. *Wasser Abwasser*, 1961: "Zur Limnologie der Speicherseen und Flußstaue," pp. 267–277, Vienna.

Wedepohl, K. H. (1953): Untersuchungen zur Geochemie des Zinks. *Geochim. Cosmochim. Acta*, 3, pp. 93–142, London.

Wedepohl, K. D. (1956): Untersuchungen zur Geochemie des Bleis. *Geochim. Cosmochim. Acta*, 10, pp. 69–148, London.

Wedepohl, K. D. (1967): *Geochem. Samml. Göschen*, 1224, 1224a, 1224b, 220 pp. Berlin: de Gruyter.

Wedepohl, K. D., Ed. (1969): *Handbook of Geochemistry*, Vol. 1. Berlin: Springer.

Wedepohl, K. D., Ed. (1978): *Handbook of Geochemistry*, Vol. 2, Parts 1–5. Berlin, Heidelberg, New York: Springer.

Wegelin, R. (1961): Über zwei eucavale Crustaceen aus dem interstitiellen Grundwasser der Leipziger Umgebung. *Int. Rev. Ges. Hydrobiol.*, 46, pp. 162–173, Berlin.

Weibel, S. R., R. B. Weidner, J. M. Cohen & A. G. Christianson (1966): Pesticides and other contaminants in rainfall and runoff. *J. Am. Water Works Assoc.*, 58, pp. 1075–1084, Baltimore.

Weisflog, D. (1968): Das Sorptionsverhalten von Niederterrassensedimenten des Rheines für Sr-85, Cs-137 und 1-131. *Dtsch. Gewässerkd Mitt.*, **12**, 5 pp. 125–134, Coblenz, Rhein.

Weisman, Y. I. (1964): On the propagation of bacterial contamination in underground waters. *Gig. Sanit.*, pp. 19–23, Moscow (in Russian).

Weithofer, K. A. (1933): Die Karlsbader Thermen und der Bergbau nebst einigen allgemeinen Bemerkungen über Mineralquellen. *Neue Jahrb. Min.*, **70-B**, pp. 116–138, Stuttgart.

Wells, A. F. (1950): *Structural Inorganic Chemistry*, 2nd ed., 727 pp. Oxford: Clarendon Press.

Weyl, P. K. (1958): The solution kinetics of calcite. *J. Geol.*, **66**, pp. 163–176, Chicago.

White, D. E. (1957a): Thermal waters of volcanic origin. *Geol. Soc. Am. Bull.*, **68**, pp. 1637–1658, New York.

White, D. E. (1957b): Magmatic, connate, and metamorphic waters. *Geol. Soc. Am. Bull.*, **68**, pp. 1659–1682, New York.

White, D. E., W. W. Brannock & K. J. Murata (1956): Silica in hot-spring waters. *Geochim. Cosmochim. Acta*, **10**, pp. 27–59, London, New York, Paris.

White, D. E., J. D. Hem & G. A. Waring (1963): Chemical composition of subsurface waters. U.S. Geological Survey Professional Paper No. 440-F, 67 pp., Washington, D.C.

Whitfield, M. (1974): Thermodynamic limitations on the use of the platinum electrode in E_h measurements. *Limnol. Oceanogr.* **19**, pp. 857–865.

Whittemore, D. O. & D. Langmuir (1975): The solubility of ferric oxyhydroxides in natural waters. *Ground Water,* **13**, pp. 360–365.

Wiberg, E. (1971): *Lehrbuch der anorganischen Chemie*, 1209 pp. Berlin: de Gruyter.

Wickman, F. E. (1944): Some notes on the geochemistry of elements in sedimentary rocks. *Ark. Kem., Mineral. Geol.*, **19B**, 2, pp. 1–7, Stockholm.

Wiegner, G. & H. Jenny (1927): Über Basenaustausch an Permutiten. *Kolloid.- Z.*, **43**, pp. 268–272, Dresden, Leipzig.

Wilcox, L. V. (1948): The quality of water for irrigation use. U.S. Department of Agriculture Technical Bulletin No. 962, 40 pp., Washington, D.C.

Wilhelm, F. (1956): Physikalisch-chemische Untersuchungen an Quellen in den bayerischen Alpen und im Alpenvorland. *Münchner Geogr. H.*, **10**, 97 pp., Regensburg.

Wilson, A. T. (1960): Sodium/potassium ratio in rain-water. *Nature*, **186**, 4726, pp. 705–706, London, New York.

Wolfskehl, O. & E. Boye (1966): Einwirkung von abgelagerter Müllasche und von Müllkompost auf das Grundwasser. *Gas- Wasserfach,* **107**, 2, pp. 36–38, Munich.

Wollny, E. (1880a): Untersuchungen über den Einflüsse der Pflanzendecke und der Beschattung auf den Kohlensäuregehalt der Bodenluft. *Forsch. Geb. Agric. Phys.*, **3**, pp. 1–14, Heidelberg.

Wollny, E. (1880b): Untersuchungen über den Kohlensäuregehalt der Bodenluft. Landw. Versuchsstn., **25**, pp. 373–391, Berlin.

Wollny, E. (1881): Untersuchungen über den Einflüsse der physikalischen Eigenschaften des Bodens auf dessen Gehalt an freier Kohlensäure. *Forsch. Geb. Agric.-Phys.*, **4**, pp. 1–24, Heidelberg.

Wollny, E. (1886): Untersuchungen über den Einflüsse der physikalischen Eigenschaften des Bodens auf dessen Gehalt an freier Kohlensäure. *Forsch. Geb. Agric.-Phys.*, **9**, pp. 165–194, Berlin.

Wollny, E. (1889): Der Einflüsse der Menge der im Boden befindlichen organische Stoffe auf den Kohlensäuregehalt der Bodenluft. *Landw. Versuchsstn.*, **36**, pp. 201–211, Berlin.

Wolter, D. (1967): Urochrome als trinkwasserhygienisches Problem. *Fortschr. Wasserchem. Grenzgeb.*, **7**, pp. 195–200, Berlin.

Wolter, R. & K. Aurand (1967): Zur radioaktiven Kontamination von Uferfiltrat und Grundwasser. *Mem. Int. Assoc. Hydrogeol. Congr. 1965,* **7,** pp. 172–175, Hannover.

Wolters, M. F. (1969): *Der Schlüssel zum Computer.* 1. Textbook, 858 + 30 pp., 2. Program, 98 pp. Düsseldorf, Vienna: Econ.

Wolters, N. & W. Schwartz (1956): Untersuchungen über Vorkommen und Verhalten von Mikroorganismen in reinen Grundwässern. *Arch. Hydrobiol.,* **51,** 4, pp. 500–541, Stuttgart.

Yanat'Eva, O. K. (1954): La solubilité de la dolomite dans l'eau en présence de gaz carbonique. *Izv. Akad. Nauk CSSR Otd. Khim. Nauk,* **16,** 6, pp. 1119–1120.

D'Yachkova, I. B. (1962): Selenium in waters from the Ural areas containing pyrite deposits. 4th *Konf. Molodykh Nauchn. Sotrudn. Inst. Mineral. Geokhim. Kristallokhim. Redk. Elem., Sb.,* pp. 40–43, Moscow (*Chem. Abstr.,* **63,** 2, 1581g, 1965).

Young, C. P. (1980): The distribution and movement of solutes derived from agricultural land in the principal aquifers of the United Kingdom, with particular reference to nitrate. *Proceedings of the International Association for Water Pollution Research,* Toronto.

Zajic, J. E. (1969): *Microbial Biogeochemistry,* 345 pp. New York, London: Academic Press.

Zartman, R. E., G. J. Wasserburg & J. H. Reynolds (1961): Helium, argon, and carbon in some natural gases. *J. Geophys. Res.,* **66,** pp. 277–306, Richmond, VA.

Zehender, F., W. Stumm & H. Fischer (1956): Freie Kohlensäure und pH von Wasser im Calciumkarbonat − Löslichkeitsgleichgewicht. *Schweiz. Ver. Gas- Wasserfach,* **36,** 11, pp. 269–275, Zurich.

Zimmermann, U., K. O. Münnich, W. Roether, W. Kreutz, U. Schubach & O. Siegel (1966): Tracers determine movement of soil moisture and evapotranspiration. *Science,* **152,** 3720, pp. 346–347, Washington, D.C.

Zimmermann, U., K. O. Münnich & W. Roether (1967a): Messung der Abwärtsbewegung der Bodenfeuchte und der Evapotranspiration mit Hilfe von Wasserstoff-Isotopentracern. *Mem. Int. Assoc. Hydrogeol.,* **7,** pp. 14–15, Hannover.

Zimmermann, U., K. O. Münnich & W. Roether (1967b): Downward movement of soil moisture traced by means of hydrogen isotopes. *Am. Geophys. Union Geophys. Monogr.,* **11,** 2 pp.

Zobell, C. E. (1946a): Studies on redox potential of marine sediments. *Bull. Am. Assoc. Petrol. Geol.,* **30,** 4, pp. 477–513, Tulsa, Ok.

Zobell, C. E. (1946b): Action of microörganisms on hydrocarbons. *Bact. Rev.,* **10,** pp. 1–49, Baltimore.

Zwittnig, L. (1964): Die Beeinflussung des Grundwassers durch Mülldeponien. *Steir. Beitr. Hydrogeol.,* N.F. **15–16,** pp. 91–106, Graz.

Index

Absorption coefficient, 20, 21, 34
Acetobacter aceti, 123
Achromobacter, 121, 128
Acid, nitric, 78, 176
Acidity, 328
Acids, organic, 60, 78, 176, 288
Actinomycetes, 120, 121, 134, 136
Activity, 31, 32, 35-37, 48, 49, 59, 61-63, 66-68
  bulk, 342, 343
Activity coefficient, 31-35, 47, 57
Adaptation, 131, 132
Adsorption, sorption, 86-105, 113, 122, 129, 133, 211, 279
Aerobacter aerogenes, 123
Aerobes, 121
Aerosols, 144, 164, 166, 170, 171
Aggressiveness (aggressivity) of CO_2, 49, 57, 85
Albite, 77
Algae, 119, 126, 134, 137
Alkali feldspar, 76, 77
Alkalinity, 245-247, 326, 328, 332
Allophane, 89
Alona guttata, 280
Aluminum, 60, 87, 88, 90, 211, 212, 261
Ammonia, 91, 107, 111, 125, 126, 131, 140, 174, 249, 250
Amphibole (hornblende), 77, 87, 282, 283
Amphibolite, 283
Anaerobes, 121, 122
Analcite, 91
Analysis:
  graphs of, 297-316
  reporting of, 291-297
Andesite, 283
Anhydrite, 26, 287
Animal husbandry, 119, 155, 156
Anorthite, 77
Antibiotics, 129, 130
Antimony, 91
Apophyllite, 91
Aquifer materials, 281-290

Argillites (hydrolysates), 212, 213, 287
Argon, 20, 274
Arsenate, 90
Arsenic, 91, 108, 154, 155, 213, 268
Arsenical waters, 334
Ascomycetes, 121
Aspergillus niger, 123
Atmosphere, 114-118
Atractides, 281
Attapulgite, 89, 90, 92
Augite (pyroxene), 77, 87

Bacilli, 120
Bacillus subtilis, 123
Bacteria, 105, 113, 120-128, 131-134, 136, 146, 154
  coliform, 134, 135, 152, 336
Bacteriophages, 134
Bank storage (bank-filtered river water), 85, 125, 126, 139, 147, 188, 204, 278
Bar charts, graphs, 299-302, 329
Barium, 25, 85, 91, 211, 212, 260, 261
Base exchange (cation exchange), 86-105, 177, 179-181
  index of, 104
Basidiomycetes, 121
Bathynella, 280
Beggiatoa, 125
Beidellite, 89
Bentonite, 8
Beryllium, 213, 258
Bicarbonate (hydrogen carbonate), 24, 35, 140, 245-248
  determination of, 247
Bicarbonate (hydrogen carbonate) water, 331-334
Biochemical oxygen demand (BOD_5), 130
Biosphere, 211
Biotite, 77
Bismuth, 91, 274
Bitumen, 107, 127

Boiler feed water, 351
Bonding of ions, 91, 92
Borate, 90
Boron, 135, 178, 213, 267
   tolerance towards, 348
Brackish water, 325
Breakdown of organic substances, 111
Brines, 26, 36, 323, 325, 334
   migration of, 323
Bromine, 144, 213, 270, 271
Brown coal (lignite), 127, 288
Buffer system, 60

Cadmium, 148, 159, 162, 213
Calcium, 24, 25, 91, 93, 104, 135, 211, 212,
   236-239
Calcium carbonate, saturation index, 49
Cancrinite, 89
Candida lipolytica, 123
Candona, 280
Carbon-14 (radiocarbon), 141, 145, 247, 248,
   344
Carbon dioxide (carbonic acid), 20, 21, 35, 36,
   45-50, 55-57, 77, 79, 85, 110-119, 127, 142,
   143, 174, 175, 245-248
   aggressive to limestone, 49, 245
   excess, 48, 49
   free, 46, 245
   free aggressive, 47, 245
Carbon isotopes, 247, 248
Carbon monoxide, 143
Carbonate, 25, 35, 37, 245-247
Carbonate determination, 247
Carbonic acid water, 331, 332
Cartographic diagrams, 317-319
Cation exchange (base exchange), 86-105,
   177, 179-181
Caustobiolites, 288-290
Cave newt (proteus anguinus), 280
Cerium-141 (radiocerium), 150
Cerium-144 (radiocerium), 103, 141, 150,
   344
Cesium, 94, 95, 213, 273
Cesium-137 (radiocesium), 96, 97, 99, 101,
   103, 141, 150, 344
Chelates, 78
Chelating agents, 78, 127
Chemical oxygen demand, 278
Chemical quality:
   effect of climate, 80-83
   variations caused by evaporation, 82, 83
   variations caused by freezing, 81, 83
Chironomides, 280
Chloride, 24, 90, 140, 144, 211, 254-256

Chloride water, 331, 332, 334
Chlorine, 140, 144, 212
Chlorine-36 (radiochlorine), 145, 256, 344
Chlorite, 77, 89, 90
Chromate pollution, 148, 262
Chromium, 92, 108, 131, 148, 159-162, 212,
   262
Chromobacter, 121
Circular diagrams, 301-304
Citric acid, 78
Cladoceras, 280
Clay membrane, 106
Clay minerals (ion exchangers), 87-94
Climate, influence on dissolved constituents,
   80-83
Coal, 111, 127, 288
Cobalt, 92, 178, 213, 262, 263
Cocci, 120, 121
Coliform bacteria, 133, 134
Coliform count, 132, 133
Colonies:
   bacterial, 121, 122, 135, 156
   microbial, 121, 156
Common ion effect, 26
Complexes, 36, 70, 77
Compressibility, 1, 15, 16
Concentration:
   units, 292, 293
   variations caused by freezing, 81, 83
Condensation, heat of, 16
Conductance:
   electrical, 18, 70, 71
   specific, 18
Connate water, 322, 323
Contact surface, influence on solution
   process, 74
Contact time, influence on solution process,
   74
Contamination, *see* Pollution
Cooling water, 351
Copepods, 134, 280
Copper, 64, 86, 91, 92, 108, 110, 131, 135,
   148, 178, 212, 263
Coprecipitation, 85, 91
Crenothrix, 125, 126
Crustacea, 280
Crystalline rocks, 283
Cyanide, 131, 148, 155
Cyclic substances, mobility, 210, 211
Cyclopoidea, 280
Cytophaga, 121

Data processing, storage, 320
Debye-Hückel theory (equation), 31

Decomposition (oxidation, breakdown of organic matter), 107, 110, 111, 115, 127-130, 146, 277
  rate of, 129, 130
  resistance to, 140
  silicates, 75-78
Deep-lying water, 322-324
Degassing, 143-145
Denitrification (nitrate reduction), 111, 112, 113, 123, 125, 126, 128, 186
Denitrifying organisms, 125
Density, 1, 2, 13-16, 105
  maximum, 2
Desulfovibrio desulfuricans, 123-125, 128
  var. aestuarii, 124, 132
  var. thermodesulfuricans, 124
Desulfurication (sulfate reduction), 69, 111, 112, 113, 123-125, 128
Detergents (surfactants), 103, 129, 131, 148
Deuterium, 3-5
Diacyclops languidoides, 280
Diagrams, 297-317
  multivariate, 307
  profile, 315-317
  Stiff, 315
  trilinear, 307-312, 330
Diatomite, 89
Diatoms, 105, 126
Dielectric constant, 1, 23, 32
Diffusion coefficient, 5, 115, 118
Dilatation (coefficient of cubical expansion), 17
Diorite, 282, 283
Dissociation, 18
  degree of, constants, 23, 31-33
Distribution coefficient, 93, 94
Dolomite, 285, 286
Dolomitization, 287
Dose equivalent, 342
Dose rate, 341-343
Dyestuffs, 92, 103, 277

Earths, rare, 150
Edaphon, 176
Eh measurement, 61-67
Eh-pH diagrams, stability field diagrams, 66, 68, 69
Electrical conductance, 18, 70-72, 215
Electrolyte, 18, 27, 29, 32, 33, 37
Electromotive force, voltage, 31, 63
Elimination of microorganisms, 131-135
Emulsion, 27
Energy:
  change, standard free, 38-44, 62, 63, 127, 128
  dose, 341, 342

free, 38
Epactophanes richardi, 280
Epilobium angustifolium (rose bay), 135
Equilibrium, pH, 48, 49, 56
Equilibrium concentration, 26
Equilibrium constants, 38, 39, 44, 59, 62, 63, 67, 68
Equivalent quantities of most important ions, 293
Escherichia coli, 132-134, 279, 336
Evaporites, 79, 80, 212, 213, 287, 322
Exchange, equations of equilibrium, 87, 88, 94
Exchange capacity, 88-92
Exchange velocity, 92

Fallout, 8, 9, 11, 141
Faraday constant, equivalent, 62, 63, 106
Fauna:
  in groundwater, 134, 135, 280
  in soils, 176
Feldspar, 75-77, 89
Feldspathoids, 89
Ferrobacillus ferrooxidans, 126
Fertilizers, 135, 140, 155, 156
  application, 103, 156
  liquid, 156
Filter effect, 196
Flagellates, 126
Flavobacterium, 121, 128
Flue gases, 144
Fluoride, 25, 269, 270
Fluorine, 145, 212, 269, 270
Force, electromotive (EMF), 31, 63
Fossil water, 321
Fresh water, 325
Fungi, 120, 121, 126, 128
Fungi imperfecti, 121
Fusarium, 121

Gabbro, 283
Gallionella ferruginea, 126
Gallium, 213, 273, 274
Gas, radioactive, 143
Gas constant, 62, 106
Gas exchange, 61, 114-118, 175
Gasoline, 145
Gay-Lussac (law), equation, 20
Geothermal gradient, 17, 203-205
Germanium, 213, 266
Germs, pathogens, 154, 279
Glauconite, 87, 89
Gneiss, 282, 283
Graetiriella unisetigera, 280

Granite, 282
Gray, 341, 342
Groundwater, 6, 10-13, 30, 60, 64, 70, 73-164,
    185-336
  dating, 10, 11, 248, 249, 321
  examples of analyses, 218-235, 298
  organisms in, 134, 135, 280
  pollution (contaminants), 96, 111, 113, 114,
    131, 132, 139-163
    indicators of, 140, 279, 336, 337
  recharge, 85, 126, 147
  temperature, 137, 197-206
  total amount of, 196, 197
  types, 321-335
Gypsum, 26, 79, 284, 287

Halacarids, 134, 280
Half-life (biological), 342, 343
Halloysite, 89, 90, 95
Hardness, trilinear diagrams, 311, 312
Hardness of water, 238, 239
Harpacticoids, 280
Heat, specific, 16
Heat of formation, 38
Heatflow, 197-199
Heavy metal ions, 91, 131
Helium, 20, 274
Henry's law, 19, 20
Hepatitis epidemica, 279, 280
Highway salting, 158, 159
Hornblende (amphibole), 77, 87, 282, 283
Hornblende-gneiss, 283
Hornblende-granite, 283
Humic acids, 60, 81, 176
Humins, 176
Humus, 81, 87, 91, 122, 129, 137, 176
Hydrachnellae, 281
Hydrate, 23
Hydration, 23
Hydration sheaths, shells, 31, 32
Hydrocarbons, 23, 107, 111, 118, 123, 124,
    127, 128, 276, 288-290
Hydrochloric acid, 60
Hydrogen ion activity, 37
Hydrogen ion concentration (pH), 18, 37,
    59-70, 75, 77, 85, 92, 107, 195
Hydrogen ions, 91
Hydrogen isotopes, 3-12
Hydrogen sulphide, 110, 111, 113, 114,
    123-125, 175
Hydrolysates, 212, 213, 287
Hydrolysis, 75-78, 83
Hydrolysis constant, 67
Hydroxide, 26

Hydroxides and oxyhydroxides of Fe, Mn,
    and Al, 87, 91

Igneous rocks, 212, 213, 281-283
Illite, 6, 89-92
Index of base exchange, 104
Indicators of groundwater pollution, 140,
    279, 336, 337
Industrial water, 350, 351
Inert (noble) gases, 274-276
Inhibitors, 130-134
Iodine, 213, 271
Iodine-131 (radioiodine), 141, 271, 344
Iodine waters, 344
Ion exchange, 86-106, 211
Ion filtration, 105, 106
Ion pair, 36
Ion product (solubility product), 23, 25, 44,
    47, 85
Ionic activity, 29-44
Ionic bonding (heteropolar bonding), 23,
    211
Ionic charge (ionic valence), 23, 210
Ionic charge number, 23
Ionic potential, 210
Ionic radius, 32, 91, 210
Ionic strength, 29, 30, 32, 33, 47-54
Ionic valence (ionic charge), 23, 210
Ionizing dose, 342
Iron, 25, 59, 64-70, 85, 107-113, 126, 212,
    240-242
Iron bacteria, 126
Irrigation water, 343-350
Isotope exchange, 6
Isotope fractionation, 4, 5
Isotope ratios, 4, 5

Juvenile water, 321, 325

Kaolinite, 6, 77, 89, 90, 92, 95
Kilo-molarity, 29, 30, 35

Lakes of internal drainage, 188-189
Langelier constant, 48
Langelier correction factor, 54
Langelier index, 45
Latent heat of fusion, 1, 4
Latent heat of vaporization, 1-3, 4
Leaching, 178
Lead, 91, 92, 108, 143, 145, 213,
    266, 267
Leucite, 89
Lignite, 127
Limestone, 285, 286

Limestone-carbonic acid equilibrium, 37, 45-59, 85
Limestone lithology, 212, 213, 285, 286
Limestone solubility, 56, 57
Limiting pressure (gas solubility), 114
Liquid fertilizers, manure, 156
Lithium, 91, 213, 258
Litre molarity, 29
Lobohalacarus weberi quadriporus, 280
Lysimeter measurements, 178, 183-187

Magmatic water, 325
Magnesium, 24, 25, 59, 91, 95, 135, 212, 239-240
Manganese, 60, 64, 86, 87, 91, 108-114, 126, 127, 140, 178, 212, 243-245
Manganese bacteria, 126, 127
Manganese pollution, Breslau, 113, 114
Marble, 284, 285
Marcasite, 107
Marine water, 324
Marl, 287
Mass action, law of, 23, 37, 38
Medicinal water, 334
Mercaptans, 177
Mercury, 92, 157, 213, 266
Metabolism, microbial, 119-123
Metamorphic rocks, 281-284
Metamorphism of groundwater, 73, 324
Metazoa, lower, 134
Methane, 116, 127, 175
Methane consumers, 127
Methane producers, 128
Methanobacterium suboxydans, 127
Methanomonas methanica, 128
Methanosarcina methanica, 127
Mica, 87
Mica-schist, 283
Microbial activity in soil, 119
Micrococcus, 121
Micromonospora, 121
Microorganisms, 87, 105, 113, 119-135, 176, 177, 279
Microspira, 132
Mine water, 107, 108
Minor and trace elements, 25, 26, 86, 135, 195, 257-276
Mixing corrosion, 85
Mixing of waters, 85
Mobility, geochemical, 210, 211
Mobility coefficient, 211
Molar volumes, 28, 106

Molarity, 29, 30
Mole fraction, 31, 106
Molecular weight of water, 1
Molybdenum, 91, 108, 135, 213, 262
Monosporium, 121
Montemorillonite, 6, 89, 90, 92, 95
Moulds, 137
Multivariate diagrams, 307, 314
Mycelia, 120, 121
Mycobacterium, 128
Mycobacterium phlei, 123
Mycorhiza fungi, 177

NaCl, tolerance of plants to, 346, 347
Natural gas, 107, 111, 124, 127, 288, 290
Nematodes, 134, 176, 280
Nernst's law, 74
Nickel, 92, 148, 212, 263
Nitrate, 24, 90, 111, 113, 122, 125, 126, 135, 136, 150, 186, 249-251
Nitrate reduction (denitrification), 111, 112, 113, 123, 125, 126, 128, 186
Nitric acid, 75, 78, 107, 174, 176
Nitrite, 113, 125, 140, 249-251
Nitrobacter, 125
Nitrococcus, 125
Nitrocrella reducta, 280
Nitrogen, 20-22, 112, 114, 125, 126, 135, 174, 213
Nitrogen compounds, 107, 125, 126, 175, 177, 186, 187, 249-251
Nitrosomas, 125
Nocardia, 120, 121, 128
Non-carbonate hardness, 103
Nonelectrolytes, 22, 23, 27
Nontronite, 89
Nosean, 89
Nyphargus, 280

Odor, odorivectors, 140, 145, 276-288
Oil field waters, brines, 6, 7, 72, 74, 84, 122, 140, 147, 148, 276, 277, 288, 290, 323
Oligochaetes, 134, 280
Olivine, 77
Organic substances, 77, 86, 89, 91, 109, 113, 114, 119, 121, 127-130, 135, 140, 175, 176, 277, 278, 288
Organochlorine, 157
Osmosis, filter effect, 105, 106
Osmotic pressure, 105, 106
Oxidation, 61, 107-114
  zone of, 113
Oxidizability, 278

Oxygen, 20-22, 62, 107-114, 117, 118, 121,
174, 175
Oxygen-18, 3-8
Oxygen isotopes, 3-8
Oxyhydroxides and hydroxides of Fe, Mn,
and Al, 87, 91

Palygorskite, 89, 90
Parabathynella, 280
Paracyclops fimbriatus, 280
Paramecium, 131
Parasites, 154
Parastenocaris fontinalis bora, 280
Pathogens, 154, 279, 336
Peat, 91, 107, 111, 127, 288
formation of, 111
Peptization, 27
Percolate, infiltration water, 172-187
Permanganate value, 140, 278
Pesticides, 130, 131, 140, 157, 158
Petroleum, 107, 111, 124, 128, 129, 288
Petroleum products, 27, 143, 144, 162, 278,
336
pH measurement, 61
pH value, 18, 37, 59-70, 75, 77, 85, 92, 93,
107, 195
Phenol, 129, 131, 135, 147, 149
Phillipsite, 91
Phosphates, 90, 267, 268
Phosphorus, 212, 267, 268
Phycomycetes, 121
Phyllite, 282, 284
Phyllognathopus, 280
Phyllognathus vigueri, 280
Physical states of water, 14, 15
Plagioclase, 77, 282
Plants, 87, 119, 135-139
Plutonium-239, 150, 345
Poliomyelitis virus, 280
Pollution:
extent of, 162
by gases, 143-145, 158
by organic liquids, 145, 146, 276, 277
spread of, 162, 163
Pollution of groundwater, 96, 111, 118, 132,
137, 139-163
Pollution from solid waste disposal, 151-155,
163
Porous rock, 281, 284, 285
Potassium, 24, 91, 94, 103, 135, 140, 211, 212,
216, 217, 236
Potassium-40 (radiopotassium), 236
Potassium (alkali) feldspar, 77
Precipitates, 212, 213

Precipitation, 57-59, 73, 78-86, 113
rain, 5, 6, 8-12, 72, 143-145, 164-172
Pressure, osmotic, 105, 106
Proactinomyces, 128
Profound water, 321
Proteus anguinus (cave newt), 280
Protozoa, 121, 126, 134, 280
Psammitic rocks, 284
Psephitic rocks, 284
Pseudomonas, 128
Pseudomonas aeruginosa, 123, 128, 130
Pumice, 89
Punched card data storage, 320
Putrefaction, 110
Pyrite, 107
Pyrophyllite, 89
Pyroxene (augite), 77, 87

Quality factor (of radiation), 342, 343
Quartz, 87, 89
Quartzitic rocks, 282

Radiation, types of, 337, 341-343
Radioactivity, dating by, 10, 247-249, 321
Radioactivity units, 341, 342
Radionuclides (radioisotopes), 93, 96-103,
141, 142, 145, 150, 258, 344, 345
Radiostrontium, 93, 103
Radium, 86, 272, 273, 345
Radon, 276, 345
Radon waters, 335
Rain water (precipitation), 5, 6, 8-12, 72,
143-145, 164-172
Rare earths, 150
Reaction:
heat of, 37, 38
rate of, 119
Redox potential, 37, 61-70, 85, 107-109, 113,
114, 126, 195
Reduction, 61, 107-114
zone of, 110, 111, 113
Regenerated water, 324
Regeneration of exchangers, 104
Relative biologic effect, 342
Residue on evaporation (dry residue), 30, 215
Resistates, 212, 213
Rhyolite, 282
River bank filtration, 85, 188. See also Bank
storage
River water analyses, 188-194
Rocks:
composition of, 212, 213, 215-217
hard, psephitic, psammitic, 284
Roentgen, 342

Root respiration, 119, 136
Rose, pie diagrams, 301-303
Rose bay willowherb (Epilobium angustifolium), 135
Rotifera, 280
Rubber, 129
Rubidium, 91, 212, 258
Rushes, plaited, 135
Ruthenium-103 (radioruthenium), 145, 150, 345
Ruthenium-106, 97-103, 150, 345

Salinity, 326, 328, 332
Salmonella, 134
Salt, cyclic, 81, 166, 168
Salt water, 266-325
Salting of highways, 158, 159
Sandstones, 212, 213, 284
Saponite, 89
Saturation concentration, 23
Saturation index, 45
Schizomycetes, 126
Scirpus lacustris, 135
Seawater, 4, 5, 30, 72, 104, 195, 196, 211-213, 323, 324
Sedimentary rocks, 284-288
Selenium, 91, 108, 213, 268, 269
Self-potential (in boreholes), 106
Semilogarithmic graphs, 316, 317
Sepiolite, 89, 90
Serpentine, 77
Serratia marcescens, 134
Settlement (human), 119, 156
Sewage, see Wastewater
Shale line, 106
Shales, 89
Silanol, 76
Silica, 75, 87, 214, 256, 257, 282
Silica gel, 89
Silicates, decomposition, weathering of, 75-78
Silicon, 135, 211, 212
Siloxan bond, 76
Silver, 25, 213, 264
Slate, 284
Slime, microbial, 87, 122, 129
Smell, threshold of, 145
Sodalite, 89
Sodium, 24, 91, 94, 135, 212, 215, 216
Sodium absorption ratio (SAR), 349, 350
Sodium quotient, 105
Soil, 103, 121, 173-177, 279
  solutions in, 178-187
Soil air, ground air, 107, 117, 118, 136, 173-175
Soil fertility, 156

Solubility:
  effect of competing ions, 26, 31-57, 94
  effect of Eh, 61-64
  effect of Eh-pH, 64-70
  effect of pH, 59-61
  gases, 19-22
  organic substances, 27, 28
  solids and liquids, 22-29
Solubility product (ion product), 23, 25, 44, 47, 85
Solution, 19, 73-75, 78-85
  colloidal, 19, 27
  ideal (infinite dilution), 30, 31
  non-ideal, 31-33
  true, 19, 22
Solution of gases, effect of pressure, 114
Sorption (in soil and near surface), 103-105
Sorption capacity, 141
Specific heat, 16
Spirilla, 120
Spring water temperature, 206-210
Stability field diagrams (Eh-pH), 64-70
Standard free energy change, 38-44, 62, 67, 112, 127, 128
Standard mean ocean water (SMOW), 4, 5
Standard potential, 61-63, 65
Standard state, 38
Stiff diagrams, 315
Streptomyces, 121, 125
Strohecker-Langelier formula, 48, 50, 56
Strontium, 25, 91-93, 135, 212, 259, 260
Strontium-89 (radiostrontium), 141, 345
Strontium-90 (radiostrontium), 91, 93, 96, 97, 99, 101-103, 141, 145, 150, 345
Sulfate, 24, 25, 69, 90, 91, 108, 111-114, 123, 136-139, 144, 147, 156, 251-254
Sulfate-reducing bacteria, 123, 124, 128
Sulfate reduction (desulfurication), 69, 111, 112, 113, 123-125, 128
Sulfate water, 333, 334
Sulfide, 25, 110-111, 113, 125
Sulfide ore deposits, 108
Sulfur, 68, 108, 113, 123-125, 135, 212
Sulfur bacteria, 110, 111, 123-125
Sulfur compounds, 107, 123-125
Sulfur-containing proteins, 110, 111, 124
Sulfur dioxide, 143, 144
Sulfur isotopes, 124, 254
Sulfur water, 334
Sulfuric acid, 60, 75, 78, 107, 174, 177
Superficial deposits, fauna of, 134, 280
Surface layers, 116, 117, 173
Surface tension, 1, 3, 4, 15, 18
Surfactants (detergents), 103, 129, 131, 148

Temperature change (by human activity), 168, 205, 206

Temperature of groundwater, 197-210

Temperature influence on solution processes, 74

Temperature measurement, 210

Temperature range, 199-205, 208, 209

Thermal conductivity, 16, 197-199, 202, 203

Thermal diffusivity, 198, 199

Thermal waters, 60, 205, 210

Thermodynamics of equilibrium systems, 37

Thiobacillus (thiobacterium), 124, 125
  denitrificans, 125
  ferroxydans, 126
  thiooxydans, 126

Thioploca, 125

Thiothrix, 125

Tillmans's constant, 47, 55

Tillmans's correction factor, 48, 50, 53, 55

Tillmans's law, 46-49

Tin, 213, 274

Titanium, 274

Titre natronique, 105

Total dissolved solids, 215

Total hardness, 103, 239

Trace elements, 86, 135

Transitional zone, 113

Trilinear diagrams, 307-312, 326

Tritium units (TU), 4

Tuff, 89, 90

Tungsten, 92, 108, 213, 273

Turbellaria, 134, 280

Typhus epidemic, 280

Uranium, 64, 108, 213, 271, 272, 345

Urochrome, 139, 336

Vanadium, 64, 92, 108, 212, 261, 262

Vapor pressure, 13-16, 34, 106

Vermiculite, 77, 89, 90, 92

Viruses, 146, 154, 279

Viscosity, 1, 13-15, 18, 105

Wastewater (sewage), 139, 140, 145-151, 157
  destruction, 149
  dilution, 148, 149
  disposal, 148, 149
  percolation, 148, 149
  radioactive, 96-101, 149-151
  sewerage, 148, 149

Water:
  connate, 322-324
  dead, stagnating, 321, 322
  fossil, 321, 322, 324
  infiltration, 321
  isotopic composition, 3-12
  juvenile, 321, 325
  marine, 324
  metamorphic, 324
  percolate, 321
  plutonic, 325
  "reduced", 111, 113
  regenerated, 324
  volcanic, 325

Water dipoles, 2, 32

Water mites, 281

Water molecules, 1-3

Water type diagrams, 325-330

Weathering:
  chemical, 73
  physical, 74

Yeasts, 120, 128

Zeolites, 89, 90, 92

Zinc, 92, 131, 135, 148, 178, 213, 264, 265

Zirconium-niobium-95, 150

Zonation, geochemical, vertical, 79, 80

Zone, isothermal, neutral, zero temperature change, 200, 203

Zones, climatic, 80-83